"Corsi's new book is a trove of information, especially about the people and motives that have driven the current climate frenzy. The discussion of climate science, especially the dominant role of the sun and clouds, is very good. There are also fascinating discussions of unfashionable and perhaps incorrect scientific theories like abiogenic hydrocarbons or alternatives to plate tectonics. These help to clarify the all-too-human nature of science, and how hard it is initially to distinguish between a paradigm shift and a mistake, like today's alarm over 'carbon pollution.'"

—**Dr. William Happer**, professor emeritus in the Department of Physics, Princeton University; chair of the university research board, 1995–2005; winner of the Thomas Alva Edison Patent Award in 2000

"Corsi, a bestselling author of books about mainstream political issues, proved to be a powerful enough mind to brilliantly deconstruct a seemingly scientific topic, the fight against climate change. He sees the fear as wildly exaggerated. Sunspots, cosmic rays, and oceans as more important for the climate than carbon dioxide. The book is not afraid of delicate scholarly topics such as chaos theory, the chemistry of Earth's core, and the extinction of dinosaurs. But for Corsi, the movement is primarily a social phenomenon that arose through the integration and evolution of several left-wing factions and ideologies including Malthusians who neglect the human intellect. The tampering of the data by the favored researchers, failed green economic policies, limitations of electric cars and other hyped technologies, as well as natural processes such as the carbon cycle are described in a way that is as detailed and sourced as it is devastating. Corsi's decision to dedicate the book to Marc Morano also proves his profound understanding of this vitally important ongoing conversation."

—**Dr. Luboš Motl**, physicist, former faculty at Harvard University

"Corsi provides an eye-opening history of the contextual background for the hysteria accompanying climate change as well as descriptions of a wide range of thinking about the history of the planet—thinking that illustrates both the insights and fallibility of the scientific enterprise."

—**Dr. Richard S. Lindzen**, professor emeritus of Atmospheric Sciences at the Massachusetts Institute of Technology; former Alfred P. Sloan Professor of Meteorology at the Massachusetts Institute of Technology

"Corsi's book is a must-read to discover just how unserious, inept, and over-certain proponents of global warming of doom are. Their 'solutions,' like battery-powered airplanes, ranging from comical to scandalous. None of them would do a thing to fix what doesn't need to be fixed."

—**Dr. William M. Briggs**, climate statistician who served on the American Meteorological Society's Probability and Statistics Committee

"Many books have been written on what I believe is a phony climate war. Dr. Corsi's treatment of this situation with across-the-board irrefutable facts and impeccable sources is a primer on how to blow away any argument on this matter. His attack seems to be on two fronts: 1) showing there is no emergency and then 2) exposing the why behind the what on the matter, the evil intentions of destroying the freedom of the individual. With twenty-five books under his belt and a man known for standing his ground when the truth is on his line, there is no better source to have than this book. I do not believe there is a climate war (it's nature and not man in control), but I do believe there is a battle raging using climate and weather to take down our way of life. I can think of no better source to have in the trenches than Dr. Corsi and his book. In fact, let's flip that around: I hope I am worthy to just be around someone like this as his writings and actions are light in a world filled with the darkness of non-truths."

—**Joe Bastardi**, chief forecaster at WeatherBELL; author of *The Climate Chronicles* and *The Weaponization of Weather in the Phony Climate War*

"Corsi has achieved the almost impossible. He shows how partisan politics and incompetent climate science have combined to distort energy economics. He describes how different views about population, hydrocarbons, capitalism, and racism have complicated the use of truth in climate science and politics. He shows how nature, not our carbon dioxide, controls the Earth's temperature. Corsi explains why we must think clearly in politics, science, and energy if we are to safeguard our economy, our freedom, and our lives."

—**Dr. Ed Berry**, physicist and American Meteorological Society Certified Consulting Meteorologist

"Corsi's work is sorely needed, exposing the climate lies and disinformation in government, university, environmental and UN official reports, and the media. With a half-century of work on climate and weather attribution studies, I find as Jerome showed incontrovertible evidence that sunspots, cosmic rays, and short and longer term cycles in the oceans are far more important for the climate than carbon dioxide. The true dangers come from the remedies being pushed, not the natural cycles in weather and climate."

—**Joseph D'Aleo**, certified consultant meteorologist; first director of meteorology at the Weather Channel and the former chairman of the American Meteorological Committee on Weather Analysis and Forecasting

Also by Jerome R. Corsi, Ph.D.

*Coup d'État: Exposing Deep State Treason
and the Plan to Re-Elect President Trump*

*Silent No More: How I Became a Political Prisoner
of Mueller's "Witch Hunt"*

*Dr. Corsi Investigates:
Why The Democratic Party Has Gone Communist*

Goodnight Obama: A Parody

THE TRUTH
ABOUT ENERGY, GLOBAL WARMING, and CLIMATE CHANGE

EXPOSING CLIMATE LIES in an AGE OF DISINFORMATION

JEROME R. CORSI, Ph.D.

A POST HILL PRESS BOOK

The Truth about Energy, Global Warming, and Climate Change:
Exposing Climate Lies in an Age of Disinformation
© 2022 by Jerome R. Corsi, Ph.D.
All Rights Reserved

ISBN: 978-1-63758-920-5
ISBN (eBook): 978-1-63758-279-4

Cover design by Mark Karis

No part of this book may be reproduced, stored in a retrieval system, or transmitted by any means without the written permission of the author and publisher.

Post Hill Press
New York • Nashville
posthillpress.com

Published in the United States of America
1 2 3 4 5 6 7 8 9 10

*Dedicated to a Contemporary Warrior for Climate Truth
Marc Morano and ClimateDepot.com,
A Project of the Committee for a Constructive Tomorrow*

Keep Fighting!

Table of Contents

Foreword by Marc Morano xi
Introduction: The Twenty-First Century "Save the Earth" Climate Delusion xiii

PART I: The Politics of Energy, Global Warming, and Climate Change 1

Chapter 1: Julian L. Simon: Eco-Sage and Natural Resources Optimist 3

Chapter 2: John P. Holdren: Eco-Malthusian Wizard Extraordinaire 19

Chapter 3: The Movement to "Reimagine Capitalism" Goes Green 45

Chapter 4: Obama Redux: The "Solyndra Syndrome" 69

PART II: The Science of Energy, Global Warming, and Climate Change 97

Chapter 5: Sun Heats Earth. 99

Chapter 6: Climategate 131

Chapter 7: Cataclysmic Climate Change 162

Chapter 8: The Chaos Theory of Climate. 196

PART III: The Economics of Energy, Global Warming, and Climate Change. . . .247

Chapter 9: Abiotic Oil. 247

Chapter 10: Renewable Energy Sad Realities 306

Conclusion: *Quo Vadimus?* (Where Are We Going?) 327

Endnotes 338
About the Author 397

Foreword

by Marc Morano

I FIRST MET JERRY CORSI back in 2004, and I was immediately struck by how he was a focused and prodigious investigative journalist. Corsi cited my investigative reporting on John Kerry's military service during the Vietnam War in his 2004 *NYT* bestseller with John O'Neill: *Unfit for Command: Swift Boat Veterans Speak out Against John Kerry.*

Corsi has been an investigative machine over the past several decades, exposing, revealing, and debunking the major news events of our time, including books on climate and energy matters.

This book, *The Truth about Energy, Global Warming, and Climate Change: Exposing Climate Lies in an Age of Disinformation on Climate and Energy*, is perhaps Jerry's career tour de force. Jerry masterfully tackles the alleged climate "crisis" and the folly of the green energy "solutions."

I was honored to find out that Jerry dedicated this book to me. I have been working as an environmental reporter since the early 1990s and have been on the climate change beat for over two decades. I battled climate hysteria, groupthink, and the meaningless "solutions" while working in the U.S. Senate Environment & Public Works Committee when Senator James Inhofe of Oklahoma was chairman. It was a pleasure to work for Senator Inhofe, who literally had the courage to stand alone against the climate establishment and oppose then-President Barack Obama's cap-and-trade climate taxes.

I founded the CFACT's Climate Depot website in 2009, and I have strived to make the website serve as a balance for the atrocious climate and energy reporting by the mainstream media. I also produced and appeared in 2016's *Climate Hustle* film and the sequel in 2020, *Climate Hustle 2.*

In a nutshell, anthropogenic climate change threats and so-called "solutions" are one of the biggest cons being imposed on the public in recent decades. It has never been about the climate, energy, or the environment. The climate agenda is about the takeover of our economy using an unscientific climate scare to achieve their ends. The climate scare is a backdoor way for progressives to impose central planning, socialism, and progressivism on the once-free West.

Everything we cherish, from our homes to the foods we eat, the vehicles we drive, our ability to travel, and our freedoms, is at stake if the manufactured climate fear campaign succeeds. Luckily, we don't face a climate emergency, but if we did and had to rely on meaningless United Nations climate pacts or the Green New Deal to save us—we would all be doomed.

It is a pleasure to be part of this book and to have Jerry's great reporting acumen once again injecting science and logic into the climate change and energy debate. Jerry spares no aspects of the climate and energy debate and takes a deep dive into the complexities of the issues. He is unafraid to reexamine controversial scientific theories that may ruffle feathers on all sides of the climate and energy debate.

The goal is to get Jerry's book into the hands of as many citizens, journalists, and policymakers as possible. Only armed with the facts can we unite and defeat the well-funded, embedded, and scientifically twisted climate change movement.

INTRODUCTION

The Twenty-First Century "Save the Earth" Climate Delusion

No longer are cap-and-trade, carbon (dioxide) taxes, and more solar and wind the promoted solutions to alleged global warming. Now we can add gender justice and defunding the police!
—**Marc Morano**, *Green Fraud*, 2021[1]

In this book, you will learn why most of what you think you know about energy—and what our kids are being taught about energy— is flat-out wrong. In one of the worst ironies of history, a frantic global movement to eliminate fossil fuels—the foundation of modern life— has achieved comprehensive power throughout the developed world at the very moment when the supply of those resources, especially in the United States, has exploded.
—**Stephen Moore** and **Kathleen Hartnett White**, *Fueling Freedom*, 2016[2]

Having first experienced and then studied the phenomenon for fifteen years, I believe that secular people are attracted to apocalyptic environmental movements because it meets some of the same psychological and spiritual needs as Judeo-Christianity and other religions. Apocalyptic environmentalism gives people a purpose: to save the world from climate change, or some other environmental disaster. It provides people with a story that casts them as heroes, which some scholars, as we will see, believe we need in order to find meaning in our lives.
—**Michael Shellenberger**, *Apocalypse Never*, 2020[3]

THE TRUTH ABOUT ENERGY

In 1895, French conservative thinker Gustave Le Bon wrote a seminal book entitled *The Crowd: A Study of the Popular Mind*.[4] In the introduction to that book, Le Bon clarified that profound changes in people's ideas are the actual cause of great upheavals that preceded civilization changes, like the fall of the Roman Empire. He explained that the transformation humanity was then experiencing had a base cause in the destruction of the religious, political, and social beliefs that grounded civilization. He felt his era was in transition and anarchy as modern scientific and industrial discoveries created new conditions of existence. He observed that the past ideas, although half destroyed, were still compelling, while the new ideas replacing them were yet in the process of transformation. Le Bon could have written that exact introduction today.

In chapter 4, "A Religious Shape Assumed by All the Conviction of Crowds," Le Bon expressed his disdain for democracy. He felt crowds whipped democracies through irrational historical moments where bizarre secular ideas assumed a religious-like popular devotion. The following paragraph from chapter 4 summarized his concerns as follows:

> We have shown that crowds do not reason, that they accept or reject ideas as a whole, that they tolerate neither discussion nor contradiction, and that the suggestions brought to bear on them invade the entire field of their understanding and tend at once to transform themselves into acts. We have shown that crowds suitably influenced are ready to sacrifice themselves for the ideal with which they have been inspired. We have also seen that they only entertain violent and extreme sentiments, that in their case sympathy becomes adoration, and antipathy almost as soon as it is aroused is transformed into hatred. These general indications furnish us already with a presentiment of the nature of the convictions of crowds.[5]

Today, the Western world is in the grip of a similar turmoil caused by the idea that we are our greatest enemy. The self-hatred extends to the belief that we are also the enemy of our mother, Earth.

This self-hatred focuses on a molecule, carbon dioxide (CO_2), which we despicable humans exhale. Even worse, we desecrate organic life itself by burning fossil fuel, releasing into the atmosphere more CO_2 the earth had preserved from living organisms that had passed away through the ages. Powering our industrial society with these hated hydrocarbon fuels, we have created an economic system, capitalism, that is inherently evil. The evil of capitalism extends social injustice to new heights as the racially privileged white race perpetuates their luxury by subjugating people of color and emitting enough CO_2 into the atmosphere to destroy the planet. The only way to save planet Earth, and in the process protect ourselves, is to decarbonize. But even that is not enough unless we also dismantle capitalism, supplanting the economics of greed with a new vision of living and working together without prejudice to sustain Earth's limited resources for the benefit of all.

In 1841, Scottish journalist Charles Mackay wrote another seminal book, *Extraordinary Popular Delusions and the Madness of Crowds*.[6] Mackay, like Le Bon, was fascinated by how crazed ideas can drive whole populations into actions motivated by a bizarre, self-destructive, mass psychosis that is hard to comprehend. In the preface to the 1852 edition of his book, Mackay boiled his thesis down to the following sentence:

> We find that whole communities suddenly fix their minds on one object, and go mad in its pursuit; that millions of people become simultaneously impressed with one delusion, and run after it, till their attention is caught by some new folly more captivating than the first.[7]

Mackay focused his attention on a series of fascinating crazes. He puzzled over the tulip craze that prompted Hollanders in the 1600s to spend fortunes on exotic roots producing color variations of the famous flower. He was amazed at the Crusades where Europeans left their homes and families to seize the Holy Land for Christianity. He was astounded by the grotesque witch mania during which those believed to be possessed by Satan were hunted down and made to suffer horrific deaths. Today, the Western world is on the precipice of abandoning the economic progress hydrocarbon energy has fueled since the Industrial Revolution to save Earth from catastrophic warmth by reducing the greenhouse gas emissions that plants depend upon for life.

In 2007, noted British columnist Christopher Booker and political analyst Dr. Richard North coauthored a book that took up these themes, entitled *Scared to Death*.[8] Booker and North marveled that Western society since the 1980s had been "in the grip of a remarkable and very dangerous psychological phenomenon."[9] Booker and North were astounded that one mysterious threat to human health and well-being after another gave rise to society-wide fear. The list of these fear crazes was extensive: salmonella in eggs, listeria in cheese, bovine spongiform encephalopathy in beef, dioxins in poultry, and so forth. In the following paragraph in their introduction to the book, Booker and North identified why these periods of psychological insanity continue to occur among today's supposedly well-educated and technologically sophisticated populations:

> Each was based on what appeared at the time to be scientific evidence that was widely accepted. Each has inspired obsessive coverage by the media. Each has then provoked a massive response from politicians and officials, imposing new laws that inflicted enormous economic and social damage. But eventually the scientific reasoning on which the panic was based has been found to be fundamentally flawed. Either the scare originated in some genuine threat that had become widely exaggerated, or the danger was found never to have existed at all.[10]

Booker went on to examine the pattern behind these scares, finding the elements in common. One was that the supposed danger had to be something universal, to which we might all be exposed, like global warming and climate change. The threat must be novel, like the assumption that the developed world is bent on warming up Earth to hazardous levels by burning more and more hydrocarbon fuels until they are all exhausted. The threat must be plausible, but there must also be a powerful element of uncertainty. The uncertainty allows alarmist speculation to run wild, imaging the damage that might result, e.g., the warming of Earth caused by anthropogenic CO_2 until Earth is hazardous to human life. Finally, society's response to the threat must be disproportionate, e.g., when the United Nation's Intergovernmental Panel on Climate Change (IPCC) demands that all governments agree to implement decarbonization schemes devised by international agreements, e.g., the Paris Climate Accords. Even when the threat is not wholly imaginary, the response eventually seemed out of proportion to reality.

The dawning of the Internet promised to bring a new era of free speech. The open access to information quickly developed into easily created blogs to express dissident views of all kinds. But ironically, the technology welcomed as a liberating tool has transformed into a tool giving totalitarian governments increased ability to suppress speech that deviates from the government-approved version of the truth. The United States justice and intelligence agencies now can monitor all electronic communications, including the keystrokes made on a laptop computer in writing a book.

Australian geologist Professor S. Warren Carey, who propounded the expanding Earth theory we will examine in chapter 8, warned that challenging orthodox beliefs in science promised no glory. In the epilogue to the 1988 book *Theories of the Earth and Universe* that he wrote as a professor emeritus, Carey warned that "the more radical the advance from the current orthodoxy, the more certain will it be scorned and rejected."[11] Carey understood this in personal terms. Carey suffered the scorn of those geologists, who were wedded to plate tectonics as their continent formation paradigm, for articulating and defending his theory of an expanding Earth. Yet, Carey had the wisdom to understand that not all challenges to orthodox thinking are necessarily correct in their views. In his last paragraph to the book, he wrote the following:

> Should we then give credence to every heretic and iconoclast with the naïveté or the zeal or persistence to challenge the established order? Of course not! Most heresy is doubtlessly false—yet latent there are the gems of the age. To discriminate unerringly within doctrine and within heresy needs a keener mind than any yet—but this must be our ever-unattainable goal.[12]

I have dedicated this book to Marc Morano, the creator of ClimateDepot.com. For decades now, Marc has challenged global warming and climate change orthodoxy. International global warming conferences have thrown Marc out and closed their doors to him. Books by global warming enthusiasts have printed the vilest denunciations of Marc's views and arguments. Marc has suffered the scorn Carey warned was inevitable for those who do not go along with what we will argue in this book is a mass delusion of Charles Mackey proportions.

Yet, Marc Morano has persisted, determined to pursue scientific truth about the climate with a purpose to prevent the Western world, and in particular the United States, from committing economic and political suicide over a scientific hoax of historic proportions. When the government mandates and subsidies run out, the fields of rotting wind turbines and rusting solar panels will be a monument to the folly of decarbonization. Should a new ice age come within our lifetimes, we will be around to lament the folly that sought to reduce atmospheric CO_2 by destroying capitalism. But should that day arrive, we fully expect the IPCC to blame the new ice age on the global warming that anthropogenic CO_2 caused.

PART I

The Politics of Energy, Global Warming, and Climate Change

CHAPTER 1

Julian L. Simon: Eco-Sage and Natural Resources Optimist

Why is there so much false bad news about the subjects of the environment, resources, and population?... An even tougher question is this one: Why do we believe so much false bad news about the environment, resources, and population?

—Julian Simon, *Hoodwinking the Nation*, 1999[1]

AFTER A CAREER AS AN ECONOMICS and business professor, Julian Simon passed away prematurely at sixty-five years old in 1998 in Chevy Chase, Maryland. At the end of his life, Simon held a position as a senior fellow at the Cato Institute in Washington, D.C., his last job after a longtime career at the University of Illinois Urbana-Champaign, followed by an academic position at the University of Maryland. Born in 1932 in Newark, New Jersey, and educated at Harvard University, Simon received his Ph.D. in business economics from the University of Chicago in 1961. Among "Green Energy" true believers, Simon has become infamous for taking a contrarian position on energy resources, arguing that our perception of scarcity is a psychological fear, one not validated by the current or historical factual record of energy abundance.

In the 1999 foreword to Simon's first book to be published posthumously, *Hoodwinking the Nation*, author Ben Joseph Wattenberg, then a senior fellow at the American Enterprise Institute in Washington, D.C., commented that Simon often felt angry that he was being ignored or ridiculed by opponents who belonged to a vast Malthusian population-environment resources conspiracy of crisis. Today, Malthusians have captured the politically correct mainstream media, rejecting Simon's contention that supplies of natural resources, including energy, are not finite and exhaustible. Simon saw the human intellect as the ultimate, infinitely renewable resource, and its potential as unlimited. He argued we would never run out of energy resources, including oil, coal, and natural gas, provided our energy resources are "mixed...with intellect."[2]

What distinguishes Simon from the Malthusians was that Simon saw human beings as the solution, not the problem. In direct contrast, Malthusians see human beings as a menace that threatens the very survival of the planet itself. Wattenberg understood this precisely, noting the attacks on Simon were often intensely personal. Simon's detractors demeaned him by stating his doctorate was "merely in business economics" and that he taught business-oriented subjects like advertising and marketing. Simon was ridiculed for starting a mail-order business and daring to write a book on how to run a successful and profitable one. "Never mind that he studied population economics for a quarter of a century and the mail-order book is still in print and in its fifth edition," Wattenberg commented.[3] Simon was perplexed that the environmental movement did not appreciate his extensive research and many publications about natural resources. What drove the "enviros" crazy, Wattenberg explained, was the following:

> But, irony again, it was Simon's knowledge of real-world commerce that gave him an edge in the intellectual wars. He knew first-hand about some things that many environmentalists of the time had only touched gingerly, like prices. If the ultimate resource was the human intellect, Simon reasoned, and the amount of human intellect was increasing both qualitatively and quantitatively, thanks to population growth, education, and technology, why, then, the supply of resources would grow, outrunning demand, pushing prices down, giving people more access to what they wanted, with more than enough left over to deal with pollution—in short, the very opposite of a crisis.[4]

Wattenberg calculated correctly that Simon's knowledge of the business world gave him an edge over the Malthusians in the intellectual wars. Suppose Simon is correct that the ultimate human resource was the human intellect. In that case, Wattenberg argued, it could also be right that our supply of natural resources would grow over time, outpacing demand, pushing prices down. Wattenberg correctly understood that Simon's vision is a severe threat to the supposed crisis in natural resources that the

Julian L. Simon: Eco-Sage and Natural Resources Optimist

Malthusians desperately want us to believe is inevitable. Simon's argument is simple: scientifically proven facts contradict the Malthusian doom-and-gloom narrative we see pervasive today in popular culture.

Appropriately, Simon titled his autobiography, published posthumously, *A Life Against the Grain: The Autobiography of an Unconventional Economist*.[5] Contrary to everything Simon argued in his numerous published writings, today's politically correct popular culture demands universal acquiesce to the proposition that human beings have created the conditions of our demise as a species. About energy resources, the politically correct popular culture requires an agreement that our wanton burning of hydrocarbon fuels has tossed so much toxic carbon dioxide into the atmosphere that we have created a greenhouse effect that will result in catastrophic climate change.

In characteristic prose, Simon began *Hoodwinking the Nation* with an essay summarizing the human history of utilizing natural resources, including energy as follows:

> Every resource economist knows that all natural resources have been getting more available rather than more scarce, as shown by their falling prices over the decades and centuries. And every demographer knows that the death rate has been falling all over the world; life expectancy almost tripled in the rich countries in the past two centuries and almost doubled in the poor countries in just the past four decades. This is the most important and amazing demographic fact—the greatest human achievement in human history. It took thousands of years to increase life expectancy at birth from just over 20 years to the high 20s about 1750. Suddenly, about 1750, life expectancy in the richest countries began to rise so that the length of a life that could be expected for a baby or an adult in the advanced countries jumped from less than 30 years to perhaps 75 years. Then starting well after World War II, the length of life that could be expected in the poor countries leaped upwards by perhaps 15 or even 20 years because of advances in agriculture, sanitation, and medicine. It is this decrease in the death rate that has caused there to be a larger world population nowadays than in former times.[6]

Today, the politically correct mainstream media would brand anyone daring to publish an argument favoring continued global use of hydrocarbon fuels as an "environmental lunatic" or possibly even an "ecological criminal." At the end of his life, Simon realized his optimism regarding the human capacity to utilize natural resources for our betterment as a species would brand him as a fringe nut case. "I was not cut out to be a Mafia boss," Simon wrote in the preface to his autobiography. "I am more like a competent and hard-working plumber or building contractor or burlesque-show baggy pants comedian, though I have more kooky ideas than most of them."[7] Yet, throughout his life, Simon insisted the results of his studies and his writings would turn out to be correct.

Over the years, I managed to acquire a student-used copy of Simon's 1981 book, *The Ultimate Resource*.[8] On the title page of the book, the student handwrote her assessment of Simon's work: "[The author is] a rich white male who has never left his office—world of graphs, equations, and charts that he bases all his theories on. Graphs, e.g., charts that are not comprehensive and only tell if population is up, if aggregate output is up, if fertility, mortality is up…but none of the other factors—environmental consequences, inequalities, humans are a resource—no limits to their abilities and innovations. Exploit the Earth and other planets if necessary to serve humans, income up…no intrinsic value in nature—only there to serve man." The polemical tone of these comments clarified that already by the 1980s, these arguments on the left were entering the realm of ideology.

Reading those comments today, I am not surprised that in this age of the neo-Marxist critical race theory, the student began her analysis of Simon's work with an ad hominem attack, pointing out that he was a white man and an academic? The student dismissed the research Simon documented in the book by insisting today's natural resource policies have produced no adverse environmental consequences and economic inequality. So, what system would the student have preferred? Would using fewer resources to preserve a more pristine environment be better, even at the cost of shortening life expectancies? Would that have been fairer to all races, all sexes, all cultures, and all religions? Today university courses rarely teach Simon's economics. Why? Because he refused to accept the orthodox conviction that we humans apply our limited intellects only to exploit, for our selfish good, the precious and scarce natural resources of our mother, Earth.

The Malthusian view has convinced millions that Earth has entered a new and final hypothesized era of geological time, the Anthropocene era. Malthusians insist that anthropogenic carbon dioxide will cause such catastrophic global warming and subsequent climate change that human activity is responsible for bringing about a coming sixth extinction. Malthusians argue that the sixth extinction will dwarf the previous "Big Five" extinctions in which nearly all life on Earth disappeared, rivaling even the giant meteorite that caused the extinction of the dinosaurs in the Late Cretaceous Period, some sixty-six million years ago. Malthusians warn us that the sixth extinction will be the last this time, and we will have no one to blame but ourselves.

Simon took a lot of abuse during his life for running against the politically correct popular culture by not adhering to Malthusian views. However, in his final analysis, Simon understood it was more important to be right about natural resources than to have a mass audience applaud his genius. That was especially true when we appreciate that the need to "decarbonize" and move to a "zero emissions" world if we are to "Save the Planet" are all views Julian Simon found hopelessly uninformed.

Why We Will Never Run Out of Oil

In his revised 1996 book, *The Ultimate Resource 2*, Julian Simon devoted chapter 11 to the question: "When Will We Run Out of Oil?" In the chapter title, Simon gave a one-word answer to his question: "Never!" Simon argued that energy is the master resource because "energy allows us to convert one material to another."[9] He argued that the low energy costs afforded by hydrocarbon fuels enable modern technological society to thrive. "On the other hand, if there were to be an absolute shortage of energy—that is, if there were no oil in the tanks, no natural gas in the pipelines, no coal to load onto the railroad cars—then the entire economy would come to a halt," he wrote. "Or, if energy were available, but at a very high price, we would produce much smaller amounts of most consumer goods and services."[10] Simon proceeded to elaborate: "The history of energy economics shows that, in spite of troubling fears in each era of running out of whatever source of energy was important at that time, energy has grown progressively less scarce, as shown by long-run falling energy prices."[11]

Simon traced fears of energy resource exhaustion back to an 1865 book published in London by W. Stanley Jevons, one of the nineteenth century's most outstanding social scientists, entitled *The Coal Question: An Inquiry Concerning the Progress of the Nation and the Probable Exhaustion of our Coal-mines*.[12] Jevons argued Great Britain's industrial progress would halt because industry would soon use all available coal. Jevons filled his book with detailed analyses of coal mines showing mine by mine the estimated amount of coal remaining, the annual consumption of that coal (depletion ratio), and the duration of the supply. He anticipated with uncanny precision the bell-shaped curve that in the next section of this chapter we will see was typical of M. King Hubbert's 1950s peak oil graphs. In his despair that the U.K. would soon run out of energy, Jevons further concluded (obviously incorrectly) that there was no chance oil would be an alternative resource able to solve the running-out-of-coal problem.

"What happened to Great Britain in 1865?" Simon asked. "Because of the perceived future need for coal and because of the potential profit in meeting that need, prospectors searched out new deposits of coal, investors discovered better ways to get coal out of the earth, and transportation engineers developed cheaper ways to move the coal," Simon explained.[13] Today, the U.K. still has thirty-three tons of economically recoverable coal reserves available at operational and legally permitted mines, plus another 344 tons at mines in planning. The use of coal in the U.K. had declined from a leading position in 1990, when coal accounted for 64.6 percent of the U.K.'s energy needs, to last place today, providing only 4.4 percent of the country's current energy needs.[14]

The reduced use of coal to produce energy in the U.K. has primarily resulted from the "Green Energy" politics there. Many coal-fired power plants in the U.K. have been closed in recent years, mainly due to the country's carbon taxes. Carbon taxes on coal-generated power plants have doubled under the U.K.'s carbon price support mechanism, which began placing punitive levies on coal-fired electrical generation in April 2013. Contrary to Jevons's expectations in 1865, the U.K. is nowhere near running out of coal after more than a century in which coal was the U.K.'s principal source of energy. Today, the use of coal for power in the U.K. is severely limited. England has not run out of coal, but neo-Marxist politics in the U.K. have focused on eliminating coal-fired electric plants as part of their unrelenting campaign to demonize the use of all hydrocarbon fuels.[15]

Similarly, Simon traced similar fears in the United States back to an 1885 U.S. Geological Survey that declared "little or no chance" of finding oil in California. In 1939, the U.S. Department of the Interior argued that U.S. oil resources would be exhausted in thirteen years. When that prediction proved a false alarm, the Interior Department revised their estimate and declared once again in 1951 that U.S. oil would be exhausted in thirteen years. All these dire oil deprecation predictions were wrong.[16]

Simon articulated many reasons why gloomy predictions about running out of coal, oil, natural gas, or any other energy resource can be presumed wrong. We summarize Simon's thinking in the following points:

- Typically, all energy resources exist on Earth in quantities much more extensive than initially estimated.
- Productivity improvements lead to more efficient use of energy resources over time.

- Advances in technology make the exploration and recovery of previously difficult-to-develop energy resources more efficient and more economically affordable.
- Innovators and entrepreneurs will always find alternative sources of energy, even while predominately used energy resources remain abundant.
- Previously dominant energy resources, such as coal, become less prevalent as more efficient energy resources, such as oil, become more understood and utilized. Simon believed liquefied natural gas would replace many oil uses, culminating in new, safer nuclear energy technologies that ultimately replace many current uses of coal, oil, and natural gas.

Simon's energy resource analysis essentially maintains that we will be running automobiles with safe miniaturized nuclear batteries (or with yet-to-be-developed safe, portable, and efficient fission technology) long before we run out of oil. Today, the U.S. Navy runs its various fleets of ships, including submarines, predominately on nuclear power. Simon wrote: "Of course nuclear power can replace coal and oil entirely, which constitutes an increase in efficiency so great that it is beyond my powers to portray the entire process on a single graph based on physical units."[17] Simon concluded this discussion by noting that while it seems impossible to keep using energy and still never begin to run out, that is the truth of what happens. He said that the "historical facts entirely contradict the commonsensical Malthusian theory that the more we use, the less there is left to use and hence the greater the scarcity." He added that in economic terms, "energy has been getting more available, rather than more scarce, as far back as we have data."[18]

Worldwide petroleum reserve statistics compiled by the Energy Information Administration (EIA) of the U.S. Department of Energy prove Simon's optimism that we will not run out of oil is well justified. According to EIA statistics, worldwide petroleum reserves totaled over 1.6 trillion barrels in 2020, factually demonstrating there are more proven crude oil reserves today worldwide than ever in recorded history, despite the worldwide consumption of oil doubling since the 1970s.[19]

M. King Hubbert and the Theory of Peak Oil

Peak oil true believers regard Shell Oil geologist M. King Hubbert as their theoretical deity. In 1956, Hubbert drew a bell-shaped curve that he said showed U.S. oil production peaking in the 1970s and declining from there until U.S. oil would be nearly depleted in 2050. Subsequently, Hubbert's adherents expanded his analysis into a worldwide prediction that we are inevitably doomed to run out of oil.

Born in San Saba, Texas, in 1936, Hubbert was too young to fight in World War I and too old to fight in World War II. Hubbert attended the University of Chicago, where he received a Ph.D. in 1937. During World War II, Hubbert served on the U.S. government's Board of Economic Warfare. In 1943, he joined Shell Oil Company, where he developed his peak oil theory. In his professional career, Hubbert worked as a highly respected geologist for oil companies while teaching geophysics at Columbia University. Upon retiring from Shell Oil in 1964, he served as a senior research geophysicist for the United States Geological Survey until his retirement in 1976. In his later years, he held positions as a professor of geology and geophysics at Stanford University and subsequently at UC Berkeley. Hubbert was well respected, and his educational background was extensive, given that his studies included advanced work in mathematics, physics, and geology.

Throughout his career, Hubbert published various professional papers in academic journals dealing with multiple aspects of Earth's crust, including studies of rock permeability as it affects underground oil and water reservoirs. Hubbert's academic publications were commonly cited in the university textbooks of the day. For instance, A. I. Levorsen, an American geologist who served as the dean of the School of Mineral Sciences at Stanford University, acknowledged Hubbert's work. Levorsen, in his 1954 college-level textbook entitled *Geology of Petroleum*, cited two academic papers Hubbert had in 1940 and then in 1953 on the subject of oil and water movements in defining oil traps in sedimentary rock.[20]

In 1956, at the spring meeting of the American Petroleum Institute in San Antonio, Texas, Hubbert presented a paper on his seminal work on peak oil, arguing oil was a finite natural resource such that oil production would peak in the United States between 1965 and 1970. It was his most famous paper called "Nuclear Energy and the Fossil Fuels," written as a consultant in general geology to the Shell Development Company, Exploration and Production Research Division in Houston, Texas.[21] Hubbert argued that oil depletion would accelerate such that eventually, the world supply of oil would be exhausted. Almost instantly, Hubbert's peak oil theory became the universally definitive oil theory among petroleum geologists.

Hubbert premised his 1956 paper by embracing the idea that hydrocarbon fuels are all organic products. He explained this as follows:

> The fossil fuels, which include coal and lignite, oil shales, and tar and asphalt, as well as petroleum and natural gas, have all had their origin from plants and animals existing upon the Earth during the last 500 million years. The energy content of these materials has been derived from that of the contemporary sunshine, a part of which has been synthesized by the plants and stored as chemical energy. Over the period of geological history extending back to the Cambrian, a small fraction of these organisms have become buried in sediments under conditions which have prevented complete deterioration, and so, after various chemical transformations, have been preserved as our present supply of fossil fuels.

He continued to explain why fossil fuels would inevitably be exhausted:

> When we consider that it has taken 500 million years of geological history to accumulate the present supplies of fossil fuels, it should be clear that, although the same geological processes are still operative, the amount of new fossil fuels that is likely to be produced during the next few thousands of years will be inconsequential. Therefore, as an essential part of our analysis, we can assume with complete assurance that the industrial exploitation of the fossil fuels will consist in the progressive exhaustion of an initially fixed supply to which there will be no significant additions during the period of our interest.[22]

For Hubbert, these conclusions were obvious. He used historical graphs to show that the production of all hydrocarbon fuels had increased over time. He then applied an integral calculus function to assume hydrocarbon production began seriously in the 1850s, starting at a zero point of production. From there, he reasoned hydrocarbon fuel production would end sometime soon. At that time, hydrocarbon fuels would be thoroughly exhausted, such that hydrocarbon fuel production would again return to zero. This analysis produced a bell-shaped curve from which Hubbert deduced there was a limit to the rate of increase in which hydrocarbon fuels could be produced. Once we reached the maximum point of oil production, hydrocarbon depletion would accelerate. The production rate would begin decreasing, finally ending up at zero production once all hydrocarbon fuels on Earth had been thoroughly mined or otherwise exploited to the point of exhaustion. By examining available estimates of known and anticipated world reserves of hydrocarbon fuels, Hubbert calculated "the culmination of world production of these products should occur within about half a century [i.e., by approximately the year 2000], while the culmination for petroleum and natural gas in both the United States and the state of Texas should occur within the next few decades."[23]

Hubbert ended his 1956 paper assuming energy from nuclear sources would begin ascending in importance starting in 1980, such that by 2060 nuclear fuel would power a world that had exhausted all petroleum resources available. On a chart that began 5,000 years ago, at the dawn of recorded history, to a point 5,000 years in the future, Hubbert ended the paper commenting that on this time scale, "the discovery, exploitation, and exhaustion of the fossil fuels will seem to be but an ephemeral event in the span of recorded history." He felt confident that nuclear fuel would be the solution, "provided mankind can solve its international problems and not destroy itself with nuclear weapons and provided the world population (which is now expanding at such a rate as to double in less than a century) can somehow be brought under control."[24]

The Demise of the Peak Oil Theory

The logical structure of Hubbert's peak oil theory is a tautology. His conclusion is nothing more than a restatement of the assumptions he postulated as his starting point. By assuming oil and all other hydrocarbon fuels are organic "fossil fuels," Hubbert had no choice but to conclude the world would eventually deplete hydrocarbon fuels to the point of exhaustion. Hubbert assumed ancient organic material in the form of plant life produced oil, not dinosaurs. But since the supply of ancient plant life was finite, hydrocarbon fuels also had to be limited.

Yet, the logical structure of the argument as a tautology remains intact. Suppose a finite amount of ancient organic material was available in geological time (regardless of whether the organic material was plant or animal). In that case, there can only be a limited amount of hydrocarbon fuels available on Earth, even if we cannot ever know for sure how many hydrocarbon fuels are yet to be discovered. Since the peak oil theory was based entirely on logic, Hubbert felt no need to prove that ancient organic material can transform into hydrocarbon fuels in sedimentary rock structures. He simply assumed hydrocarbon fuels were "fossil fuels." Nor did he feel he had to know for sure the exact amount of hydrocarbon fuels truly available today and in the future. He just assumed we would run out of hydrocarbon fuels because there were only so many plants and animals on Earth in geological time. For those who believe in peak oil, there is no way to refute the argument. When hydrocarbon production fails to peak at the predicted time, adherents of the peak oil theory simply revise their predictions to move the depletion dates out to a more distant time in the future. The point is that the logic that hydrocarbon fuels come from prehistoric organic material demands we conclude the quantity of hydrocarbon fuels available on Earth has to be limited.

We should also appreciate that the peak oil theory developed by Hubbert as a logical tautology has a psychological impact. Once we accept that hydrocarbon fuels, including coal, oil, and natural gas, are organic products of prehistoric time, we lock ourselves into a Malthusian fear that inevitably we must run out, if not today, then tomorrow. The complete psychological impact is that as population increases and the world becomes more dependent on burning hydrocarbon fuels, we become the cause of our demise. We are doomed because we have locked ourselves into the conclusion that the tautology demands: namely, that hydrocarbon fuels of ancient organic origin are, by definition, not renewable.

To both Julian Simon and M. King Hubbert, nuclear fuels were the ultimate energy solution because, again, by definition, nuclear fuels are renewable. The psychological rub is that for Malthusians, the nuclear energy solution is not psychologically satisfying. Malthusians view nuclear energy as inherently dangerous because it involves hazardous, radioactive energy. The Malthusian "solution" to their hypothesized exhaustion-of-natural-resources doomsday scenario demands finding a limit to the human experience.

"Hubbert's Peak" was the label that peak oil advocates derived Hubbert's famous bell curves. One of the more interesting critics of Hubbert's Peak logic was the prominent oil and gas analyst Michael C. Lynch, known for his record of producing long-term oil and natural gas market forecasts.[25] In a 2010 published paper entitled "The New Pessimism about Petroleum Resources: Debunking the Hubbert Model (and Hubbert Modelers)," Lynch argued that Hubbert's initial analysis was anything but rigorous or scientifically formal, even though Hubbert documented his 1956 paper with numerous graphs and equations:

> The initial theory behind what is now known as the Hubbert curve was very simplistic. Hubbert was simply trying to estimate approximate resource levels, and for the lower-48 US, he thought a bell-curve would be the most appropriate form. It was only later that the Hubbert curve came to be seen as explanatory in and of itself, that is, geology requires that production *should* follow such a curve. Indeed, for many years, Hubbert

himself published no equations for deriving the curve, and it appears that he only used a rough estimation initially. In his 1956 paper, in fact, he noted that production often did not follow a bell curve. In later years, however, he seems to have accepted the curve as explanatory.[26]

One of those who agreed with Lynch was Kenneth Deffeyes, who went to work at Shell Oil's research lab in 1958 when Hubbert was the top dog. Despite his admiration for Hubbert, Deffeyes had to acknowledge that Hubbert's peak oil argument had the feel of a "back-of-the-envelope" drawing. In his 2001 book, *Hubbert's Peak: The Impending World Oil Shortage*, Deffeyes, then a professor at Princeton, recalled Hubbert at Shell Oil as follows:

> The numerical methods that Hubbert used to make his predictions are not crystal clear. Today, 44 years later, my guess is that Hubbert, like everybody else, reached his conclusion first and then searched for raw data and methods to support his conclusion. (Despite sharing roughly 100 lunches and several long discussions with Hubbert, I never had the guts to cross-examine him about the earliest roots of his prediction. Lunch discussions were more cheerful when Hubbert chose the topic.) Guessing the answer first and then searching for supporting arguments is a common scientific procedure; it is not cheating. Hubbert had a message; he packaged his message in a format that he found convincing.[27]

Yet, despite any reservations he may have had, Deffeyes could not resist the stampede as Hubbert's peak oil theory became dogma among mainstream professional geologists working in the oil industry. Even when Hubbert's original prediction that oil depletion would begin between 1965 and 1970 was proved wrong, adherents like Deffeyes just kept moving the goalposts further out in time. For instance, in the first paragraph of his 2005 book, *Beyond Oil: The View from Hubbert's Peak*, Deffeyes boldly predicted that world production of crude oil would peak on Thanksgiving Day 2005. He wrote:

> The supply of oil in the ground is not infinite. Someday, annual world crude production has to reach a peak and start to decline. It is my opinion that the peak will occur in late 2005 or in the first months of 2006. I nominate Thanksgiving Day, November 24, 2005, as World Oil Peak Day. There is a reason for selecting Thanksgiving. We can pause and give thanks for the years from 1901 to 2005 when abundant oil and natural gas fueled enormous changes in our society. At the same time, we have to

face up to reality: World oil production is going to decline, slowly at first then more rapidly.[28]

But the critical point here is that Deffeyes was wrong. World oil production did not reach a zenith on Thanksgiving Day 2005, nor anytime soon after that.

Peak Oil Theorists Fold Their Tent

Another prominent peak oil adherent is the British petroleum geologist Colin J. Campbell. He was an internationally respected petroleum geologist born in 1931 and received a Ph.D. in geology from Oxford. Among his most influential papers was an article he coauthored with Jean Laherrère entitled "The End of Cheap Oil" that *Scientific American* published in 1998.[29] Like Campbell, Laherrère had also spent more than forty years working in the oil industry. In 2000, Campbell founded the Association for the Study of Peak Oil and Gas (ASPO). At its height, ASPO published a newsletter (mainly authored by Campbell) that published one hundred issues, with the last one published in 2009.

In 2019, Ugo Bardi, a professor of chemistry at the University of Florence in Italy, published a paper in *Energy Research & Social Science*. He credited Campbell with proposing the term *peak oil* for the highest global oil production level. In his article, Bardi explained why Campbell, his theory of peak oil, and ASPO have folded their tents and essentially disappeared. In his article, Bardi reported the following:

> The expected world peak has not arrived, at least in terms of a reduction of the global supply of liquid fuels and, in general, the concept of peak oil has faded from the mainstream discussion as well as from the scientific literature. ASPO international seems to have disappeared as an active association around 2012-2013, although some national branches of the organization still exist. The generally accepted explanation for the fading interest in the concept attributes it to "wrong predictions" of the date of the peak and, from there, most mainstream reports tend to define the whole concept as wrong and misleading.[30]

Under President Donald Trump's strong support of hydrocarbon fuels, the United States defied all peak oil predictions, becoming once again a net exporter of oil and a world leader in the production of oil and natural gas.

On August 20, 2019, the Energy Information Administration (EIA), the statistical agency of the U.S. Department of Energy, announced that the United States established new production records, with U.S. petroleum and natural gas production increasing in 2018 by 16 percent and 12 percent, respectively. The EIA further announced that the United States surpassed Russia in 2011 to become the world's largest natural gas producer and surpassed Saudi Arabia in 2018 to become the world's largest petroleum producer. The EIA report noted that the 2018 increase in the United States, which boomed under the Trump administration, constituted "one of the largest absolute petroleum and natural gas production increases from a single country in history."[31]

On October 9, 2020, the EIA further reported that the United States led the world by holding 22 percent of the world's proven coal reserves. Russia was in second place with 15 percent of the world's proven coal reserves.[32] On April 13, 2021, the EIA reported that in 2020, the last year of the Trump administration, the United States became an annual net exporter of petroleum, a position the peak oil advocates never imagined would ever again be possible. The same EIA report noted that in 2020, the United States imported the least amount of petroleum since 1991. "After generally increasing every year from 1954 through 2005, U.S. total gross and net petroleum imports peaked in 2005," the EIA noted. "Increases in domestic petroleum production and in petroleum exports helped to reduce total annual petroleum net imports every year except one since 2005. In 2020, annual petroleum net imports were negative (at -0.65 million barrels/day), the first time this occurred since 1949."[33]

Conclusion

These U.S. achievements in producing hydrocarbon fuels again validate Julian Simon's analysis that despite peak oil predictions, the U.S. was not running out of oil, natural gas, and coal. As Simon would have pointed out (had he lived to see the U.S. resurgence in hydrocarbon fuel production): first, the U.S. had more reserves than were previously estimated; and second, technological advances, including fracking and hydraulic drilling, have made it economically feasible to obtain energy resources that were once largely unexplored and typically underdeveloped.

The peak oil theory has been debunked by the resurgence of the U.S. petroleum industry under President Trump.[34] In 2019, the Energy Information Administration published five charts that showed how wrong doomsday projections had been. The forecast in 2010 was that by 2019, we would have 5.8 thousand metric tons of CO_2 emissions in the atmosphere; the actual figure was 5.3 thousand metric tons. The projection in

2010 was that U.S. natural gas in 2019 would be 19.8 trillion cubic feet: the U.S. natural gas production in 2019 was 30.6 trillion cubic feet. The projection in 2010 was that the U.S. in 2019 would produce 6.1 million barrels of oil per day; the U.S. oil production in September 2019 was 12.5 million barrels per day.

Yet, the failure of oil to be completely depleted, as the peak oil theory predicted, has done nothing to destroy the organic theory of the origin of oil—the premise upon which petroleum doomsayers like Colin Campbell based their peak oil proclamations. In its heyday, the adherents of peak oil were highly opinioned, dismissive of any critic who dared challenge their rigid beliefs that the world had to be running out of oil. In truth, the peak oil theory was nothing more than an ideology based on a tautology, a secular, religious-like belief that tolerated no discussion or criticism. Yet despite the demise of the peak oil theory, conventional wisdom still continues to insist coal, oil, and natural gas are fossil fuels. In other words, while coal, oil, and natural gas appear as abundant as Julian Simon predicted, conventional wisdom has not challenged the idea that the origin of hydrocarbon fuels must be more abundant than accounted for by the limited amount of organic material that did not decompose into constituent chemicals. Given how hard the peak oil theory died, the demise of the fossil fuel theory will undoubtedly be another prolonged and brutal battle. Surprisingly, even today, conventional wisdom proceeds uncritically on the largely unquestioned assumption that running out of oil is inevitable.

The Biden administration has set out on a course that should prove the reality that the amount of hydrocarbon fuel produced worldwide and the price of hydrocarbon fuels has much more to do with politics than it has to do with the principles of economics or with petroleum geology being understood. The likelihood is that if the production of petroleum ever peaks worldwide, the cause will be politics and ideologically driven arguments that demonize the use of hydrocarbon fuels precisely because hydrocarbon fuels are abundant, powerful, easy to use, relatively cheap, and central to the economic abundance the capitalist system produces. As we will see in the next chapter, ecologists and environmentalists demonize hydrocarbon fuels because they view population growth as threatening. At heart, ecologists and environmentalists tend to be miscreants and misanthropes. Why is it that anyone would see human society as better off for abandoning hydrocarbon fuels? But then, why would anyone start with the view that Earth would be a fine place if only we could get rid of people? Or that our lives would be more fulfilled if we destroyed the capitalist system?

As one of its first acts in office, the Biden administration reversed a key Trump administration policy and canceled the Keystone Pipeline. The cancellation of this pipeline evidenced the Biden administration's decision to move away from U.S. reliance on hydrocarbon fuels as part of the administration's support of the Green New Deal's push toward solar and wind power. On June 17, 2021, three days short of five months in the White House, the Biden administration had to acknowledge that oil imports from Russia set a record high in March 2021. According to the International Energy Agency (IAE), U.S. imports of crude oil and petroleum products from Russia reached 22.9 million barrels in March 2021, despite strained relations between Washington and Moscow.[35]

On July 6, 2021, the Biden administration implored the Organization of the Petroleum Exporting Countries (OPEC) oil cartel to pump more oil so U.S. gasoline prices would stop rising on Biden's watch.[36] In June and July 2021, after OPEC and Russia disagreed on increasing production quotas, the cost of gasoline at the pump in the United States began exceeding three dollars a gallon nationwide and with the price of crude oil surpassing seventy dollars per barrel on world markets. Wall Street investors began taking seriously the concern that the Biden administration's plan to cut U.S. oil production and to engage in a $7 trillion deficit-spending spree would trigger a wave of global inflation not seen since the 1970s presidency of Democrat Jimmy Carter.

Peak oil will once again have its day in court under the Biden administration. Under this administration, neo-Marxist ecologists and left-leaning environmentalists will convince the world that we are climate criminals. A significant indictment will be against those who want to use Earth's abundant hydrocarbons for our economic betterment, understanding that hydrocarbon fuels are the most potent form of available and affordable energy used by humans in the history of the world to date.

CHAPTER 2

John P. Holdren: Eco-Malthusian Wizard Extraordinaire

My personal opinion is that we have to keep geoengineering on the table. We have to look at it very carefully because we might get desperate enough to want to use it.

—John Holdren, Associated Press interview, April 8, 2009[1]

IN DIRECT CONTRAST TO JULIAN SIMON'S experience of elite academics and mainstream media marginalizing him, John Holdren has been a celebrated academic and a mass media favorite for decades. Holdren was born in 1944, twelve years after Julian Simon, with his birthplace in Sewickley, a small town in Allegheny County, Pennsylvania, that sits on the Ohio River northwest of Pittsburgh. Holdren's early studies were in aeronautics, astronautics, and plasma physics. Holden received his bachelor's degree at MIT in 1965, where he was drawn in his early years to study plasma physics. Plasma in nuclear physics is considered the fourth state of matter, comparable to solids, liquids, and gases.

Plasma is superheated gas that, in its highly electrified/magnetized state, causes atoms to dissociate into positively charged ions and negatively charged electrons.[2] The sun's mass, making up over 99.85 percent of the solar system, consists of a plasma state in which hydrogen and helium gas are completely ionized into hydrogen and helium ions. After graduating from MIT, Holdren went to Stanford University. He received a Ph.D. in 1970, studying under Oscar Buneman, a genius in plasma physics who played a role during World War II in developing radar. Had Holdren stayed with plasma physics as his professional career, he might have played a role in developing nuclear fusion, an elusive technology based on the nuclear reactions that power the sun. While yet unrealized, nuclear fusion can create on Earth nearly limitless quantities of carbon-emission-free energy.

Harrison Brown: Why Holdren Abandoned Plasma Physics

Holdren records two turning points in his early life that determine why he abandoned nuclear physics to pursue a career in ecology and environmental studies. The first occurred while in high school. He read Harrison Brown's 1954 book *The Challenge of Man's Future*.[3] In 1986, more than three decades after reading that book, Holdren helped edit a volume of essays dedicated to Harrison Brown's lifetime of work. "By the time I read it [Brown's book] as a high school student," Holdren noted in his foreword, "the book had been widely acclaimed as a monumental survey of the human prospect, illuminated through analysis of the interaction of population, technology, and the resources of the physical world."[4]

Holdren acknowledged that reading Brown's book played a crucial role in shaping his subsequent professional career. "I knew even before high school that science and technology held a special interest for me, and I suppose I had some prior interest in the larger human condition. But *The Challenge of Man's Future* pulled these interests together for me in a way that transformed my thinking about the world and about the sort of career I wanted to pursue."[5]

Brown's impact on Holdren was truly formative. Regardless of the topic, whether it be vital statistics, food, or natural resources in general, Brown was cautious about how much we could accomplish. He warned there were "fundamental physical limitations to man's future development and of the hazards which will confront him in the years and centuries ahead."[6] In *The Challenge of Man's Future*, Brown wrote that "both Malthus's reasoning and the principles he enunciated were sound."[7] In writing about Brown in 1986, Holdren praised him for understanding that "the problems of population, the rich-poor gap, and the prospects for war and peace" are interrelated issues.[8] "Thirty years after Harrison Brown elaborated these positions, it remains difficult to improve on them as a coherent depiction of the perils and challenges we face," Holdren stressed.[9] He commented that he includes himself among those "who have been restating his [Brown's] points (usually less eloquently) in the three decades since he first made them."[10]

John P. Holdren: Eco-Malthusian Wizard Extraordinaire

As recently as February 15, 2007, Holdren gave a presidential address entitled "Science and Technology for Sustainable Well-Being" to the American Association for the Advancement of Science (AAAS). Holdren headed the AAAS from February 2006 to February 2008.[11] In that address, Holdren singled out Brown as one of the "several late mentors" to whom he was thankful for "insight and inspiration." Holdren's embrace of Brown as a mentor could not have been more complete. "My preoccupation with the great problems at the intersection of science and technology with the human condition—and with the interconnectedness of these problems with each other—began when I read *The Challenge of Man's Future* in high school. I later worked with Harrison Brown at Caltech,"[12] he acknowledged in a slide that accompanied his presentation. Brown was a professor of geochemistry at the California Institute of Technology (Caltech) from 1951 to 1977, coinciding with Holdren's years as a graduate student at Stanford.

In 1995, Holdren delivered the Nobel Peace Prize lecture as chair of the executive committee of the Pugwash Council on behalf of the Pugwash Conferences on Science and World Affairs, a group dedicated to pursuing nuclear disarmament. Holdren acknowledged Brown and Russian physicist Andrei Sakharov in the lecture, noting both were Pugwash participants. Holdren shared Brown's Malthusian views on a variety of subjects, including natural resource limitations and overpopulation, but also on nuclear weapons.[13]

Harrison Brown Urges World Government to End Overpopulation

Holdren's selection of Harrison Brown as his mentor gives us a good hint of how radical Holdren himself was as a young man. Consider the type of solutions Brown proposed in *The Challenge of Man's Future*. Brown proposed using eugenics to "prevent the long-range degeneration of human stock." He recommended implementing the "science of human genetics" in a two-step process designed to improve the species. Brown wrote the following:

> First, man can discourage unfit persons from breeding. Second, he can encourage breeding by those persons who are judged fit on the basis of physical and mental testing and examinations of the records of their ancestors. A small start has been made in this direction in the cases of childless couples where the male is sterile and artificial insemination is utilized to impregnate the female. It is quite likely that artificial insemination will be

used with increasing frequency during the coming decades, and increasing care will be taken to insure the genetic soundness of the sperm.[14]

Brown doubted humans had the intelligence today needed to breed the desired characteristics of a "super-race."[15] "We can carry out selection processes satisfactorily with sheep, cows, horses, and dogs, for in all cases we are able to examine the animals objectively and decide upon desirable characteristics," he wrote. "Unfortunately man's knowledge of human genetics is too meager at the present time to permit him to be a really successful pruner," he lamented. "The science of human genetics is not very old, and reliable facts and figures which enable one to differentiate satisfactorily between genetic effects and environmental effects are few and far between." Still, he looked forward to a day when our knowledge would advance. "And it is quite possible that by the time another ten or fifteen generations have passed, understanding of human genetics will be sufficient to permit man to do a respectable job of slowing down the deterioration of the species," he wrote.[16] These passages suggest that Brown would be very enthusiastic about eugenics applied to developing a race of superior humans, given the advances in understanding DNA so far as to experiment with gene-splicing technology.

Still, while we were waiting for these scientific advancements in genetics, Brown felt we had even then "sufficient information to permit man to make a start toward pruning, however small it may be."[17] Brown was determined to begin the process of "pruning" the human race, even though our understanding in the 1950s of genetics was limited:

> Is there anything that can be done to prevent the long-range degeneration of human stock? Unfortunately, at the present time there is little, other than to prevent breeding in persons who present glaring deficiencies clearly dangerous to society and which are known to be of a hereditary nature. Thus we could sterilize or in other ways discourage the mating of the feeble-minded. We could go further and systematically attempt to prune from society, by prohibiting them from breeding, persons suffering from serious inheritable forms of physical defects, such as congenital deafness, dumbness, blindness, or absence of limbs. But all these steps would be negligible when compared with the ruthless pruning of man that was done by nature.[18]

Here Brown left no doubt he endorsed Darwin's evolutionary principles to suggest we could perfect nature and advance the process of natural selection by engaging in eugenics once our scientific understanding of genetics had advanced.

"A 'super-race' of men or a panel of gods could examine us objectively and plan a wise pattern," he continued. "But in the absence of either, we will probably remain pretty much as we are for hundreds of thousands of years."[19] Still, Brown remained concerned that "it does appear that the feeble-minded, the morons, the dull and backward, and the lower-than-average persons in our society are outbreeding the superior ones at the present time. Indeed, it has been estimated that the average Intelligence Quotient of Western population as a whole is probably decreasing with each succeeding generation."[20]

Earlier on in the book, Brown had recommended controlling overpopulation by a combination of the following methods:

1. Restriction of sexual intercourse;
2. Abortion;
3. Sterilization; and
4. Fertility control, "either through the practice of coitus interruptus or through the use of chemicals or devices designed to prevent contraception."[21]

Despite expressing concerns that humans possessed the foresight and intelligence to apply eugenics to shape a "super-race," in the conclusion to his 1954 book Brown appeared to be enthusiastic about the project. "A broad eugenics program would have to be formulated which would aid in the establishment of policies that would encourage able and healthy persons to have several offspring and discourage the unfit from breeding at excessive rates," he commented.[22] "Precise control of population can never be made completely compatible with the concept of a free society; on the other hand, neither can the automobile, the machine gun, or the atomic bomb," he continued. "Just as we have rules designed to keep us from killing one another with our automobiles, so there must be rules that keep us from killing one another with our fluctuating breeding habits and with our lack of attention to the soundness of our individual genetic stock."[23]

Brown concluded that "population stabilization and a world composed of completely independent sovereign states are incompatible." He insisted that "population stabilization" is a goal "with which a world government must necessarily concern itself." Brown called for world government authorities to set "maximum permissible population levels" for all world regions. Each world region could then be self-sufficient in both agriculture and industry.[24] Brown even contemplated that infanticide is an acceptable solution to overpopulation in extreme situations. He insisted that "if we cared little for human emotions and were willing to introduce a procedure which most of us would consider being reprehensible in the extreme, all excess children could be disposed of much as excess puppies and kittens are disposed of at the present time."[25]

Brown suggested "a substantial fraction of humanity" was reproducing as if "it would not rest content until the earth is covered completely and to a considerable depth with a writhing mass of human beings, much as a dead cow is covered with a pulsating mass of maggots."[26] He believed there are "physical limitations of some sort which will determine the maximum number of human beings who can live on the earth's surface."[27] Brown regretted "there can be no escaping the fact that if starvation is to be eliminated, if the average child who is born is to stand a reasonable chance of living out the normal life span with which he is endowed at birth, family sizes must be limited."[28]

As far as Brown was concerned, government-mandated population control was necessary to prevent overpopulation. He cautioned: "Either population-control measures must be both widely and wisely used, or we must reconcile ourselves to a world where starvation is everywhere, where life expectancy at birth is less than 30 years, where infants stand a better chance of dying than living during the first year following birth, where women are little more than machines for breeding, pumping child after child into an inhospitable world, spending the greater part of their adult lives in a state of pregnancy."[29] Ultimately, Brown accepted limiting human freedom as a necessary condition of entrusting a world government to "stabilize" population. Brown concluded, "it is difficult to see how the achievement of stability and the maintenance of human liberty can be made compatible."[30]

Brown proposed a rule government officials would have to follow to mandate birth control measures. "Let us suppose that in a given year the birth rate exceeds the death rate by a certain amount, thus resulting in a population increase," he postulated. "During the following year the number of permitted inseminations is decreased, and the number of permitted abortions is increased, in such a way that the birth rate is lowered by the requisite amount." Next, Brown insisted that in a year where the death rate exceeds the birth rate, "the number of permitted inseminations would be increased while the number of abortions would be decreased." Brown formulated his rule as follows: "The number of abortions and artificial inseminations permitted in a given year would be determined completely by the difference between the number of deaths and the number of births in the year previous."[31] His ideal solution was dystopian. "If all babies were born from test tubes, as in Aldous Huxley's *Brave New World*, the solution would be fairly simple: The number of babies produced on the production line each year could be made to equal the number of deaths," he concluded.[32]

Nothing on the record indicates John Holdren ever abandoned his admiration for Harrison Brown or rejected Brown's ideas.

Harrison Brown Agrees: The World Is Running Out of Fossil Fuels

In *The Challenge of Man's Future*, Brown fully embraced the organic theory of the origin of petroleum. Brown explained that "oil, like coal, was formed slowly over long periods of time from living matter that became trapped in sediments." He suggested that "the evidence strongly supports the theory that marine organisms living in shallow coastal waters were the basic stuff from which most petroleum deposits evolved." He believed petroleum traced back to "dead organisms carried into the oceans by rivers and streams were, along with the mud of the ocean floor, subsequently compacted into shale."[33]

While his explanation of the organic origin of petroleum is simplistic even for petroleum geology at that time, he fully embraced the idea that oil was a fossil fuel:

> Later the folding of mountain ranges elevated the shale, and the organic material, in the liquid form of petroleum, flowed, under the influence of gravitation, gas pressure, and the circulation of water into pools or pockets. Thus, the locations of oil deposits are determined by the locations of ancient shorelines, mountain upheavals, and overlying rocks suitable for the trapping of pools of displaced oil.[34]

Two years before Hubbert expounded his famous bell-curve peak, Brown had concluded that since oil was a fossil fuel, humans would eventually run out of hydrocarbon fuels. He expressed this concern as follows:

> Consumption of the earth's stores of fossil fuels has barely started; yet we can see the end. In a length of time which is extremely short when compared with the span of human history, and insignificant when compared with the length of time during which man has inhabited the earth, fossil fuels

will have been discovered, utilized, and completely consumed. The "age of fossil fuels" will be over, not to be repeated for perhaps another 100 million years. Will its passing mark the end of civilization and perhaps the beginning of the downward path to man's extinction? Or can we expect other sources of energy to fulfill the need?[35]

Brown concluded the last available source of energy would be the sun converted into solar energy. He despaired that atomic energy would be sufficient for the seven billion people he estimated would be on Earth by 2000. Why? The answer was predictable. Brown felt we would eventually run out of uranium, just as he figured we would exhaust all available hydrocarbon fuels, as well as nearly every other natural resource he discussed.

Harrison Brown Urges World Government to Prevent Atomic War

Like Holdren, Brown was a scientist, specifically a nuclear chemist, by academic training. During World War II, he worked at the Manhattan Project's Metallurgical Laboratory at the University of Chicago. Nobel Prize chemist Glenn Theodore Seaborg recruited Brown to the Manhattan Project as Brown's first job after leaving Johns Hopkins University, where he had received his Ph.D. Seaborg met Brown in 1938 when Brown, then an undergraduate at the University of California, Berkeley, attended a Wednesday evening seminar that Seaborg was giving at the university. At that time Seaborg was a research assistant. Brown joined Seaborg's chemistry group, where he did lead work conceiving and developing the chemical processes required for the isolation of plutonium produced in what at the time was called "large nuclear piles." Seaborg later commented that Brown's work developing the plutonium processes into the Fat Man atomic bomb dropped on Nagasaki was seminal to Brown's later pacifism. Seaborg commented as follows: "And, I believe, his [Brown's] work on the processes for producing the nuclear weapon, even though for wartime use against a dangerous enemy, colored his outlook and furnished the background of experience that led later to a lifetime dedicated to the achievement of arms control and arms limitation measures."[36] Brown's first book, published in 1945, entitled *Must Destruction be Our Destiny? A Scientist Speaks as a Citizen*,[37] warning us of the dangers of nuclear war.

Brown concluded the creation of the atomic bomb made the creation of a world government even more urgent. He felt a world government was necessary for the course of human history to proceed favorably, but not inevitable given the human propensity to pursue selfish goals. "The atomic bomb did not create the need for world government: the need existed long before uranium fission was discovered," Brown wrote in his 1945 book. "The possibility that technological achievements might ultimately precipitate disaster in the world has become increasingly apparent during the last fifty years. The atomic bomb is but the latest in a series of developments that make political nationalism a senseless concept, and there is every reason to believe that these developments will continue at an ever-increasing pace."[38]

Even when contemplating a one-world government, Holdren saw eye-to-eye with his mentor, Brown.

Holdren Meets Paul and Anne Ehrlich, Becomes an Ecologist

The second seminal event in his young life occurred when Holdren met Paul Ehrlich.

Harrison Brown's thinking on overpopulation molded Holdren's thinking on the subject. But Brown was of a different generation. Born in 1917, toward the end of World War I, Brown experienced World War II as an adult. When Holdren was born in 1944, too young to remember World War II, Brown was already working on the Manhattan Project. Yet, for John Holdren, Paul Ehrlich was the perfect match to be a potential colleague. When Holdren got his Ph.D. from Stanford in 1970, Brown was teaching at Caltech, where he worked with Holdren as a postgraduate student.

Paul R. Ehrlich was born in 1932 in Philadelphia, the same year Julian Simon was born. Both Ehrlich and Simon were too young to experience the Great Depression as an adult or fight in World War II. Both were too old to have fought in Korea. Ehrlich received his Ph.D. in biology from the University of Kansas in 1957, where he studied etymology and published his dissertation on Lepidoptera (i.e., butterflies). In 1959, he joined the Stanford University faculty and was appointed as a professor of biology in 1966. While in graduate school, Ehrlich met Anne Howland, his future wife, a University of Kansas undergraduate French major, one year younger than Paul. Their first book together, in 1961, was entitled *How to Know the Butterflies*.[39]

As a graduate student at Stanford, Holdren not only met Brown, but he also met Paul Ehrlich. Brown may have viewed Holdren as a promising young scholar, but Ehrlich (some twenty-seven years younger than Brown) considered Holdren might be useful as a coauthor. Just twelve years older than Holdren, Ehrlich was on track for making Brown's thinking about global overpopulation an enduring pillar of popular culture. With the publication of his 1968 book *The Population Bomb*, Ehrlich achieved international celebrity status.[40]

During the 1965–1966 academic year in which Paul and Anne had moved to Australia on a National Science Foundation (NSF) fellowship, the two spent a few weeks touring India on a trip that included visits to Thailand and Cambodia. Though officially looking to collect butterfly specimens as part of their NSF grant, the Ehrliches were greatly impressed by the poverty and crowds they saw in Delhi.[41] *The Population Bomb* embraced the original fears expressed by Thomas Malthus in 1989 that population growth outstripping the ability to produce food would be the ultimate undoing of humankind. In a section of chapter 1 subtitled "Too Many People," Ehrlich wrote:

> Americans are beginning to realize that the underdeveloped countries of the world face an inevitable population-food crisis. Each year food production in these countries falls a bit further behind burgeoning population growth, and people go to bed a little bit hungrier. While there are temporary or local reversals of this trend, it now seems inevitable that it will continue to its logical conclusion: mass starvation. The rich may continue to get richer, but the more numerous poor are going to get poorer. Of these poor, a minimum of three and one-half million will starve to death this year, mostly children. But this is a mere handful compared to the numbers that will be starving in a decade or so. And it is now too late to take action to save many of these people.[42]

Ehrlich was serious: his primary conclusion was that there were too many people in the world. "The battle to feed all of humanity is over," Ehrlich explained in *The Population Bomb*. "In the 1970s the world will undergo famines—hundreds of millions of people are going to starve to death in spite of any crash programs embarked upon for now." Nothing could save the world, Ehrlich argued, except population control, defined as "the conscious regulation of the numbers of human beings to meet the needs, not just of individual families, but society as a whole."[43] In the "Prologue" to *The Population Bomb*, Ehrlich signaled that Brown was also his mentor: "The birth rate must be brought into balance with the death rate or mankind will breed itself into oblivion." This sentence appeared lifted almost word-for-word from Brown's conclusion in *The Challenge of*

Man's Future, published fifteen years earlier. "We can no longer afford to treat the symptoms of the cancer of population growth; the cancer itself must be cut out," Ehrlich continued. "Population control is the only answer."[44]

As a result of writing *The Population Bomb*, Ehrlich appeared more than twenty times on NBC's *The Tonight Show Starring Johnny Carson*.[45] In an appearance on Johnny Carson's show that aired on January 31, 1980, Ehrlich left no doubt about his Malthusian positions on oil and population. "Do we really want to threaten to blow up the world over a resource which we know damn well is going to be gone in twenty or thirty years anyway?" Ehrlich asked regarding the Middle East conflict. "In other words, it's not like we are not sucking that oil out of the ground. All we're doing is moving the timetable up."[46] Ehrlich's "running out of oil" by 2000 or 2010 predictions were wrong.

Regarding people, Ehrlich explained to Carson that every country is now overpopulated. "If U.S. population started to decline slowly, if we got our death rate a little above our birthrate and held it there for a couple hundred years, the fact that we were moving back to 150 or 100 million people would not scare me a bit. Everybody's lives would get better in the long run because there would be more of everything to go around. There is a finite pie. The more mice you have nibbling at it, the smaller every mouse's share."[47] When he met Ehrlich, Holdren dropped any idea of pursuing a career in plasma physics. Even while he was finishing his graduate studies, Holdren began publishing with Ehrlich. If achieving notoriety was Holdren's goal for judging career success, he made the right decision.

In December 1969, Holdren's first publication with Paul Ehrlich was a paper entitled "Population and Panaceas: A Technological Perspective," published in an academic journal, *BioScience*.[48] By working with Holdren, Ehrlich added a coauthor who was about to be a Ph.D. with an impressive scientific pedigree, advanced plasma physics. Holdren's academic credentials were a considerable enhancement to the pedigree of Ehrlich's Ph.D. in biology, studying butterflies. By working with Ehrlich, Holdren realized an opportunity to change careers. He leapfrogged the interest in overpopulation that he had developed from Brown into a working collaboration with Ehrlich.

Predicting doomsday disasters with Paul and Anne Ehrlich as his coauthors, Holdren launched a dazzling career in academics and politics. Consider his job credits: professor of environmental policy at the Kennedy

School of Government at Harvard University; director in the Science, Technology, and Public Policy Program at the Kennedy School of Government at Harvard University; professor of energy and resources emeritus at the University of California, Berkeley; director at the prestigious Woods Hole Research Center; and "science czar" director in the Office of Science and Technological Policy during the Obama administration. Compare this to the academic disdain Julian Simon endured in an academic career that ended mainly in obscurity. As any successful Hollywood director knows, disaster stories sell, while the type of reassuring "feel good" stories Simon told, even if true, tended to languish on the vine of public interest. For a moment, please consider how Simon, in his autobiography, characterized his academic career:

> What others don't notice when they kindly praise me is that in my entire professional life, spanning more than three decades, I have never (I mean literally *never*, truly an amazing statistic) had a single standard mark of professional respect (let alone honor) in my academic professions. I've never held an office in a professional association (not even membership on the committee that nominates *other* people for offices and honors or on the committee that counts the ballots for candidates); never been offered a prestigious teaching job; never been asked to give a to-be-published paper at an annual economics association meeting. Any scholar must be amazed that anyone could publish as huge a pile of books and papers as I have over many years and never be asked to referee a paper for the major "official" journal in my three fields of economics, demography, and statistics, or with two or three exceptions in 30 years, by almost any other top journal. Shake your heads at that, brothers and sisters. You can hardly get any less formally distinguished than that.[49]

Simon again was right. False bad news about the environment, natural resources, population, and the climate sells. Since the 1950s, Harrison Brown, Paul Ehrlich, and John Holdren prove the path to writing best-selling books, achieving academic success, and achieving international fame remains Malthusian, whether the underlying arguments are valid or not, or whether the doomsday predictions happen or not.

A Radical Textbook Proposing Planetary Population Control

In 1977, John Holdren teamed up with Paul and Anne Ehrlich to coauthor a revised college textbook entitled *Ecoscience: Population, Resources, Environments*.[50] Paul and Anne Ehrlich originally published the book in

1970 without Holdren under a different title: *Population, Resources, Environment: Issues in Human Ecology*.[51] The new edition of the book was expanded from 383 pages to 1,051 pages. Today, this revised version is extremely hard to obtain. Along with the new "ecoscience" perspective, Holdren introduced his radical solutions to many environmental problems he saw human beings as causing. The emphasis in the revised textbook on the need for an authoritarian world government to aggressively limit population growth became politically inconvenient when Holdren faced Senate confirmation to become Obama's "science czar."

The debt the 1977 coauthors felt they owed to Brown was made clear on page 1 of the revised first chapter now called "Population, Resources, Environment: Dimensions of the Human Predicament." It was headed by the following quote from *The Challenge of Man's Future*:

> It is clear that the future course of history will be determined by the rates at which people breed and die, by the rapidity with which nonrenewable resources are consumed, by the extent and speed with which agricultural production can be improved, by the rate at which the underdeveloped areas can industrialize, by the rapidity with which we are able to develop new resources, as well as by the extent to which we succeed in avoiding future wars. All of these factors are interlocked.[52]

Holdren's contribution with Paul and Anne Ehrlich went beyond Brown to conceptualize Earth's environment in system theory terms as an ecosystem. Holdren's recasting of the population concern into ecosystem language added an air of scientific necessity to the solutions to the "population bomb" problem. Holdren's perspective was that overpopulation risked destroying Earth's ecosystem. Holdren was not sure in 1977 whether we faced global cooling or global warming. But he was certain we were going to threaten our own survival one way or the other.

In the textbook, Holdren and his coauthors came to the same conclusions reached by Brown. Namely, the only solution involved massive population control enforced by global governance. The *Ecoscience* authors urgently called to create a "Planetary Regime" empowered to act as an "international superagency for population, resources, and environment." Among the various responsibilities of the Planetary Regime would be population control. "The Planetary Regime might be given responsibility for determining the optimum population for the world and for each region and for arbitrating various countries' shares within their regional limits," the authors argued. "Control of population size might remain the responsibility of each government, but the Regime should have some power to enforce the agreed limits."[53]

The authors argued involuntary birth control measures, including forced sterilization, may be necessary and morally acceptable under extreme conditions, such as widespread famine brought about by climate change. One way to discourage illegitimate childbearing "might be to insist that all illegitimate babies be put up for adoption—especially those born to minors who generally are not capable of caring properly for a child alone." Alternatively, the authors suggested unwed mothers might place their babies up for adoption, writing: "If a single mother really wished to keep her baby, she might be obliged to go through adoption proceedings and demonstrate her ability to support and care for it."[54]

Yet, the authors added the following caveat: "If some individuals contribute to general social deterioration by overproducing children, and if the need is compelling, they can be required by law to exercise reproductive responsibility—just as they can be required to exercise responsibility in their resource-consumption patterns—*providing they are not denied equal protection*."[55] However, the equal protection caveat was essentially meaningless as long as the government equally treated all persons who are overproducing children severely. The critical point here is that the authors failed to see any constitutional protection for the right to bear children. "Some people—respected legislators, judges, and lawyers included—have viewed the right to have children as a fundamental and inalienable right," the authors continued. "Yet neither the Declaration of Independence nor the Constitution mentions a right to reproduce."[56]

Similarly, the authors argued that a right to privacy did not extend to an unlimited right to have children, elaborating as follows: "Where the society has a 'compelling, subordinating interest' in regulating population size, the right of the individual may be curtailed. If society's survival depended on having more children, women could be required to bear children, just as men can constitutionally be required to serve in the armed forces. Similarly, given a crisis caused by overpopulation, reasonably necessary laws to control excessive reproduction could be enacted."[57] The point was that family size and composition was no longer the responsibility of the parents in the "ideal society."[58]

Among the efforts to control the population, the authors thought steps in India to vasectomize all fathers of three or more children and efforts in China to sterilize mothers after their third child were considered justified. The authors also advocated developing a long-term sterilizing capsule implanted under the skin. They discussed that the government could issue a license to entitle a woman to a specified, limited number of children. They went so far as to consider the need to add a sterilant to drinking water or staple foods. "Compulsory control of family size is an unpalatable idea, but the alternatives may be much more horrifying," the authors concluded. "As those alternatives become clearer to an increasing number of people in the 1980s, they may begin *demanding* such control."[59]

Thus, the pioneering work Brown did in the 1950s to popularize the Malthusian overpopulation nightmare into a likely post–World War II global future reality was advanced in the 1970s by the team of John Holdren working with Paul and Anne Ehrlich. What emerged from these new efforts was the justification for a neo-Marxist dystopian future. An all-powerful, global governance structure would impose strict population control measures, including forced abortions and mandatory sterilization to prevent the "population bomb" from causing ecological/environmental disaster.

In so doing, neo-Marxist "brave new world" enthusiasts found that blaming a hypothesized future ecological/environmental catastrophe resulting from overpopulation was a brilliant social control argument. A frightened global population would willingly consent to planetary government totalitarianism to save humanity from destroying itself. Holdren understood that in a neo-Marxist world, hypothecated ecological/environmental disasters attributed to anthropogenic causation were a potent psychological tool to gain control over humanity on a global scale. A compliant population would abandon even fundamental human rights if an all-powerful dystopian government could save the planet. Under Holdren's direction, neo-Marxist power-seekers learned that postulating anthropogenic ecological/environmental disasters could serve their ideological goals.

Holdren and Climate Change

A new focus on climate change emerged in the work of Paul and Anne Ehrlich with the addition of Holdren. In the 1970 textbook, Paul and Anne Ehrlich were concerned that overpopulation would soon exhaust the natural resources needed on Earth to sustain human life. But by adding "ecoscience," Holdren broadened the environmental perspective to the climate. Today, the concepts of ecosystems and ecoscience have taken over academic America to the point where even computer hardware and software talk about "operating systems" as "computer ecosystems." The concerns Paul and Anne Ehrlich wrote about in their 1970 textbook were virtually identical to the Malthusian population concerns Paul Ehrlich expressed in his blockbuster book published two years earlier, *The Population Bomb*. The first sentence of chapter 1 in the 1970 textbook, entitled "The Crisis," reads as follows: "The explosive growth of the human population is the most significant terrestrial event of the past million millennia."[60] Chapter 4, entitled "Limits of the Earth," asks the authors' key question: "What is the capacity of the Earth to support people?"[61]

The 1977 edition makes climate change a genuine concern. Yet, Earth was experiencing a bout of global cooling at that time, and the environmental left was preoccupied with the possible coming of a new ice age. Thus, the 1977 collaborative author team could only equivocate whether the anticipated climate disaster was global warming or global cooling in the form of a new ice age. Still, in the 1970 edition (where the climate was not a primary concern), as in the 1977 revision (where climate change, whether global warming or the coming of a new ice age, was a real threat to human life on Earth), we human beings are the culprit causing our destruction. The point is that Paul and Anne Ehrlich, in the 1970 edition without Holdren, fail to enumerate global warming in the litany of ways humans are self-destructive. Consider the following quotation, taken from the first sentences from the first chapter of the 1970 edition:

> The explosive growth of the human population is the most significant terrestrial event of the past millennia. Three and one-half billion people now inhabit the Earth, and every year this number increases by 70 million. Armed with weapons as diverse as thermonuclear bombs and DDT, this mass of humanity now threatens to destroy most of the life on the planet. Mankind itself may stand on the brink of extinction; in its death throes it could take with it most of the other passengers of Spaceship Earth. No geological event in a billion years—not the emergence of mighty mountain ranges, nor the submergence of entire subcontinents, nor the occurrence of periodic glacial ages—has posed a threat to terrestrial life comparable to that of human overpopulation.[62]

In the 1970 edition of the textbook, Paul and Anne Ehrlich still focused on overpopulation as the Malthusian threat, the same villain of *The Population Bomb*—a villain the Ehrlich team inherited from Harrison Brown.

While the 1970 textbook edition mentions CO_2 as a greenhouse gas added to the atmosphere by burning hydrocarbon fuels, the authors limited the discussion to about four pages. Again, they were ambivalent as to what effect the added CO_2 would have. The following paragraph at the heart of the 1970 edition's discussion on "Pollution and Climate" could never be written by today's neo-Marxist, global-warming Malthusians:

> Unfortunately it is impossible to predict exactly what will happen to the overall temperature of the Earth over the next few decades, or what the local effects of changes will be. It is not even known whether the amount of radiation produced by the sun is a constant—and that is essential information if changes in the heat budget of the planet are ever to be predicted. As a result, although we can be certain that man is affecting the climate (and probably accelerating change), we cannot yet isolate man's contribution to the changes we observe.[63]

Paul and Anne Ehrlich appear to have written nothing claiming that there were too many butterflies. Nor in the 1970 edition of their textbook was climate change a genuine concern. But with the addition of Holdren to the team, a new preoccupation with climate change moved center stage, along with the transposition of mere problems with the environment to the higher systems-thinking, scientific-sounding level of "ecoscience."

A New Ice Age or Global Warming?

The publication of a 1977 blockbuster book entitled *The Weather Conspiracy: The Coming of a New Ice Age*,[64] allegedly based on CIA climate studies, capped off a wave of widespread concern that global cooling, not global warming, was going to kill us all. A close look at Harrison Brown's 1954 book clarifies that he was not worried about global warming. Arguing that carbon dioxide produces a greenhouse effect, Brown strongly advocated increasing carbon dioxide in the air. He wanted to increase global warming because plants thrive on carbon dioxide. Growing more food was necessary to stave off the overpopulation crisis hypothecated. Brown lamented that Earth's atmosphere contains only a minute percentage of carbon dioxide, only about 0.03 percent of the total atmosphere. He worried that "in the absence of winds and breezes the air can become depleted locally of carbon dioxide, and the growth rate is lessened." He advised that a tripling of carbon dioxide in the air would double the growth rate of tomatoes, alfalfa, and sugar beets.[65]

Brown argued that "controlled atmospheres enriched in carbon dioxide" would be an essential component of the enormous greenhouses built to grow plants in nutrient-rich solutions that would be needed to feed an overpopulated world. His answer was to pump more carbon dioxide into the world. "It would perhaps be easier to adopt methods which would increase the carbon-dioxide concentration in the atmosphere as a whole than to attempt to build elaborate greenhouses to confine the enriched air." Thus, he concluded: "If, in some manner, the carbon-dioxide content of the atmosphere could be increased threefold, world food consumption might be doubled." Brown envisioned "on a world scale, huge carbon-dioxide generators pouring the gas into the atmosphere."[66]

Brown was clear that world governments should cooperate to generate more carbon dioxide in the air, not reduce human-generated carbon dioxide from the atmosphere. He went so far as to recommend burning more coal to generate electricity, precisely because burning coal emitted carbon dioxide. "There are between 18 and 20 tons of carbon dioxide over every acre of the Earth's surface," he noted. "To double the amount in the atmosphere, at least 500 billion tons of coal would have to be burned—an amount six times greater than that which has been consumed during all of human history." As an alternative, Brown recommended producing the needed carbon dioxide from limestone: "In the absence of coal, the equivalent in energy would have to be provided from some other source so that the carbon dioxide could be produced by heating limestone."[67]

In his 1986 book entitled *The Machinery of Nature*, Paul Ehrlich clarified that John Holdren introduced the climate change concern. Though initially, Holdren's ideas on climate change were confused. On page 274 of *The Machinery of Nature*, Paul Ehrlich wrote: "As University of California physicist John Holdren has said, it is possible that carbon-dioxide climate-induced famines could kill as many as a billion people before the year 2020." Holdren based his prediction on a bizarre theory that human emissions of carbon dioxide would produce a climate catastrophe in which global warming would cause global cooling with a resultant reduction in agricultural production that in turn would cause a widespread disaster. Ehrlich also explained Holdren's theory by arguing "some localities will probably become *colder* as the warmer atmosphere drives the climactic engine faster, causing streams of frigid air to move more rapidly away from the poles." The movement of the frigid air from the poles caused by global warming "could reduce agricultural yields for decades or more—a sure recipe for disaster in an increasingly overpopulated world," Ehrlich wrote.[68]

In 1971, Holdren and Ehrlich edited a book of readings entitled *Global Ecology*.[69] In the book they coauthored a paper, "Overpopulation and the Potential for Ecocide," in which they predicted the likelihood of a new ice age caused by human activity. They wrote that since 1940, urban air pollution had reduced the incoming light in the atmosphere. They claimed aerosols, agricultural air pollution due to dust, and volcanic ash had created a "screening phenomenon...responsible for the present world cooling trend—a total of about .2°C in the world mean surface temperature over the past quarter century." They commented that this number seems small "until it is realized that a decrease of only 4°C would probably be sufficient to start another ice age." They concluded that "a final push in the cooling direction comes from man-made changes in the direct reflectivity of the earth's surface (albedo) through urbanization, deforestation, and the enlargement of deserts."[70] They summed up the argument as follows:

> The effects of a new ice age on agriculture and the supportability of large human populations scarcely need elaboration here. Even more dramatic results are possible, however; for instance, a sudden outward slumping in the Antarctic ice cap, induced by added weight, could generate a tidal wave of proportions unprecedented in human history.[71]

Equivocating whether human-caused global warming or global cooling was the more likely future trend, the authors concluded that either way would produce an eco-disaster. Any immediate shift in climate, regardless of whether toward global warming or global cooling, would have hazardous effects upon agriculture and food production. Remarkably, in this paper, Ehrlich and Holdren discounted that global warming would come from the "greenhouse effect" of emitting carbon dioxide from burning hydrocarbon fuels. Instead, they argued that global warming could be initiated simply by human-caused excess heat generation from burning fossil and nuclear fuels. Here is Ehrlich and Holdren's summary paragraph on climate change in this 1986 paper:

> If man survives the comparatively short-term threat of making the planet too cold, there is every indication he is quite capable of making it too warm not long thereafter. For the remaining major means of interference with the global heat balance is the release of energy from fossil and nuclear fuels. As pointed out previously, all this energy is ultimately degraded to heat. What are today scattered local effects of its disposition will in time, with the continued growth of the population and energy consumption, give way to global warming. The present rate of increase in energy use, if continued, will bring us in about a century to the point where our heat input could have drastic global consequences. Again, the exact form such consequences might take is unknown; the melting of the icecaps with a concomitant 150 foot increase in sea level might be one of them.[72]

By 1986, Holdren appears convinced the ultimate ecoscience catastrophe will be climate change, regardless of whether the global climate change is cooling or warming. Holdren's definition of the resulting catastrophe is relatively undeveloped compared to Al Gore's 2006 film, *Inconvenient Truth*. However, by 1986, Holdren had already identified the melting of the ice caps and rising sea levels as the global warming catastrophe. But note, Holdren said the ice caps would melt, and the seas would rise from the heat resulting from burning hydrocarbon and nuclear fuels, not from CO_2 emissions.

Paul Ehrlich Loses "Bet" with Julian Simon

In the early 1980s, Julian Simon's irritation grew because Paul Ehrlich had attracted such tremendous international attention even though his predictions on global overpopulation had failed. Finally, in June 1982, Simon published an article in the *Social Science Quarterly* entitled "Paul Ehrlich Saying It Is So Doesn't Make It So." Simon wrote:

> His [Paul Ehrlich's] predictions of the last decade or so about increasing scarcities and a worsening environment in the United States have proved wrong almost without exception—a track record of poor predictive validity which should lead one to have little confidence in his present predictions.[73]

In the article, Simon announced that Ehrlich had accepted his offer to bet on whether resource prices would go up or come down in the future. Designed to test Ehrlich's resource exhaustion thesis, Simon agreed to bet on what the 1990 price would be of any five metals Ehrlich picked. With the assistance of Holdren, Ehrlich picked the following metals: copper, chrome, nickel, tin, and tungsten. The point of the bet was to see if the 1990 value, adjusted for inflation, exceeded or fell below the metal's value in 1980. Ehrlich bet that the metals would become scarcer in the decade and the prices would go up. Both Ehrlich and Simon agreed that the use of each metal would increase in the coming decade. Simon bet each metal would be cheaper in 1990 than it had been ten years earlier. Simon won the bet. To Ehrlich's amazement and embarrassment, all five metals cost relatively less in 1990 than they had in 1980.[74]

Early in his career, Simon would have agreed with Ehrlich about overpopulation and the scarcity of natural resources. To explain his turnaround, Simon cited an epiphany he had in 1969 while visiting the Iwo Jima Memorial in Washington, D.C. While there, Simon recalled a famous eulogy the Jewish chaplain Rabbi Roland Gittelsohn had given at Iwo Jima for a dedication ceremony after the war for the 5th Marine Division's soldiers who had fought and died there. Gittelsohn's sermon has become a Marine Corps legend. His eulogy began as follows:

> This is perhaps the grimmest, and surely the holiest task we have faced since D-Day. Here before us lie the bodies of comrades and friends. Men who until yesterday or last week laughed with us, joked with us, trained with us. Men who were on the same ships with us, and went over the sides with us, as we prepared to hit the beaches of this island. Men who fought with us and feared with us. Somewhere in this plot of ground there may lie the individual who could have discovered the cure for cancer. Under one of these Christian crosses, or beneath a Jewish Star of David, there may rest now an individual who was destined to be a great prophet to find the way, perhaps, for all to live in plenty, without poverty and hardship for none. Now they lie here silently in this sacred soil, and we gather to consecrate this earth to their memory.[75]

The eulogy made an essential impact on Simon. He asked himself an important question: "What business do I have trying to help arrange it that fewer human beings will be born, each one of whom might be a Mozart or a Michelangelo or an Einstein—or simply a joy to his or her family and community and a person who will enjoy life?"[76] Soon after, on Earth Day 1970, Simon gave a speech at the University of Illinois questioning his original premise that overpopulation posed a scientifically provable threat to the future of humankind on this planet.[77] The realization accounts primarily for Simon's understanding that human intelligence is the "ultimate resource" capable of overcoming perceived natural resource shortages and limitations.

Simon always distinguished that he considered natural resources, including oil, to be "not finite," but he did not think natural resources were "infinite," i.e., available without limit. In an appearance with William Buckley on Buckley's PBS show *Firing Line* broadcasted on November 8, 1981, Simon explained the distinction as follows:

> It seems to me that the notion of something being finite is very much a matter of how we look at it and what we choose to do about it. That is, the food in your larder is finite right now and if you have twice as many guests over tonight, you may crowd it to the limit and you may exhaust it; and you may think of what's in there now as being finite, but because

you know there's a supermarket down the street, you know that you can replenish the larder and therefore you don't think of it in another way as being finite and limited.

Simon continued:

In the same way we tend to think about many of the other resources that we deal with, whether it be copper or wheat or oil. We tend to think of the supply as being fixed at a given moment because of any of the many preconceptions that we have about it, and then it seems finite and we think about running out; but if we think instead about our capacities to increase that supply by finding substitutes or by finding better ways to get more of it or by replacing it, just as we in fact grow oil in Illinois, then we begin no longer to think about the supply of oil or copper as being finite.

Simon summed up his position as follows:

I do think there is a distinction between looking at it from the point of view of the word "finite" and the word "infinite." What I want to do is dispose of the word "finite" and not to bring in the word "infinite" and to argue to you that the supply of anything is infinite. What I want to suggest to you is that we can indeed think of what's important to us with respect to copper and that is, services that we get from copper, as not being finite in any way, and in fact the whole history of mankind with respect to copper has been one of cooper getting more and more abundant each year.[78]

Understanding this distinction is key to understanding the fundamental difference between an optimist on natural resources, such as Julian Simon, and a Malthusian pessimist like Harrison Brown, Paul and Anne Ehrlich, and John Holdren.

The Folly of Geoengineering

In contrast to Julian Simon's commonsense approach to natural resources, John Holdren as Obama's science czar endorsed several bizarre geoengineering projects. As was made clear by the quotation that started this chapter, Holdren, in a 2009 interview with the Associated Press, made clear he was not about to take off the table exploring geoengineering to save the planet. Under the rubric of geoengineering, several harebrained projects gained prominence. One idea involved placing mirrors in space to reflect sunlight from Earth. Another wanted to fertilize the ocean with iron to encourage the growth of algae to soak up atmospheric CO_2. Possibly seeding clouds and the upper atmosphere with metal strips or pollution particles would bounce the sun's rays back into space, so they do not warm Earth's surface.[79]

In the March/April 2009 issue of the Council on Foreign Relations magazine *Foreign Affairs*, a group of five authors led by David Victor, a professor at Stanford Law School, published an article entitled "The Geoengineering Option: A Last Resort Against Global Warming?"[80] In the paper, Victor and his coauthors described their geoengineering strategies as "deploying systems on a planetary scale, such as launching reflective particles into the atmosphere or positioning sunshades to cool the earth." The article's fundamental premise was that by increasing the reflectivity of the atmosphere, more of the sun's rays would reflect into space. "Increasing the reflectivity of the planet (known as the albedo) by about one percentage point could have an effect on the climate system large enough to offset the gross increase in warming that is likely over the next century as a result of a doubling of the amount of carbon dioxide in the atmosphere," Victors and his coauthors argued.

Another scheme involved launching sulfur particles and other reflective materials into the upper stratosphere using high-flying aircraft, naval guns, or giant balloons. Alternatively, a plan was to shoot air pollution in microscopic particles into the upper atmosphere to reflect the sun's rays into outer space. As a result, Earth would absorb less carbon dioxide. One of the more bizarre of the proposed geoengineering schemes involved using 1,900 wind-powered "cloud ships" (alternatively called "albedo yachts"). The ships would sail the world's oceans to suck up seawater and spray it out in minuscule droplets through tall funnels designed to create large white clouds.[81]

Despite the slim chances any of these schemes might work, Victor and his coauthors still felt geoengineering was essential to prevent climate warming. "The highly uncertain but possible disastrous side effects of geoengineering interventions are difficult to compare to the dangers of unchecked global climate change,"[82] Victor and his team insisted. In truth, the magic of geoengineering intrigued Holdren, despite the almost silly nature of the schemes.

Geoengineering schemes returned to fashion as the Green New Deal gained support among neo-Marxist ecologists/environmentalists in the Biden administration. Among the new round of geoengineering is the idea to grind up olivine, a volcanic rock that jewelers know as peridot. By depositing this pea-colored sand offshore along 2 percent of the world's coastlines, the idea would be to capture 100 percent of CO_2 emissions from the ocean. Another brainstorm, "marine cloud-brightening," calls for spraying a fine mist of seawater into the clouds, so the salt makes them brighter and therefore more reflective of the sun's heat.[83]

John Holdren: The Intensity of a True Believer

As Obama's science czar, Holdren's pronouncements on global warming became increasingly dogmatic. By 2009, Holdren had no doubt global warming was the climate change catastrophe and anthropogenic CO_2 was the culprit. Consider the following congressional testimony Holdren gave in 2009, at the beginning of his term as Obama's science czar:

> We now know that climate is changing all across the globe. The air and the oceans are warming, mountain glaciers are disappearing, sea ice is shrinking, permafrost is thawing, the great land ice sheets on Greenland and Antarctica are showing signs of instability, and sea level is rising. And the consequences for human well-being are already being felt: more heat waves, floods, droughts, and wildfires; tropical diseases reaching into the temperate zones; vast areas of forest destroyed by pest outbreaks linked to warming; alterations in patterns of rainfall on which agriculture depends; and coastal property increasingly at risk from the surging seas.

Having articulated this grim scenario of catastrophic climate change, Holdren next explained to Congress why climate change was happening. Consider his next paragraph:

> We know the primary cause of these perils beyond any reasonable doubt. It is the emission of carbon dioxide (CO_2) and other heat-trapping pollutants from our factories, our vehicles, and our power plants, and from use of our land in ways that move carbon from soils and vegetation into the atmosphere in the form of CO_2. We also know that failure to curb these emissions will bring far bigger impacts from global climate change in the future than those experienced so far. Devastating increases in the power of the strongest hurricanes, sharp drops in the productivity of farms and ocean fisheries, a dramatic acceleration of species extinctions, and inundation of low-lying areas by rising sea level are among the possible outcomes.[84]

There is no hesitation or doubt in any of these statements. As Holdren expressed it, the truth that Earth was warming because humans were addicted to hydrocarbon fuels demanded no further scientific proof and permitted no room for serious questioning.

A bestselling book published in 1951 when Holdren was six years old was Eric Hoffer's *The True Believer: Thoughts on the Nature of Mass Movements*. The book grew out of the era's fascination that destructive personalities like Hitler and Mussolini had given birth to ideologies that captured the imagination of millions who were, in turn, motivated to create a war that killed more than sixty million people. Hoffer described true believers who adhere to mass movements as having "a proclivity for united action." He stressed that true believers "breed fanaticism, enthusiasm, fervent hope, hatred and intolerance."[85] As we will see in the next chapter, the global warming/climate change movement has morphed into becoming a secular religion among the political left worldwide.

In a sense, Holdren gave up trying to harness the sun's energy when he abandoned plasma physics. In becoming a Harrison Brown/Paul Ehrlich true believer, Holdren embraced their enchantment with eugenics, abortion, forced sterilization, and a "planetary government" on a mission to rid humanity of degenerates who did not (in their imaginations) deserve to live. These destructive impulses were the root that led to Holdren's almost religious-like devotion to the concept that hydrocarbons—the most plentiful, simple, and yet powerful energy ever discovered by humans—must be banned, or we all will most certainly die from global warming.

Conclusion

Modern fearmongers who predict that the population will ultimately outstrip our ability to produce food fail mainly for the same reason Thomas Malthus failed in his famous 1798 book, *An Essay on the Principle of Population*. Malthus based his calculations on failed mathematics. As a law of nature, he proposed that population growth proceeds at a geometric rate (i.e., 2, 4, 8, 16, 32, 64, etc.). But food production moves at an arithmetic rate (i.e., 1, 2, 3, 4, 5, 6, etc.). Thus, if not restrained, population growth must inevitably result in disaster, ultimately affecting the most successful populations that failed to be trimmed by war, natural disasters, or disease. In 1800, the world population was just topping one billion people. In 2021, the world population will be approaching eight billion people.

While famines have occurred, the population crisis Malthus thought was inevitable has failed to materialize. Likewise failed was Holdren's prediction that one billion people would die by 2020 in a new ice age. Yet, despite a string of failed predictions—despite a history in which human populations have continued to multiply and grow—Malthusian doomsayers continue to command the stage.

As we have seen, Julian Simon saw more people as positive for Earth since he felt the planet could abundantly produce given the expansion of human talent—the "ultimate resource"—that would be available in a world populated by eight billion or more people. For Harrison Brown, Paul and Anne Ehrlich, and John Holdren, the overpopulation problem caused by "too many" people must be stopped even if it requires a global totalitarian government denying human rights to accomplish the goal. Malthusian totalitarians understand that fear of survival will motivate people to accept a totalitarian regime because the doomsdayer authoritarian promises to eliminate the threat if people will just obey.

People who would fight to preserve liberty and freedom of expression switch and become intolerant once they become confident that governmental restrictions are necessary to survive as a species. People become intolerant of those wanting to protect the unborn because they come to see "anti-abortion activists" as failing to understand government-imposed population control as a necessity for the human species to survive. People who are vilified as "climate deniers" are those who treasure human ingenuity and believe burning hydrocarbon fuels is essential to providing economic prosperity, confident CO_2 creates a green Earth that benefits all life. Julian Simon has to be considered "Climate Denier Number One." Simon failed to be fooled by the anti-people Malthusian logic behind today's dominant narrative that we must abandon hydrocarbon fuels and decarbonize if we are determined to do all we can to "Save the Earth." In the next chapter, we shall see how neo-Marxists have coopted the global warming movement to advance their anti-capitalist agenda.

CHAPTER 3

The Movement to "Reimagine Capitalism" Goes Green

*Mass movements can rise and spread without belief in a god,
but never without belief in a devil.*
—Eric Hoffer, *The True Believer*, 1951[1]

*Nature took about a million years to lay down the amount of fossil
fuel that we now burn worldwide every year—and in doing so
it seems that we are causing rapid change of the Earth's climate.
Such a level of exploitation is clearly not in balance,
not harmonious and not sustainable.*
—John Houghton, *Global Warming: The Complete Briefing*,
Third Edition, 2004[2]

AS WE NOTED AT THE CONCLUSION to the previous chapter, hydrocarbon energy is the devil the green movement has chosen to vilify in a global consensus. The theme that we must decarbonize to save the planet can be traced back to the creation of the Intergovernmental Panel on Climate Change (IPCC) in 1988. Formed by the World Meteorological Organization (WMO) and the United Nations Environment Program (UNEP), the IPCC, headquartered in Geneva, Switzerland, comprises 195 member states with the full endorsement of the United Nations General Assembly. The IPCC's First Assessment Report (1990) influenced the formation of the United Nations Framework Convention on Climate Change (UNFCCC), an international treaty to reduce global warming and cope with the consequences of climate change.

The Fifth Assessment Report completed in 2015 provided the scientific input for the landmark Paris Climate Accords agreed to at the United Nations Climate Change Conference in 2015.[3] In 2018, the IPCC issued an alarming special report insisting that we must limit global temperatures to no more than 1.5°C above preindustrial levels within the next thirty years to prevent catastrophic global warming. Since its formation, the IPCC has led the global scientific effort to document and develop a policy advancing the argument that anthropogenic carbon dioxide (CO_2) emissions, resulting from burning hydrocarbon fuels, are in the process of causing that catastrophic global warming.[4]

The IPCC's 2018 special report warned that by 2017 the world had already warmed by 1°C, which meant the world would have to cut 2010-level carbon emissions by at least 45 percent by 2030. Cutting CO_2 emissions this drastically was the only hope we had to achieve net zero emissions (NZE) by 2050. The IPCC cautioned that if we failed to make significant cuts in CO_2 emissions, the world would register the 1.5°C increase above preindustrial levels between 2030 and 2050. Failure to meet these targets would produce catastrophic climate change consequences. They specified that global temperature increases between 1.5°C and 2.0°C would cause sea levels to rise, generate unbearably high temperatures in inhabited regions, and cause heavy precipitation or severe droughts in various geographic areas. When the 2018 IPCC report came out, Ove Hoegh-Guldberg, director of the Global Change Institute at the University of Queensland in St. Lucia, Australia, said that the world would become an almost impossible place to habitat without aggressive action. "As we go to the end of the century, we have to get this right," he demanded.[5]

Additional adverse consequences of the world warming another 1.5°C to 2°C, the IPCC warned, included disastrous impacts on biodiversity and ecosystems, including species loss and extinction. The panel projected that an increase with global warming of 1.5°C would cause climate-related risks to health, livelihoods, food security, water supply, human security, and economic growth. Particularly at risk would be disadvantaged and vulnerable populations, indigenous peoples, and local communities dependent on agriculture or coastal livelihoods. The environmental risk would be disproportionately higher to Arctic ecosystems, dryland regions, small island developing states, and Least Developed Countries (LDCs). The panel also warned of increases in poverty and negative impacts on human health with anticipated increases in heat-related morbidity and mortality.

They insisted that global warming of 2°C would destroy ecosystems on 13 percent of the world's landmasses. The report detailed that global warming of 2°C would cause the Arctic to experience ice-free summers, cause coral reefs to disappear, produce extreme hot days in midlatitudes, and generate severe storms in high-elevation regions of eastern Asia and eastern North America.

With the target of preventing Earth's temperatures from rising another 1.5°C, the IPCC report gave a specific, actionable target to the global warming alarm that they first sounded in their initial 1990 Climate Assessment Report. In 1988, James Hansen, a climatologist at the National Aeronautics and Space Administration (NASA) Goddard Institute for Space Studies in New York, was among the first to sound the climate alarm publicly. In his testimony to the U.S. Senate Energy and Natural Resources Committee, on June 23, 1988, Hanson asserted the greenhouse effect of anthropogenic CO_2 had warmed Earth to unprecedented highs:

> Number one, the earth is warmer in 1988 than at any time in the history of instrumental measurements. Number two, the global warming is already large enough that we can ascribe with a high degree of confidence a cause-and-effect relationship to the greenhouse effect. And number three, our computer climate simulations indicate that the greenhouse effect is large enough to begin to affect the probability of extreme events such as summer heat waves.... It is changing our climate now.[6]

Hanson told the Senate committee there was only a 1 percent chance that he was wrong on blaming rising temperatures worldwide on the buildup of greenhouse gases in the atmosphere.[7]

In 2018, three decades after Hanson's testimony, the IPCC reported clear specific goals for CO_2 reduction, setting in stone targets that various nongovernmental organizations (NGOs), regional governments including the European Union (EU), and nation-states could follow to implement action proposals to decarbonize.

While the IPCC has been demonizing coal, oil, and natural gas since its first 1990 climate assessment, the 2018 special report added urgency to the plea to decarbonize. With the 2018 report, the IPCC issued specific and well-articulated numerical global temperature goals, demanding a cut in using hydrocarbon fuels sufficient to make sure Earth's temperature would not rise 1.5°C above preindustrial levels. Historically, hydrocarbon fuels have provided cheap and readily available energy to propel our vehicles, heat and cool our homes, and power our offices, factories, hospitals, and schools. But now, the message was clear: Keep using coal, oil, and natural gas if you all want to die. Switch to renewable fuels if you care to save the planet for yourselves and for future generations.

The fear of overpopulation central to Harrison Brown in the 1950s and Paul Ehrlich in the 1960s gave way by the 1990s to John Holdren's climate change fear. The problem by 1990 was not just that there were too many people, but that with so many people burning hydrocarbon fuels, the resulting CO_2 emissions would trigger a greenhouse effect. The result would be rapidly increasing global warming ending up in catastrophic climate change. Under Brown and Ehrlich's scenario, overpopulation and starvation were certain to kill off a large proportion of humans on Earth today, but it would take time. Under the IPCC's 2018 warning, we have less than thirty years left to deal with global warming. By 2050, the IPCC warned catastrophic climate change was certain to have become irreversible. Harrison Brown, Paul Ehrlich, and John Holdren argued that stopping overpopulation would require an authoritarian global government capable of extreme measures, including forced sterilization. The IPCC's 2018 warning made clear that overpopulation plus the resulting continued increased global use of hydrocarbon fuels to power developed and developing economies threatened our survival on the planet.

Like overpopulation, the IPCC characterizes global warming/climate change as an anthropogenic crisis. We have nobody to blame but ourselves for the wanton burning of readily available, cheap hydrocarbon fuels. The CO_2 emitted as a consequence would certainly cause vast areas of Earth to become uninhabitable from the effects of unbearable heat. The global warming/climate change crisis would predictably trigger a sixth (and possibly final) extinction that would kill us all. Like overpopulation, combating global warming would require coordinated action by NGOs and governments acting with a worldwide consciousness. If ever authoritarian governmental action was required, the 2018 IPCC issued the mandate, specifying both the climate goal that we must reach (i.e., preventing Earth from heating another 1.5°C) and the target date (2050) for action.

The 2018 IPCC special report had for ecological/environmental neo-Marxists the impact of gospel truth. The political left worldwide quickly translated the IPCC's specific climate targets as action points that demanded we reexamine capitalism itself.

Green Activists Take Over the "Reimagining Capitalism" Movement

The original movement to "reimagine capitalism" can be traced back to the rise of the U.S. labor movement in the 1920s and the 1930s. In 1935, with the passage of the National Labor Relations Act (generally known as the Wagner Act), U.S. Congress established the right of collective bargaining, legitimating a labor movement that acknowledged management's responsibility to share "earnings" with workers. In the 1960s and '70s, Harvard sociologist Daniel Bell, with books like *The End of Ideology* (1960)[8], *The Coming of Post-Industrial Society* (1973)[9], and *The Cultural Contradictions of Capitalism* (1976)[10], introduced the notion that capitalism itself had become outmoded beyond reform. Bell's post-capitalist vision insisted corporations have a moral responsibility beyond profits and beyond the welfare of the workers employed to address the economic interests of those less fortunate in society as a whole.

In the wake of the 2007–2008 global economic crisis caused by the collapse of the subprime real estate market in the United States, progressive academics jumped on Bell's bandwagon. Suddenly, a new language emerged in the universities. Never fond of praising capitalism for creating unprecedented wealth, leftist intellectuals suddenly began talking about "democratizing corporations." A democratized corporation would think beyond profits. Such a corporation would take into account the interests of "stakeholders" (i.e., those affected by a corporation's actions) to move beyond the profit motives of "shareholders" (i.e., stockholders in particular, plus others like investment bankers who have an economic interest in corporate management). A democratized corporation would seek to redistribute wealth gained by profit to benefit victims of social injustice, including the disadvantaged poor.

After academic progressives sounded the call to democratize capitalism, the green movement insisted the campaign to "reimagine capitalism" had to take responsibility for implementing the IPCC's 2018 clarion call to achieve net zero emissions globally by 2050. The green movement's embrace of reimagining capitalism with the mission to combat global warming can also be traced back to the 2007–2008 global economic collapse. For a year from 2007 to 2008, British economist Ann Pettifor met with a group of British economists, ecologists, and environmentalists in her London flat to "set out to draft a plan for the transformation of the global economy away from its addiction to fossil fuels."[11] Pettifor and her group began calling their plan the Green New Deal (GND) "to echo the transformational financial and environmental policies of the 1933-1945 Roosevelt administration in the US."[12] Pettifor conceptualized their project as "based on the understanding that the economy and the ecosystem are tightly integrated—and that to protect the ecosystem we need to radically transform today's rapacious capitalism."[13]

In 2018, Pettifor and her Green New Deal group met with political activists Justice Democrats (JusticeDemocrats.com) who were visiting London. On their return to the U.S., the Justice Democrats urged Alexandria Ocasio-Cortez to adopt the GND as the basis of her congressional campaign that was challenging a sitting Wall Street Democrat in New York.[14] In drafting their version of the GND, the Justice Democrats drew heavily on the work of Mariana Mazzucato, an economist with dual Italian-U.S. citizenship who is a professor of economics of innovation and public value at University College London.[15]

In her 2019 book, *The Case for the Green New Deal*, Pettifor made the case that climate concerns demanded massive and immediate government intervention in the economy. "Both the US and the UK GNDs are based on the understanding that because climate breakdown is a security threat to the nation as a whole, the state has a major role to play in the transformation—just as if the nation were facing the threat of war," she wrote.[16] Pettifor argued that the Green New Deal is a demand for a revolution. She continued with the following declaration:

> The Green New Deal is a demand for a revolution in international financial relationships, the globalized economy, and humanity's relationship to nature. We demand an end to the imperialism of the dollar. An end to the toxic ideology and institutions of capitalism, based on extreme individualism, greed, consumption and competition—and fueled by spiraling levels of unregulated credit. Instead we insist and will uphold the boundaries and limits imposed by the capacities of both the ecosystem and the economy. We regard it as an urgent priority that the top 20 per cent of the world's big emitters, responsible for 70 per cent of global emissions, are made to radically reduce their carbon use. Carbon equity—between North and South, taking existing stocks of carbon into account—is fundamental to the Green New Deal. Finally, we demand—and will build—an economy based on social and economic justice, one that celebrates the altruism, cooperation and collective responsibility that is a characteristic of human nature.[17]

The British GND departed from the U.S. version in that the British GND sought to combat globalism with localism. As defined by Colin Hines, a former coordinator of Greenpeace International's Economics Unit and a member of the British GND, localism "would ensure that all goods, finance, and services that can reasonably be provided locally should be."[18] Still, Pettifor leaves no doubt she intended the British GND to be transformative, demanding "total decarbonization and a commitment to an economy based on fairness and social justice."[19]

Decarbonization Becomes Central to "Reimagining Capitalism"

Mariana Mazzucato, in her coauthored introduction to a 2016 book of essays entitled *Rethinking Capitalism: Economics and Policy for Sustainable and Inclusive Growth*, wrote that "the performance of Western Capitalism in recent decades has been deeply problematic."[20] Mazzucato's specific criticisms included her conclusion that the financial crash of 2008 and the prolonged recession and slow recovery that followed "have provided the most obvious evidence that Western capitalism is no longer generating strong or stable growth."[21] She noted that private investment in the Western capitalist countries has fallen as a percentage of gross domestic product (GDP). Stagnant economic growth was a consequence, with real median household income in the United States barely higher in 2014 than in 1990, even though GDP had increased by 78 percent in the same period. Meanwhile, the gap between the rich and the poor has grown larger. The richest one-hundredth of the U.S. population realized approximately 91 percent of the income gains in the years after the 2008 crash. Between the late 1990s and the late 2000s, the proportion of low-paid workers increased in most advanced economies. Unemployment has remained stubbornly high, particularly among young people. In 2010, the top 10 percent of the U.S. population owned 70 percent of all wealth.[22]

Mazzucato argued that underlying these economic trends in modern capitalism, there is another, deeper trend: "That is that of rising greenhouse gas emissions, which have put the world at severe risk of catastrophic climate change."[23] She argued that throughout capitalism's economic history, "economic growth has been accompanied by environmental damage, from the pollution of the air, water, and land to the loss of habitats and species, a constant subtraction from its successes in increasing welfare." Mazzucato insisted that "two hundred years of fossil fuel use in the developed world, now compounded by rapid growth in the emerging economies, means that, unless current emissions are drastically reduced, the world faces serious damage."[24] She cited the IPCC estimate that an increase of 2°C means that "we can expect a much higher incidence of extreme weather events (such as flooding, storm surges and droughts), which may lead to a breakdown of infrastructure networks and critical services, particularly in coastal regions and cities; lower agriculture productivity, increasing the risk of food insecurity and the breakdown of food systems; increased ill-health and mortality from extreme heat events and diseases; greater risks of displacement of peoples and conflicts; and faster loss of ecosystems and species."[25]

She lays this climate disaster at the doorstep of modern capitalism, commenting on the evidence that the coming CO_2 crisis has been known for a quarter of a century. Still, we have done very little to avoid catastrophe. Why? "The major reason is that the production of greenhouse gas emissions—particularly carbon dioxide—is so embedded in capitalism's historic systems of production and consumption, which have been built on the use of fossil fuels."[26] She noted that some 80 percent of the world's energy still comes from coal, oil, and natural gas. She concludes as follows: "Modern capitalism has in effect been storing up profound risks to its own future prosperity and security."[27]

Mazzucato's solution was to call for governments to refocus their efforts in a more systems-oriented approach to public policy, emulating how the U.S. government organized in the 1960s with NASA to send a human-crewed spaceflight to the moon and back. Specifically, Mazzucato called for "a moonshot" approach to changing capitalism.[28] She wanted a bold governmental effort to achieve policy results across the board. She had in mind a long list of government-funded social welfare projects. These projects ranged from providing affordable quality health care to all, improving our educational systems, cleaning the oceans, and reducing the "digital divide" between those who have ready access to the Internet and those who do not. Mazzucato was equally enthusiastic about applying a moon-mission, government-involved, systems approach to solving global warming and climate change. She wrote:

> Imagine if we were to bring the courage, spirit of experimentation and willpower of the moonshot to bear on the greatest problem of our time: the climate emergency. Imagine having leaders who proudly declare: "We choose to fight climate change in this decade not because it is easy, but because it is hard, because that goal will serve to organize and measure the

best of our energies and skills, because that challenge is one that we are willing to accept, one we are unwilling to postpone, and one which we intend to win."[29]

The "moon mission" concerning global warming would involve a commitment to implement the Green New Deal fully. Once again, Mazzucato called for an ambitious plan. She had the following goals in mind: upgrading all existing buildings in the United States for energy efficiency; working with farmers to eliminate pollution and greenhouse gas emissions as much as is technologically feasible; overhauling transportation systems to reduce emissions, including expanding electric car manufacturing and building charging stations to be readily available; and expanding high-speed rail. To this list, she added the following: "On top of that, the mission has social goals, including a guaranteed job with a family-sustaining wage, adequate family and medical leave, paid vacations and retirement security, and 'high-quality health care' for all Americans."[30]

With that statement, Mazzucato just went "woke" on the efforts she felt necessary to achieve decarbonization. As Alexandria Ocasio-Cortez has argued in defense of her Green New Deal program, achieving social justice goals is an integral part of creating the more livable "democratic" world she envisions once we eliminate the use of hydrocarbon fuels. In February 2019, together with Senator Edward Markey of Massachusetts, Representative Ocasio-Cortez unveiled the Green New Deal. She insisted that social justice goals were essential to achieving a sustainable future for everyone. She reasoned that global warming is an issue of equality since climate change hurts poor people the most while threatening to widen further the wealth gap between upper and lower classes in the United States.[31]

Naomi Klein Freaks Out

In her 2014 book, *This Changes Everything: Capitalism vs. the Climate*, Canadian author and political activist Naomi Klein confessed that she did not take the global warming issue seriously for many years. "I denied climate change for longer than I care to admit," she wrote. "I knew it was happening, sure." She continued, insisting she was not like "Donald Trump and Tea Partiers going on about how the existence of winter proves it's all a hoax."[32] Once she decided to pay attention, Klein realized how the global warming issue completely fit into her extensive leftist agenda. "And through conversations with others in the growing climate justice movement, I began to see all kinds of ways that climate change could become a catalyzing force for positive change—how it could be the best argument progressives have ever had to demand the rebuilding and reviving of local economies; to reclaim our democracies from corrosive corporate influence; to block harmful new free trade deals and rewrite old ones; to invest in starving public infrastructure like mass transit and affordable housing; to take back ownership of essential services like energy and water; to remake our sick agricultural system into something much healthier; to open borders to migrants whose displacement is linked to climate impacts; to finally respect Indigenous land rights—all of which would help to end grotesque levels of inequality within our nations and between them."[33]

In her enthusiasm as a recent true believer convert to the secular ideology of global warming leading to catastrophic climate change, Klein somehow forgot how her conversion reaffirmed the central proposition of her career. "I have spent the last fifteen years immersed in research about societies undergoing extreme shocks—caused by economic meltdowns, natural disasters, terrorist attacks, and wars," she acknowledged. Ironically, Klein did not seem to appreciate how she was falling into the same pattern. Wasn't Klein proposing to exploit a climate crisis to force social changes she desired? Klein detested the shock doctrine when corporate interests used the tactic. She wrote the following:

> As I discussed in my last book, *The Shock Doctrine*, over the past four decades corporate interests have systematically exploited these various forms of crisis to ram through policies that enrich a small elite—by lifting regulations, cutting social spending, and forcing large-scale privatizations of the public sphere. They have also been the excuse for extreme crackdowns of civil liberties and chilling human rights violations. [34]

But Klein did not appreciate that the shock doctrine psychology could apply to herself as well. Klein viewed the global warming cause as justified, so she considers the shock doctrine tactics completely acceptable. Her apparently hypocritical argument is reconciled only because with respect to the climate crisis, she represented herself as an unquestioned paragon of truth and justice.

After her conversion experience, Klein admitted she began pursuing global warming science with the ecstatic enthusiasm of a true believer. Suddenly, she found she "no longer feared immersing myself in the scientific reality of the climate threat." She became voracious in her pursuit of global warming/climate change "truths," acknowledging that she "stopped avoiding the articles and the scientific studies and read everything I could find."³⁵ She had a purpose once she saw how she could use the shock doctrine of "global warming will kill us all" as an instrumentality to achieve the social justice rectifications that were her true goal. She felt she was motivated by noble goals—saving people, fighting off evil global-warming capitalists. Hence, she saw no hypocrisy in her motives.

Here's how Klein explained her enthusiasm to embrace global warming causes and climate change theories. She wrote:

> And I started to see signs—new coalitions and fresh arguments—hinting at how, if these various connections were more widely understood, the urgency of the climate crisis could form the basis of a powerful mass movement, one that would weave all these seemingly disparate issues into a coherent narrative about how to protect humanity from the ravages of both a savagely unjust economic system and a destabilized climate system. I have written this book because I came to the conclusion that climate action could provide just such a rare catalyst.³⁶

In 2019, Klein's next book, *On Fire: The (Burning) Case for a Green New Deal*, displays page after page of her raging enthusiasm for Alexandria Ocasio-Ortez's FDR-like bold maneuver to save Earth. The book begins with nearly poetic descriptions of the March 2019 global School Strike for Climate. She gushed over the 100,000 "bodies" in Milan, 40,000 in Paris, 150,000 in Montreal all pouring out of schools into the streets like "rivulets were rushing rivers." She gushed over the "signs bobbed above the surf of humanity" as she typed out their messages in all capital letters: "THERE IS NO PLANET B! DON'T BURN OUR FUTURE. THE HOUSE IS ON FIRE!"³⁷ Klein detailed how the Puerto Rican community in New York came out onto the streets because of their friends and relatives who were "still suffering in the aftermath of Hurricane Maria."³⁸ She adored the student strikers in Delhi, India, who "braved the ever-present air pollution (often the worst in the world) to shout through their white medical masks, 'You sold our future, just for profit!'"³⁹ She lauded the generations of Australians distraught that "half the Great Reef Barrier, the world's largest natural structure made up of living creatures, had turned into a rotting underwater mass grave."⁴⁰

Klein probably did not realize that she had fallen into a classic psychological trap. Each climate change event claimed was proof of global warming resulting from CO_2 emissions in her uncritical embrace. In psychology, confirmation bias refers to a tendency to process information to confirm preexisting expectations. Those experiencing confirmation bias tend to pursue supporting evidence while dismissing or failing to follow contradictory evidence.[41]

So, hurricanes, air pollution, and coral reefs dying are all seen by Klein as proof positive that CO_2 emissions are killing the planet. In the extreme, global warming true believers interpret every extreme climate event as evidence of global warming. Even snow blizzards and long stretches of subzero temperature become proof of climate change caused by global warming. Under confirmation bias, every climate event that happens out of the ordinary, especially if the weather event is dramatic or has catastrophic consequences, is yet more evidence. Brush fires in California during the summer confirm for true believers that global warming is happening; so too with floods in Europe or China, and volcanoes in South America, or tsunamis in the South Pacific. Yet, commonsense logic parts ways when true believers insist events that look like global cooling, snow blizzards, for instance, are attributed to being caused by global warming. To understand the true believers' logic, we must appreciate how and why "global warming" morphed into "climate change."

Going back to the previous chapter, we pointed out that in the 1970s, climate fear centered on the possible coming of a new ice age. When global temperatures shifted to warming again, true believers switched the theme from global cooling to warming. Either way, the fear was climate change. Confirmation bias then kicks in to see every extreme weather event as caused by human action. Because we burn hydrocarbon fuels, we release CO_2 into the atmosphere. The CO_2 is seen as harmful whether it produces global cooling or global warming. Either way, anthropogenic CO_2 causes climate change. When global cooling was in vogue, global cooling caused extreme weather events. Now that global warming is in favor, global warming causes extreme weather events. But either way, human beings, by burning hydrocarbons and releasing CO_2 into the atmosphere, have damaged the planet by changing the climate. Thus, we are at fault because climate change is the reason polar ice caps melt, the reason the seas rise, and the reason hurricanes and tornadoes are so violent. Hysterical true believers will just assume the future global cooling resulted as a climate change artifact caused by global warming. The concept "anthropogenic climate change" becomes ideologically driven, assuming the status of a religious-like belief. The term "climate change" is now capable of morphing into either a future period of global warming or global cooling, with CO_2 remaining the culprit. If Earth's temperature goes into another ice age, climate hysterics could quickly begin claiming today's global warming produced tomorrow's global cooling.

Toward the end of her book, *On Fire*, Naomi Klein began weaving critical race theory into her concerns over global warming and climate change. She wrote the following:

> When it comes to climate action, it's abundantly clear that we will not build the power necessary to win unless we embed justice—particularly racial but also gender and economic justice—at the center of our low-carbon politics. *Intersectionality*, the term coined by black feminist legal scholar Kimberlé Crenshaw, is the only path forward. We cannot play "my crisis is more urgent than your "crisis"—war trumps climate; climate trumps class; class trumps gender; gender trumps race. That trumping game, my friends, is how you end up with Trump.[42]

Of course, Klein's critical race theory that was mapped onto global warming ideology ends up predictably hating Donald Trump. "See, it's Trump's fault," Q.E.D., the argument is over, and Klein wins. With this statement, Klein clarified that the green movement is more about leftist politics than about ecological or environmental science. For today's left, trained in critical race theory, everything—including capitalism and the use of hydrocarbon fuels—boils down to racism. When everything is "racist," nothing is racist (except possibly the person making the accusation).

The Sunrise Movement: "We Are the Revolution!"

The youth-led Sunrise Movement is a Green New Deal progressive activist organization employing leftist organizational techniques and demonstration tactics that date back to the 1960s to advance the global warming ideological agenda. The group came together in 2017 to impact the 2018 midterm elections. The Sunrise Movement came to prominence on November 13, 2018, a week following the midterm elections, when a group of some 250 Sunrise Movement protestors, joined by Rep. Alexandria Ocasio-Cortez and members of the Justice Democrats, staged a sit-in at the congressional D.C. office of House Speaker Nancy Pelosi. When U.S. Capitol Police arrested fifty-one of the demonstrators, the Sunrise Movement's urge to Pelosi to take action in the House to pass the Green New Deal became national news.[43]

On July 29, 2021, Sunrise Movement protestors chained themselves to the gate entrances of the White House, effectively shutting it down, as part of a protest against President Joe Biden for allegedly gutting the infrastructure bill before Congress by trimming the spending proposed for green energy projects. Protestors were also demonstrating against two new oil pipeline proposals. The pipelines involved in the protest were the Line 3 pipeline, which was planned to export oil from Alberta, Canada, through Minnesota to Wisconsin while crossing indigenous land, and the Mountain Valley Pipeline from southern Virginia to northwestern West Virginia.[44] With protest actions such as the Pelosi sit-in and the White House gate-chaining, the Sunrise Movement has emerged as the political action arm of the Green New Deal initiative.

Yet, if the Sunrise Movement was protesting the Biden administration while the "woke" activists within the Democratic Party were setting the plan in the White House and Congress, there was a disconnect. Despite all the noise over decades of blaming greenhouse gases for global warming, voters had no political urgency to enact a hugely expensive Green New Deal. From this perspective, the climate change alarmists had failed to meet the *Shock Doctrine* threshold of convincing the American people that massive government intervention to decarbonize was an urgent action item to be put on the nation's political agenda. In contrast, during the 1960s, Students for a Democratic Society (SDS)—one of the models for the Sunrise Movement—convinced thousands of draft-age Americans that fighting in Vietnam was not a cause worthy of risking life and limb.

The Sunrise Movement has incorporated the left's current "woke" ideology into its Green New Deal political action strategy. The green movement in the U.S. political left has come full circle, abandoning its sole focus on environmental issues. The result is that the Sunshine Movement has become a neo-Marxist political organization manipulating climate fears to advance the radical left's social justice and critical race plan to transform the United States into a government-controlled socialist state. In the introduction to their 2020 book on the Sunrise Movement entitled *Winning the Green New Deal: Why We Must, How We Can*, Varshini Prakash, the executive director of the Sunrise Movement, and Guido Girgenti, a founding director of the Sunrise Movement, explained why the group has incorporated "woke" ideology into its climate change, green political agenda. Prakash and Girgenti explained they were worried about avoiding a reaction like a backlash that developed in France, where a fuel tax hike led to the formation of the Yellow Vest movement in protest. As a result, the workers' movement that developed threatened the stability of the French government. Prakash and Girgenti elaborated as follows:

The Movement to "Reimagine Capitalism" Goes Green

That's why the Green New Deal recognizes the government's duty to guarantee fundamental human rights throughout our response to the climate crisis. It's the right thing to do, and it's the only way to get real climate policy implemented. The right to clean water, already violated in communities like Flint, will be made even more tenuous amid climate-induced drought, so water infrastructure must be at the top of our priority list. The right to quality health care through private employment could be imperiled as millions shift jobs due to climate policy; thus we must establish Medicare for All. The right to economic security can be met through a federal "job guarantee" policy offering dignified work to all who want it. And we must protect against institutional racism in all these programs, to guarantee that *all* people, black, brown, and white, benefit equally.[45]

Ian Haney López, the Chief Justice Earl Warren Professor of Public Law at the University of California, Berkeley, wrote an article entitled "Averting Climate Collapse Requires Confronting Racism" for the Sunrise Movement book. In this article, López made clear that social justice warriors fighting the decarbonization battle faced capitalism as their primary obstacle. López explained what stands in the way of government intervening on a massive scale to prevent global warming as follows:

> The primary culprit seems to be the reigning free market ideology that slanders government as serfdom's handmaiden and instead heralds loosely regulated capitalism as the surest route to liberty. This ideology prevents the government from acting forcefully on behalf of most Americans, thereby condemning the vast majority to chronic economic, health, and environmental jeopardy while it concentrates wealth in corporations and family dynasties. Yet it continues to hold sway not just in the halls of power but among broad swaths of voters.[46]

López asked what stood in the way of the government acting forcefully to defeat the free-market ideology. Here is his answer:

> Racism—or more particularly, the Right's strategic manipulation of racial resentment. The intense concentration of society's wealth generates widespread social misery. What can justify awarding the ultrarich so much while most struggle to get by on so little? Stories about free markets are not nearly sufficient to this task. They are far too abstract. Instead, the Right deploys visceral racist narratives about who "we" are, who threatens us, and who protects us, all told in coded language to hide the racism

while nevertheless triggering racial resentment and breaking public confidence in government for the collective good.

Decades of right-wing narratives linking people of color, hostility toward government, and class war have culminated in a strong connection between racial resentment and climate denial.[47]

Later in the article, López fingered conservative philanthropist and investor Charles Koch, noting that Koch's "immense fortune...rests on a huge petrochemical and industrial conglomerate." López charged that Charles Koch and his brother David "spent barrels of cash protecting their bottom line against efforts to protect the environment." López expressed outrage at what he believed were efforts by the Koch brothers to fund studies casting doubt about "climate science" while buying cooperation from "pliant politicians."[48]

With this proclamation, López grafts white privilege accusations and insinuations of white supremacy into the Sunrise Movement's green argument by suggesting in his conclusion that the real culprits are capitalism and the "planet-destroying" money-making schemes of "fossil fuel millionaires."[49] López went so far as to suggest the real political mission of the Green New Deal youth movement behind the Sunrise Movement was not to end global warming. The real purpose, López claimed, was to showcase to the American public how the wealthy used racism to divide people to protect and multiply their hydrocarbon-obtained riches. López closed by insisting the Green New Deal "provides a vehicle to help build a progressive, multiracial wave."[50]

Why Not a Climate Lockdown to Stop Global Warming?

Economist Mariana Mazzucato did not hesitate to advocate the idea. In September 2020, she wrote: "As COVID-19 spread earlier this year, governments introduced lockdowns to prevent a public-health emergency from spinning out of control. In the near future, the world may need to resort to lockdowns again—this time to tackle a climate emergency."[51]

Clearly, with people around the world confined for over a year in what amounted to house arrest, with businesses and central business districts closed by governments around the globe, less CO_2 was being emitted into the atmosphere. As early as April 2020, scientists reported that CO_2 emissions had dropped by as much as 40 percent, according to satellite readings. "As motorways cleared and factories closed, dirty brown pollution belts shrunk over cities and industrial centres in country after country within days of lockdown," the *Guardian* in London reported, noting that the improving air quality reduced the risks of asthma, heart attacks, and lung disease. "For many experts, it is a glimpse of what the world might look like without fossil fuels."[52] What the COVID-19 lockdown proved was that the climate change/global warming crisis, precisely like the overpopulation crisis, became reduced to a conclusion that there are just too many people on Earth. The overpopulation concern added to the climate change fear precisely fit the ideology of Harrison Brown, Paul Ehrlich, and John Holdren: the world would be a much better place if only most of the people were gone. For the Green New Deal movement, cities without vehicles were a lovely sight—just what the doctor ordered to end global warming and cure the problem of climate change.

The *Guardian* continued to report that road traffic fell in the U.K. by more than 70 percent, "to levels seen last when the Beatles were in shorts." Jonathan Watts, the *Guardian*'s global environmental editor, lamented the climate relief was likely temporary, lasting only until the COVID-19 lockdown ended. Still, the *Guardian* reported that "key environmental indices, which have steadily deteriorated for more than a half a century, have paused or improved." Watts continued:

> In China, the world's biggest source of carbon, emissions were down about 18% between early February [2020] and mid-March [2020]—a cut of 250m tonnes, equivalent to more than half the UK's annual output. Europe is forecast to see a reduction of around 390m tonnes. Significant falls can also be expected in the US, where passenger vehicle traffic—its major source of CO_2—has fallen by nearly 40%. Even assuming a bounce back once the lockdown is lifted, the planet is expected to see its first fall in global emissions since the 2008–9 financial crisis.[53]

The *Guardian* noted with apparent approval that the lockdowns were also affecting the fossil fuel industry with fewer drivers on the road and airplanes in the air. Watts commented that car sales had fallen by 44 percent in March, with motorway traffic down 83 percent. "So many more people are learning to teleconference from home that the head of the Automobile Association in the UK advised the government to switch infrastructure investment from building new roads to widening internet bandwidth," Watts continued. "This is potentially good news for the climate because oil is the biggest source of carbon emissions that are heating the planet and disrupting weather systems. Some analysts believe it could mark the start of a prolonged downward trend in emissions and the beginning of the end for oil."[54]

In an article published in the July 2020 issue of *Nature Climate Change*, a group of international scientists sympathetic with climate change activists presented a more realistic view of the COVID-19 lockdowns.[55] Corinne Le Quéré at the School of Environmental Studies at the University of East Anglia in the U.K. led the international scientific team that favored the idea of climate lockdowns. Still, the researchers concluded climate lockdowns were insufficient to meet the CO_2 emission-reduction goals set by international groups like the UN's IPCC. The researchers concluded that while the COVID-19 lockdowns had decreased global CO_2 outputs by 17 percent by early April 2020, the long-term impact depended on the duration of the confinement. They gave a high estimate of a 7 percent reduction in global CO_2 emissions if the COVID-19 lockdown restrictions lasted until the end of 2020.[56] The researchers noted that declines in CO_2 emissions in the range of one to two billion metric tons per year were necessary to meet the "safe worldwide temperature range" as defined by the Paris Climate Accords. The scientists estimated that the dramatic drop in CO_2 witnessed during the early days of the pandemic and global shutdowns would need to be matched by repeated pandemic lockdowns every two years for the rest of the decade to meet the goals of the Paris Climate Accords. Yet, in an article published in *Nature Climate Change* a year later, Corinne Le Quéré and her colleagues cautioned that lockdowns would not yield lasting reductions in CO_2 emissions because "a fossil fuel-based infrastructure" still sustains the global economy.[57] The researchers concluded the only way to get a steady drop in global CO_2 emissions would be for world governments to invest in green energy and divest from fossil fuels in the years after the pandemic. In their article published in 2021, Corinne Le Quéré and her colleagues cautioned disappointedly that most COVID-19 recovery plans were turning out to be "in direct contradiction with the countries' climate commitments."[58]

Let's return to why Mazzucato felt climate lockdowns were a good idea. She argued that under a climate lockdown, governments "would limit private-vehicle use, ban the consumption of red meat, and impose extreme energy-savings measures, while fossil-fuel companies would have to stop drilling." Mazzucato argued that COVID-19 was itself "a consequence of environmental degradation" that could rightly be termed "the disease of the Anthropocene." Mazzucato completed her analysis by returning to her familiar themes of "reorienting corporate governance, finance, policy, and energy systems toward a green economic transformation."[59] She wrote:

> Corporate governance must now reflect stakeholders' needs instead of shareholders' whims. Building an inclusive, sustainable economy depends on productive cooperation among the public and private sectors and civil society. This means firms need to listen to trade unions and workers' collectives, community groups, consumer advocates and others.[60]

She continued:

> Likewise, government assistance to business must be less about subsidies, guarantees, and bailouts, and more about building partnerships. This means attaching strict conditions to any corporate bailouts to ensure that taxpayer money is put to productive use and generates long-term public value, not short-term profit.[61]

Writing in the *Guardian* in March 2020 at the start of the global pandemic lockdown, Mazzucato urged that COVID-19 was a crisis that gave the world a chance to "do capitalism differently."[62] Echoing Obama's associate Rahm Emanuel that no good crisis should be allowed to go to waste, Mazzucato argued that the COVID-19 crisis gave governments the chance to do the right thing. Under the health crisis, governments could require that companies receiving government bailouts plan to lower carbon emissions. The government could also demand that corporations receiving pandemic bailouts make serious efforts to invest in workers, shifting from a shareholder orientation to a stakeholder orientation. "But we now have an opportunity to use this crisis as a way to understand how to do capitalism differently," she insisted. "This requires a rethink of what governments are for: rather than simply fixing market failures when they arise, they should move towards actively shaping and creating markets that deliver sustainable and inclusive growth."[63]

Klaus Schwab, COVID-19, and the "Great Reset"

While the political left worldwide saw the COVID-19 pandemic lockdown as a way to end global dependence on fossil fuels, Klaus Schwab, founder of the World Economic Forum headquartered in Cologny, Switzerland, saw COVID-19 as an opportunity to perpetuate globalism by implementing what he called the Great Reset. Schwab was not willing to be left behind by an ecology/environment-led global leftist movement that was demanding a green orientation to reimagine capitalism. Schwab declared climate change as an opportunity for forward-thinking multinational corporations to pledge the implementation of big-dollar green initiatives. Schwab noted that General Motors (GM), America's largest car manufacturer, had promised to go carbon neutral in its global products and operations by 2040. Apple committed to being 100 percent carbon neutral in its supply chain and products by 2030. Schwab announced he is "really excited" about these changes and believes the trend of a more stakeholder-centric view of the world is ahead. Schwab announced he has "a new mindset" about the need for climate action. He was confident the World Economic Forum could assist forward-thinking globalist corporations to "get it right."[64]

In a book published in June 2020 entitled *COVID-19: The Great Reset*,[65] Schwab laid out an aggressive plan for governments and corporations to rethink capitalism in a post-COVID-19 world where we are vulnerable not only to pandemics but also to the ecological/environmental dangers of global warming and climate change. In an article he published on the World Economic Forum's website in the same month his book was published, Schwab gave a succinct summary of his vision.[66] In that article, Schwab defined his Great Reset plan as having three components: (1) governments must steer markets "toward fairer outcomes," including upgrading international trade agreements and creating "the conditions for a 'stakeholder economy'"; (2) governments must develop policies that promote equitable outcomes, implementing wealth taxes, the withdrawal of fossil-fuel subsidies, and new rules regarding intellectual property, trade, and competition; and (3) corporations must "harness the innovations of the Fourth Industrial Revolution to support the public good, especially by addressing health and social challenges," working with universities to develop vaccines and health testing centers in anticipation of the next pandemic.[67] With these directives, Schwab has morphed his long-term interest into establishing a global economic system dominated by multinational corporations into a plan where those corporations must include green initiatives and operations. At the same time, governments must prepare to deal with future pandemics and lockdowns. Schwab encouraged governments to use government relief funds not to build corporate profits but to build "'green' urban infrastructure and to create incentives for industries to improve their track record on environmental, social, and governance (ESG) metrics."[68]

In his book, Schwab made clear that the coronavirus pandemic plunged the world into a crisis with "no parallel in modern history."[69] He

The Movement to "Reimagine Capitalism" Goes Green

characterized the COVID-19 crisis as changing everything. In the introduction to the book, Schwab wrote the following:

> We cannot be accused of hyperbole when we say it is plunging our world in its entirety and each of us individually into the most challenging times we've faced in generations. It is our defining moment—we will be dealing with its fallout for years, and many things will change forever. It is bringing economic disruption of monumental proportions, creating a dangerous and volatile period on multiple fronts—politically, socially, geographically—raising deep concerns about the environment and also extending the reach (pernicious or otherwise) of technology into our lives. No industry or business will be spared from the impact of these changes. Millions of companies risk disappearing and many industries face an uncertain future; a few will thrive. On an individual basis, for many, life as they've always known it is unravelling at an alarming speed. But deep, existential crises also favor inspection and can harbor the potential for transformation. The fault lines of the world—most notably social divides, lack of fairness, absence of cooperation, failure of global governance and leadership—now lie exposed as never before, and people feel the time for reinvention has come.[70]

Schwab correctly understands that the Overton window has shifted to demand government and corporate policies must accept the IPCC conclusions that the world faces a global crisis requiring decarbonization to prevent global warming that causes catastrophic climate change. For his multinational corporations to continue with their globalist objectives, Schwab insisted that they learn to adapt and operate in a global, politically correct environment that embraces and promotes the green agenda. To be clear, the Overton window is a conceptual construct identifying politically safe ideas that are deemed perfectly acceptable for public discussion. Views outside the Overton window are too radical for the public to accept and, hence, too dangerous to be principles of public policy.[71]

The point is that the politically correct public policy debate worldwide has shifted to accept global warming and climate change ideology as indisputably genuine. Schwab understands that the World Economic Forum and multinational corporations must establish corporate policies that advocate reducing CO_2 emissions. Put simply, global warming and climate change have become the only views considered orthodox, mainstream, or correct. To complete the analysis, Schwab constructed his Great Reset to make sure multinational corporations will continue to play a central role in dominating the emerging globalist economy, even when globalism itself goes green. Given this perspective, it becomes clear that Schwab's Great Reset embraces green operating policies. To oppose decarbonization orthodoxy would be suicidal. In other words, Schwab appears to have concluded that if multinational corporations cannot stop a green consciousness from dominating global ideology, he has no choice but to embrace a green future for the worldwide economy.

In his 2020 book, Schwab acknowledged the COVID-19 pandemic could be a setback for the green movement. He acknowledged that governments wanting to achieve recovery would be under pressure to "pursue growth" at any cost to cushion the impact on unemployment. He understood that companies would be under such pressure to increase revenues that "sustainability in general and climate considerations in particular will become secondary." He appreciated that low oil prices if maintained after the pandemic ends could drive both consumers and businesses "to rely even more on carbon-intensive energy."[72] Yet, Schwab concluded that "scattered factual evidence" convinced him that "the future would be greener than we commonly assume."[73] The enduring crisis, Schwab wrote, would be the climate crisis, not the pandemic risk. In conclusion, he stressed the following:

> Hopefully, the threat from the COVID-19 won't last. One day, it will be behind us. By contrast, the threat from the climate crisis and its associated extreme weather events will be with us for the foreseeable future and beyond. The climate risk is unfolding more slowly than the pandemic did, but it will have even more severe consequences.[74]

Skeptics of the Great Reset fear a Schwab-created future that ends up in Orwellian globalism. Under the Great Reset, multinational corporations will intensify efforts to work with governments to monitor people. With artificial intelligence (AI) Internet technology, a partnership between global corporations, international NGOs, and government could exploit pandemics for social control. An Orwellian future can require vaccine passports to maintain social compliance. In a subsequent iteration, AI-created and -held social acceptability scores can be used to determine who gets a job, who gets a bank account, and even what personal freedoms are permitted.

Those doubting Schwab's intentions should visit the World Economic Forum webpages specifically lauding Texas as an example of the "energy transition" to a decarbonized energy world that Schwab now considers inevitable, provided the World Economic Forum remains in control.[75] The Texas State Legislature has passed laws mandating various percentages of green energies, including solar and wind, that must constitute a specified percentage of the electricity generated in the state. At the same time, as energy companies in Texas have moved to comply, using green renewable fuels to generate electricity, and increased brownouts and blackouts have resulted.[76] Texas households are also beginning to understand that energy companies are implementing control measures to reduce brownouts and blackouts. For instance, Texas utilities are installing technological measures such as "'smart' thermostats" that will enable the energy companies to increase the home temperature to lower air conditioning during hot spells and raise temperatures during cold periods without the homeowner's consent.[77]

Conclusion

The ecological/environmental left has joined with politically motivated neo-Marxists to transform the Green New Deal into an attack on capitalism. For Green New Deal radical activists, the mixture of fighting for decarbonization as an energy policy and advocating a broad socialist agenda is an easy transition. Today, decarbonizing has morphed into neo-Marxism. Climate change activists also want to fight inequality, impose social justice, fight racism, and create government-funded, minimum annual income, along with government-guaranteed employment.

Green New Deal activists have further radicalized to embrace 1960s-style movement organization tactics.

Four accomplished progressive authors—Kate Aronoff, Alyssa Battistoni, Daniel Aldana Cohen, and Thea Riofrancos—wrote a 2019 book entitled *A Planet to Win: Why We Need a Green New Deal* advocating movement tactics. Consider the following excerpt from the book:

> Tackling the climate crisis will require action from unions, social movements, Indigenous peoples, racial justice groups, and others to take back power from the elites who've presided over the climate emergency. That's why the Green New Deal must combine climate action with attacks on social inequalities. Only then can we build enough public support and grassroots organizing to break the stranglehold of the status quo, and give people reasons to keep fighting for more. For all its flaws, the original New Deal excelled in creating a positive feed-back loop between public spending on collective goods and mass mobilization, thus overcoming anti-socialist hostility from the business class and political elites. A Green New Deal would likewise have to make climate action viscerally beneficial, turning

victories into organizing tools for yet greater political mobilization—and for ongoing liberation. Done right, investments in climate action could facilitate real freedom for everyone, the kind that only economic security for all makes possible.[78]

Moving beyond John Holdren, Alexandria Ocasio-Cortez has found in decarbonization an "end of world" fear powerful enough to unite progressives into an ideological battle against capitalism itself. Reluctant to characterize the movement as neo-Marxist, the Green New Deal movement has preferred identifying FDR and the social welfare programs his administration crafted during the Great Depression of the 1930s.

Blind adherence to doctrinal ideology is key to the fanaticism necessary to create a successful social or political movement. Eric Hoffer explained that active mass movements strive "to interpose a fact-proof screen between the faithful and the realities of the world."[79] He pointed out that the facts on which a true believer bases his conclusions "must not be derived from his [i.e., a true believer's] experience or observation."[80] For true believers, "the ultimate and absolute truth is already embodied in their doctrine, and there is no truth nor certitude outside it," Hoffer insisted.[81] "To rely on the evidence of the senses and of reason is heresy and treason. It is startling to realize how much unbelief is necessary to make belief possible. What we know as blind faith is sustained by innumerable unbeliefs."[82]

Hoffer concluded his comments on the importance of ideology (that he calls doctrine) for true believers by commenting in his 1951 book that the fanatical Communist refuses to believe any unfavorable report or evidence about Russia. The militant Communist refuses to see "with his own eyes the cruel misery inside the Soviet promised land."[83] The same holds for progressives who transform a supposedly scientific argument about CO_2 emissions into the political fight against capitalism. The Green New Deal movement insists we must move to solar energy and wind power only because these are not hydrocarbon fuels. The Green New Deal assumes solar energy and wind power are essential to a future committed to social justice. But, as we will see in the next chapter, the experience of the Obama administration to implement a green plan raises serious questions. Will the world these neo-Marxists control be any more productive, more livable, or more abundant in the decarbonized future they plan to create?

CHAPTER 4

Obama Redux: The "Solyndra Syndrome"

> *If executed strategically, our response to climate change can create more than 10 million well-paying jobs in the United States that will grow a stronger, more inclusive middle class enjoyed by communities across the country, not just in cities along the coasts.*
>
> —Senator Joe Biden 2020 campaign promise[1]

> *As a candidate, President Biden promised his "Build Back Better Recovery Plan" would create 10 million jobs—including millions in the resilient infrastructure and clean energy fields. These bold words echo promises made by President Obama and then-Vice President Biden in response to the "Great Recession," in 2009. The partisan American Recovery and Reinvestment Act included $90 billion for green jobs and billions more for the failed-Cash for Clunkers program. After funds were distributed, companies like Solyndra, A123 Systems, Beacon Power, and others went belly up and billions of taxpayer dollars were wasted.*
>
> —Senator John Barrasso, R-WY, Ranking Member, U.S. Senate Committee on Energy and Natural Resources, 2021.[2]

ON MAY 5, 2021, WYOMING Republican Senator John Barrasso, M.D., the ranking member of the Senate's Energy and Natural Resources Committee, gave a speech on the Senate floor. He called for the release of an investigative report that charged the Biden administration with repeating the Obama administration's failure to create the five million new clean energy jobs that Barack Obama had promised in his 2008 presidential campaign.

Barrasso called the phenomenon the "Solyndra Syndrome," charging that Biden was heading down the same path that the Obama administration proved would not work. Throwing billions of taxpayer jobs at various clean energy ventures had not worked for Obama, and Barrasso argued it would not work for Biden either. But like Obama, Biden was also putting a stranglehold on the oil and gas industries, where jobs were being created. A summary of Barrasso's report noted the following:

> President Biden and his administration seem determined to double down on the Obama administration's failed policies, while also punishing the oil and gas sector. He wants to invest hundreds of billions of taxpayer dollars in new green job training programs, new green energy financing, increased high speed rail, and new electric vehicle subsidy programs. At the same time, the president has taken executive action to damage the oil and gas industry. He signed an executive order to stop the construction of the Keystone Pipeline, ending the prospect for roughly 11,000 American jobs in 2021 alone. He has also implemented a moratorium on new oil and gas production on public lands that if made permanent could cost one million jobs and jeopardize the nearly $10 billion in revenue.[3]

Barrasso charged that President Biden was "doubling down" on failure, repeating the mistake the Obama administration had made frittering "away billions of taxpayer dollars on green gambles like Solyndra, while taxpayers got fleeced." Incredibly, the Democrats, once the party had control of the White House and Congress with Biden as president, decided to repeat Obama's failures with renewable solar and wind.[4]

Ironically, on September 4, 2009, then-Vice President Joe Biden was the one who announced that the Department of Energy (DOE) had just finalized a $535 million loan guarantee for Solyndra, LLC. This green energy company manufactured "innovative cylindrical solar photovoltaic panels that provide clean, renewable energy."[5] Biden enthusiastically noted the DOE loan guarantee aimed to finance the construction of Solyndra's manufacturing plant. He also bragged that the annual production of solar panels from the first phase of Solyndra's plans would provide the energy equivalent to powering 24,000 homes a year for over half a million homes during the project's lifetime.

In 2009, Solyndra estimated the new plant would initially create 3,000 construction jobs and lead to 1,000 jobs once the facility opens. Solyndra also expected hundreds of new jobs would be created as Solyndra's solar panels were installed on rooftops across the country. "This announcement today is part of the unprecedented investment this Administration is making in renewable energy and exactly what the Recovery Act is all about," Vice President Biden said. "By investing in the infrastructure and technology of the future, we are not only creating jobs today, but laying the foundation for long-term growth in the 21st-century economy." The Obama White House insisted that the first-phase financing of the Solyndra facility would manufacture up to seven gigawatts of solar panels capable of generating electricity equivalent to three or four coal-fired plants.[6] At its height, the *MIT Technology Review* touted Solyndra as one of the fifty most innovative companies globally.[7] President Obama personally visited Solyndra's solar panel manufacturing plant at the company's headquarters in Fremont, California, on May 26, 2010.[8]

The Solyndra Debacle

On September 6, 2011, Solyndra filed for bankruptcy, suspended operations at its headquarters, and laid off 1,100 workers. Solyndra went bankrupt despite $535 million in federal loan guarantees and more than $700 million in venture capital funding. [9]

The U.S. Department of Energy blamed the Solyndra bankruptcy on the Chinese, claiming the China Development Bank offered more than $30 billion in financing to Chinese solar manufacturers, "about 20 times more than U.S.-backed loans to solar manufacturers."[10] DOE charged that Chinese government funding of Chinese competitors allowed Chinese solar manufacturers to capture market share by undercutting prices. In 1995, the U.S. produced 40 percent of the world's solar panels, compared to 5 percent in 2011. DOE noted that at the same time, China's market share in the manufacturing of the world's solar panels had grown from 6 percent to 61 percent. However, DOE stayed on message, reassuring the American taxpayer that solar power would be producing a quarter of the world's electricity within four decades. The DOE report insisted that by then, more than $3 trillion worth of solar panels would be manufactured, creating "a vast economic and employment opportunity to be seized by countries that succeed in this sector."[11]

The real story appears to have been that, yes, the Chinese low-cost manufacturing was a factor in Solyndra's demise. Still, the truth seems to have been that the demand in the U.S. for solar panels on rooftops was nowhere near what the Obama administration wanted us to believe. At the time of the Solyndra bankruptcy, Axiom Capital Management's solar power analyst, Gordon Johnson, told *Bloomberg* that the supply of photovoltaic panels exceeded market demand. Johnson claimed that the supply in 2011 had climbed to almost triple the level of demand, crashing prices in the industry. "It could be Armageddon," Johnson said. "Demand is about to fall at a time when you're going to have a significant increase in supply. In a commoditized industry, that is a formula for disaster."[12]

On August 24, 2015, Gregory H. Friedman, DOE inspector general (IG), issued a special report blaming the bankruptcy on Solyndra's management. In the report, Friedman characterized the actions of Solyndra officials as "at best, reckless and irresponsible or, at worst an orchestrated effort to knowingly and intentionally deceive and mislead the Department."[13] The IG concluded that Solyndra had provided DOE "with statements, assertions, and certifications that were inaccurate and misleading." He charged that Solyndra management had "misrepresented known facts, and, in some instances, omitted information that was highly relevant to key decisions in the process to award and execute the $535 million loan guarantee."[14] Solyndra had represented to DOE that it had four sales contracts executed worth over $1.4 billion over the next five years but failed to disclose Solyndra had offered these customers price concessions not reflected in the sales figure. Solyndra's actual sales data was at considerably lower volumes and below-contract prices. Yet, to get the federal loan guarantee, Solyndra management had submitted to DOE a proforma spreadsheet estimate of sales with sales prices listed at the original, higher contract levels. The IG's report concluded the "Solyndra ordeal resulted in a loss to U.S. taxpayers likely to exceed $500 million and a corresponding loss of confidence in the loan guaranteed program."

On December 25, 2011, the *Washington Post* reported on its study of Obama's entire $80 billion clean energy technology program that had involved an analysis of thousands of memos, company records, and internal emails.[15] The conclusion was highly critical. "Meant to create jobs and cut reliance on foreign oil, Obama's green-technology program was infused with politics at every level."[16] With regard to Solyndra, the company records examined by the newspaper found that even when Solyndra warned that financial disaster might lie ahead, the Obama administration "remained steadfast in its support for Solyndra." The *Washington Post* concluded that "as Solyndra tottered, officials discussed the political fallout from its troubles, the 'optics' in Washington and the impact that the company's failure could have on the president's prospects for a second term. Rarely, if ever, was there discussion of the impact that Solyndra's collapse would have on laid-off workers or the development of clean-energy technology."[17]

The same *Washington Post* article further concluded that the Obama administration gave preferred access to investors in Solyndra who had donated to Obama's 2008 presidential campaign. Some of these preferred investors took jobs in the administration and helped manage the clean energy program. "Documents show that senior officials pushed career bureaucrats to rush their decision on the [Solyndra] loan so Vice President Biden could announce it during a trip to California," the *Washington Post* commented.[18] The newspaper also noted that Obama's May 2010 stop at Solyndra's headquarters, "like most presidential appearances," was "closely managed political theater." Additionally, the newspaper noted that Solyndra's strongest political connection was George Kaiser, a Democratic fundraiser and oil industry billionaire who happened to be an Obama campaign bundler in 2008. Kaiser had hosted Obama at his home. The *Washington Post* noted that Kaiser's family foundation owned more than a third of Solyndra, and Kaiser "took a direct interest in its [Solyndra's] operations."[19]

Peter Schweizer, head of the Government Accountability Institute, reported that 80 percent of the money spent in Obama's 2009 Recovery Act on green energy companies went to companies owned by individuals who had sat on Obama's finance committee for his 2008 presidential campaign. Given the number of influential donors in Biden's 2020 presidential campaign that have considerable financial stakes in green energy companies, Schweizer predicts Biden's "Build Back Better" green energy program amounts to nothing more than "a wealth transfer to Biden's biggest bundlers."[20]

Obama's Green Energy Investments Crashed and Burned

By 2015, the Obama administration had subsidized solar and other renewable energy in the United States with taxpayer money averaging $39 billion per year over five years. The five year federal subsidy for developing renewable energy technologies amounted to nearly $200 billion, with the dismal result that this massive investment in renewable energy resulted in less than 1 percent of additional electrical generation.[21] In total, the Obama administration financed some thirty-four faltering or bankrupt green energy companies, including the following: solar panel manufacturers Solyndra LLC ($535 million loss in federal loan guarantees) and Abound Solar Manufacturing, LLC ($400 million loss); Fisker Automotive ($529 million), a green vehicles program; and green energy storage companies Beacon Power ($43 million) and A123 Systems ($132 million).[22]

The Obama administration's experience with developing renewable energy technologies raises essential questions over the prospects of the Green New Deal under the Biden administration. Suppose green energy technologies were technologically feasible and economically profitable on the nationwide scale demanded by the Green New Deal. Why did the Obama administration lose so many billions of dollars on the initiative? If green energy technologies worked and were profitable, the Obama administration's determined efforts in over three dozen renewable fuel ventures with public funding on a scale never attempted would have been successful. Senator Barrasso noted that as a further insult to the American taxpayer, China's Wanxiang Group, a prominent Chinese auto parts maker, bought the assets (including the intellectual property) of A123 Systems and Fisker Automotive at a deeply discounted value after the two companies declared bankruptcy.[23]

Obama's Green Energy Boondoggle Extended to Spain

Obama's green energy fiascos were not limited to the United States. Consider the case of Abengoa, a Spanish multinational company headquartered in Seville.

Abengoa was a renewable energy company that perfectly scripted the Obama administration's renewable energy agenda in Europe. The company was a European counterpart to the U.S.-based Solyndra. On November 25, 2015, the *Washington Times* reported Abengoa had received at least $2.7 billion in federal loan guarantees since 2010 to build several large-scale solar power projects in the United States, with no certainty the company could pay back the government loans.[24] An earlier exposé, published in *Townhall* on August 4, 2012, made transparent the U.S. government's funding provided to Abengoa. The then estimated $2.8 billion Abengoa received in U.S. federal grants and loans made the company the second-largest recipient of the $16 billion doled out through the U.S. Department of Energy's Section 1705 loan guarantee program,[25] the same DOE program that had funded Solyndra.[26]

In November 2015, Abengoa began insolvency proceedings that resulted in a divestment plan announced in April 2016. The divestment plans involved selling off four renewable energy plants at a collective value of €57.2 million (USD 65.13 million). The divestment revenue represented a debt reduction of €50.3 million (USD 57.26 million) and a net cash flow of €12.2 million (USD 13.9 million), helping the company meet its debt restructuring targets set out in its feasibility plan.[27] The asset sale announced in April 2016 came after the renewables giant sold its 20 percent share in the 100MV Shams-1 concentrated solar power (CSP) plant in the United Arab Emirates in February 2016 to the Abu Dhabi–based renewable energy company Masdar.[28]

The bankruptcy, the largest in Spain's history, was triggered after Gonvarri, an arm of Spain's industrial group Gestamp, decided in November 2015 against a plan to invest around €350 million (USD 371 million) into the company.[29] On November 25, 2015, after the Abengoa bankruptcy was public, Reuters reported the company's bonds were "virtually worthless," as the company's share price plummeted 54 percent in a single day.[30] In April 2016, in a separate move, a local court in Mexico ordered the seizure of all Abengoa assets in Mexico to settle an action against the company brought to Mexican courts by bondholders seeking to prevent Abengoa from selling the Mexican assets without paying the bondholders.[31]

Al Gore, Hillary Clinton, and the Export-Import Bank Get Involved with Obama's "Spanish Solyndra"

Prominent Democrats joined Obama in the Abengoa disaster to involve the Export-Import Bank to get more U.S. taxpayer money into the Spanish venture.

The fascination of Democratic Party politicos with Abengoa began in 2007 when Al Gore's U.K. Generation Investment Management (GIM) bought a stake in Abengoa, a company Gore touted as "the largest solar platform in Europe."[32] GIM was started in 2004 by Al Gore and several Goldman Sachs executives, including David Blood, Mark Ferguson, and Peter Harris. In November 2015, Goldman Sachs announced plans to invest $150 billion in renewable energy projects, including solar and wind farms, energy efficiency upgrades for buildings, and power grid infrastructure.[33]

In her 2016 presidential campaign, Hillary Clinton argued for the reauthorization of the Export-Import Bank, insisting she wanted to be "the small business president." On June 8, 2015, Breitbart reported that under the Obama administration, Export-Import Bank lending had increased to 248 percent, with U.S. taxpayers now holding nearly $140 billion in Export-Import Bank exposure.[34] The same article noted that Abengoa had obligations of over $225 million in Export-Import Bank support. The report further disclosed that Bill Richardson, appointed by President Bill Clinton to serve as U.S. Secretary of Energy from 1998-2001, was involved in the Export-Import Bank affair. Richardson was both an advisory board member to the Export-Import Bank and a member of the Abengoa advisory board when the Export-Import Bank made the loan commitments to the Spanish-based renewable energy company.

On January 6, 2013, the *Washington Free Beacon* reported that the Export-Import Bank had approved a $78.6 million direct loan to Abengoa in December 2012 and a $73.6 million direct loan to a wind farm owned by Abengoa in Uruguay, noting Richardson's conflict of interest.[35] The *Washington Free Beacon* made clear the Export-Import Bank was in the business of extending taxpayer-backed loans to foreign buyers of U.S. exports. The Bank claimed the loans to Abengoa would generate 510 American jobs. "These two transactions demonstrate the strength of American energy technology and highlight the importance of this growing sector," Export-Import Bank Chairman and President Fred P. Hochberg said in a statement, as reported by the *Washington Free Beacon*. "In order for the U.S. to compete globally, our companies must continue to produce cutting-edge energy technology," the statement continued. "President Obama set an ambitious goal of doubling U.S. exports in five years, and these types of projects will help us meet that goal in 2015."[36]

Wall Street received Obama's green energy ambitions enthusiastically, realizing how many millions of dollars Wall Street investment bankers would make structuring and financing green energy ventures, whether the companies themselves succeeded or failed. When the companies did fail, the Abengoa case study illustrates how creative Wall Street investment bankers were to make sure Wall Street did not lose, even if taxpayers lost money in bankruptcies.

As Abengoa began to face solvency problems, the company spun off a "yieldco" under the name "Abengoa Yield plc," as a NASDAQ-listed company (NASDAQ: ABY). This Wall Street maneuver sought to create "sufficient separateness provisions" to insulate Abengoa Yield plc from the parent company's bankruptcy, Spain's Abengoa SA.[37] A "yielding company" or "yieldco" was, in 2016, a relatively new Wall Street innovation. The concept behind a yielding company is that the parent company developing renewable energy resources faces high risk, including insolvency. Still, once in operation, renewable energies should produce low-risk cash flows provided government subsidies remain in place. Goldman Sachs strongly supported the spin-off, given that Goldman Sachs was one of the 113 institutional shareholders owning Abengoa Yield shares, at that time valued at more than $45 million. After the Abengoa demise, these previous

Democratic Party champions, their minions in the mainstream media, and Hillary-leaning Goldman Sachs buried any further discussion of Abengoa. This company derisively became known on Wall Street as the "Spanish Solyndra."

Majority Speaker Harry Reid Partnered with China in Nevada

Under Obama's leadership, prominent Democrats in Congress realized the administration's green energy push offered a unique opportunity to cash in and get wealthy working with the Chinese. In 2008, smart money bet China would capture the solar panel simply because China's lack of moral inhibitions, which would have prevented them from utilizing slave or near-slave labor, would inevitably make China the world's lowest-cost manufacturer.

On April 3, 2012, *Bloomberg* reported Chinese billionaire Wang Yusuo, one of China's wealthiest citizens and the founder of Chinese energy giant ENN Group based in Langfang, China, had teamed up with Senate Majority Leader Harry Reid.[38] ENN sought to win incentives involving some 9,000 acres in Laughlin, Nevada, an unincorporated resort town in Clark County, about 113 miles southeast of Las Vegas on the Colorado River near the California border. ENN sought to purchase that land for $4.5 million, less than one-eighth of the land's $38.6 million assessment value. ENN's plans for Laughlin involved investing approximately $5 billion to construct a solar power station and a million-square-foot solar panel farm. *Bloomberg* described Laughlin, Nevada, as "pockmarked with foreclosed properties and the skeleton of a 14-story resort abandoned."[39] *Bloomberg* explained ENN intended to manufacture solar panels in Nevada despite the nearly 50 percent plunge in solar panel prices globally in the previous fifteen months that led to the bankruptcy of solar equipment maker Solyndra. ENN saw in Nevada an opportunity to avoid the 4.73 percent tariff the Obama administration had just placed on solar equipment imported from China after U.S. solar panel manufacturers complained about China's unfair competition.

What had China done to win Harry Reid's favor? *Bloomberg* documented that ENN had contributed $40,650 individually and through its political action committee to Senator Reid over three election cycles. They also reported that the ENN project produced legal work for Reid's son, Rory, a lawyer at the Las Vegas law firm Lionel Sawyer & Collins. That law firm, the largest in Nevada, was founded in 1967 by attorney Samuel S. Lionel and former Nevada governor Grant Sawyer. They were Democrats, like Reid, known for representing in Nevada top mining, energy, property development, and casino interests. The law firm had donated to Reid more than $40,000 in campaign contributions in the last three election cycles. By September 2012, ENN had appointed Rory Reid as a company representative. They used Rory Reid to front the Chinese company's proposal to build a $5 billion solar panel plant on the 9,000-acre desert plot in Nevada.[40] A Reuters report published on August 31, 2012, documenting that Wang Yusuo personally recruited Reid when he escorted Reid and a delegation of nine other U.S. senators on a tour of the ENN energy operations in Langfang, China.[41]

While he was Senate majority leader, Reid had ambitious plans to enrich himself and his family by transforming thousands of open range and ranch acres in Nevada into solar energy projects involving his Chinese partners. At the same time, Reid was campaigning hard against coal-powered plants in the state. Nor was Reid shy about positioning family members to benefit from his various business deals in Nevada. Years earlier, on June 23, 2003, the *Los Angeles Times* reported on Reid's ties with the Howard Hughes Corporation.[42] The Hughes Corporation paid $300,000 "to the tiny Washington consulting firm of son-in-law Steven Barringer to push a provision allowing the company to acquire 998 acres of federal land ripe for development in the exploding Las Vegas metropolitan area." The *Los Angeles Times* further reported the legal work for that deal was done by the law firm Lionel Sawyer & Collins when four of Reid's sons—Reid, then age forty; Leif, then thirty-five; Josh, then thirty-one; and Key, then twenty-eight—all worked for the law firm.

On July 10, 2013, environmental journalist Chris Clarke published an article about ENN on the website of KCET television news in Los Angeles.[43] Clarke noted that the Laughlin site on which ENN had planned to build its solar project was the home to the 1,580-megawatt coal-fired Mojave Generating Station, owned by Southern California Edison, which Reid and other Democrats in Congress forced to close in 2005. Demolition began in 2009 because the coal-burning facility, using coal mined on Navajo and Hopi reservations, arguably polluted the southwest desert air.

After the Mojave Generating Station had been closed and demolished, Reid contacted Southern California Edison, the majority owner of the destroyed coal-fueled power plant, to ask permission to use the site for a solar power plant. "When the plant closed down, the local communities and Native American tribes lost valuable jobs," Reid said in a statement published by his office. "I am urging Southern California Edison to convert the plant and its assets into a vibrant solar power producer to give the area an economic boost." Reid further suggested to Southern California Edison that industrial plants could sell sulfur dioxide credits, rather than continue to operate and spew out pollution, and the company could use that money to build a solar power plant.[44]

Ironically, the Chinese canceled the ENN project planned for Laughlin, Nevada, when NV Energy, a major public utility generating electricity in Nevada, explained the company had no need for new solar power projects for the foreseeable future. To comply with requirements under Nevada law to use renewable energy sources, NV Energy already had more than enough solar power projects in progress or under planning. On June 17, 2013, the Associated Press affirmed that ENN Group had terminated its agreement to purchase 9,000 acres in Laughlin. "ENN Mojave Energy LLC has informed Clark County officials that it's terminating its agreement to purchase 9,000 acres after it was unable to find customers for the power that would have been generated there," the AP reported. "The company, a subsidiary of ENN Group, says it was unable to sign the necessary power purchase agreements to sell the energy to utilities in Nevada or neighboring states."[45] Clark County Commissioner Steve Sisolak said it had been a great pleasure working with ENN, but "when it all came down to the end, they just couldn't sell the power." Sisolak explained why the ENN project in Nevada failed. "Alternative energies are still more expensive than fossil fuels and they [ENN] couldn't get (the costs) down to a point where they could sell any of the power," he told the *Las Vegas Sun*. "Even if we had given them an extension for a year or two, it wouldn't have made a difference."[46]

Democrats Attempted to Profit from Carbon Emissions Tax Scheme

One more incident involving a Democratic operative's apparent attempt to cash in on green energy involves Franklin Raines, a former Clinton administration budget director and a housing advisor to Senator Barack Obama in his 2008 presidential campaign. Raines earned $90 million in his five years as CEO of the government mortgage giant Fannie Mae, from 1999 to 2004, only to resign in disgrace. Raines was one of several Democratic operatives who directed Fannie Mae to finance subprime loans to minorities. The political left in the 1990s insisted that banks were denying loans based on race to low-income potential homeowners. The political left charged this in a discriminatory practice that was called "red lining."

The collapse of the subprime real estate market in 2007–2008 caused the global recession that followed—a recession that, as we have seen, leftist economists today use to argue that capitalism is a flawed system.

The public record documents that Raines and his partners began filing U.S. patents while Raines was head of Fannie Mae. The patents gave Raines and his partners ownership and control over any carbon "cap and trade" taxing scheme the Obama administration might implement. The patents covered Environmental Protection Agency regulations under the Clean Air Act and any new legislation Congress might pass. Consider the following proof of these statements:

- The first patent in question is U.S. Patent #6904336, entitled "System and Method for Residential Emissions Trading," applied for by Franklin Raines and his associates on November 8, 2002, while Raines was CEO of Fannie Mae. On June 7, 2005, this first patent was issued.[47]
- In three separate assignments made in April and July 2004, Raines and his associates assigned this first patent #6904336 to two entities: Fannie Mae and CantorCO2e.[48] Carlton Bartels, the chief executive of Wall Street trading and investment firm Cantor Fitzgerald and head of the spin-off CantorCO2e organization, was one of Raines's partners listed as an inventor and co-owner of patent #6904336. In August 2011, CantorCO2e became BCG Environmental Brokerage Services following the acquisition of CantorCO2e's North American business by BGC Partners, Inc. (Nasdaq: BGCP), a leading global brokerage company primarily serving wholesale financial markets. BGCP has offices in twenty-five cities around the world. Since 1992, emissions trading services has been the core business of BCG Environmental Brokerage Services.[49]
- On April 28, 2005, Raines and his partners, including Carlton Bartels, applied for a second carbon emissions patent under the same name, "System and Method for Residential Emissions Trading," issued on November 7, 2006, as U.S. Patent # 7133750.[50]
- A close examination of this second patent clarifies that it was nearly identical to the first, with only a few sentences in the claims modified such that the substantive meaning and intent of the second patent were not different from the first. The U.S. Patent and Trademark Office records clarify that this second patent, U.S. Patent #7133750, was never assigned to any other party, with the result

that Raines and the other individuals listed on the patent as inventors retained all ownership rights.

What was the point for Raines and Fannie Mae of applying for these carbon emission patents? The idea appears to have been for Fannie Mae to create what would have amounted to a type of securities investment that would be called collateralized carbon obligations, or CCOs. The idea was to emulate a similar methodology Fannie Mae utilized in the heyday of Fannie Mae's subprime real estate lending. In those years, collateralized loan obligations, or CLOs, combined individual home loans into investment securities. The collapse of the collateralized loan obligation market was one of the precipitating causes in 2007–2008 of the financial failure that led to the subsequent global recession.

These carbon emission patents gave Raines and his associates ownership of the methodology for creating collateralized carbon obligations. Here, carbon emission reduction payments on a proposed carbon exchange would become investment securities packaged by Fannie Mae as CCOs. In effect, these patents gave Raines and his associates control over a future carbon-exchange market. With this methodology under U.S. Patent control, Raines and Fannie Mae could demand payment from Wall Street firms seeking to create collateralized carbon obligations.

The patents described a methodology where regulators could measure ongoing energy use by installing new meters in homes. The language of the patents authorized Fannie Mae to create "a computer-implemented method" to convert household energy savings into tradable credits. Among many suggestions, the patents recommended replacing older appliances with more energy-efficient appliances. The patents specifically suggested a natural gas hot water heating system could easily replace an oil-fired boiler. The patents also instructed homeowners participating in the program to consider installing insulation in attics and exterior walls as well as more efficient windows.

CantorCO2e was originally a spin-off subsidiary of the Wall Street investment firm Cantor Fitzgerald. The patents listed CantorCO2e's CEO Carlton Bartels as a co-inventor of the household carbon-emissions investment-packaging methodology. Bartels died tragically when one of the hijacked airplanes hit the Cantor Fitzgerald offices in the Twin Towers during the 9/11 terrorist attacks. After his death, his wife, Jane Bartels, took legal ownership of his property interests in the patents. CantorCO2e at that time was also listed as an offset aggregator on the Chicago Climate

Exchange. This listing suggested CantorCO2e planned to work in creating CCO securities. By utilizing the Raines-filed patents, CantorCO2e and Fannie Mae could package the household energy credits traded as CCO securities on a planned future carbon-credit exchange.

In a letter dated May 25, 2010, after these patents were made public, Fannie Mae general counsel Alfred M. Pollard attempted to explain to Reps. Darrell Issa (R-California) and Jason Chaffetz (R-Utah) of the U.S. House Committee on Oversight and Government Reform, that the actions taken by Raines and Fannie Mae to file these patents were perfectly acceptable. Pollard wrote the following:

> Residential emission trading, if developed in the market, would be conducted by others in the financial industry or other subject matter experts. Similarly, Fannie Mae did not pursue a patent out of a desire for potential royalties, but instead with the hope it could help facilitate the implementation by others of its original concept that residential builders could leverage their investments in building energy efficient houses.[51]

Although Raines has consistently denied he had any personal profit motive in filing these patents, the history would suggest Democratic operatives close to the Obama administration saw green energy as a way to make money. Those controlling these carbon patents could reap enormous profits by managing the envisioned Chicago Climate Exchange (CCX) and packaging collateralized carbon obligations on the CCX. Despite repeated efforts during Obama's first term in office, climate environmentalists failed to get any carbon emissions cap-and-trade legislation passed by Congress.[52]

In 2004, Franklin Raines's career at Fannie Mae came to an end. Fannie Mae's regulator, the Office of Federal Housing Enterprise Oversight (OFHEO), and the Security and Exchange Commission's (SEC) top account issued reports that Fannie Mae had misstated under Raines's stewardship earnings for three and a half years. The $9 billion restatements of earnings the OFHEO and SEC required ended up wiping out 40 percent of Fannie Mae's originally stated profits from 2001 to mid-2004.[53] Raines resigned from Fannie Mae in December 2004 with a $19 million severance package.[54]

Raines continued playing the victim until April 2008, when Raines and two other Fannie Mae top executives were ordered in a civil lawsuit to pay nearly $31.4 million for their roles in what amounted to an Enron-like accounting scandal. Raines and the other Fannie Mae executives were also accused in the civil suit of manipulating Fannie Mae's books to manufacture earnings over six years that stretched from 1998 through 2004 to trigger for themselves millions of dollars in otherwise unearned bonuses.[55] Raines was also forced to give up his Fannie Mae stock options in the final settlement, then valued at $15.6 million. None of those accused admitted wrongdoing in the matter.

Obama's Green Jobs Hoax

Campaigning for president in 2008, Obama promised that $150 billion in government spending on green energy projects would create five million green-collar jobs in ten years. Again, this promise was a flop. The millions of jobs that Obama's green energy plans promised never materialized.

On March 12, 2012, Obama visited the Copper Mountain Solar Project in Boulder City, Nevada. White House press aides arranged for press photographers to stage Obama's podium against an impressive background of solar panels stretched to the distant horizon. The Copper Mountain Solar Project, built on public land, planned to use one million solar panels to provide solar energy to 17,000 homes. "Three weeks ago, President Barack Obama stood in front of a sea of gleaming solar panels in Boulder City, Nevada, to celebrate his administration's efforts to promote 'green energy,'" Andy Sullivan reported for Reuters on April 13, 2021.[56] "But the millions of 'green jobs' Obama promised have been slow to sprout, disappointing many who had hoped the $90 billion earmarked for clean-energy efforts in the recession-fighting federal stimulus package would ease unemployment—still above 8 percent in March." Yet, Sullivan noted, the Copper Mountain Solar facility employed only ten people. According to the American Wind Energy Association, Sullivan further reported that the wind industry had shed 10,000 jobs since 2009. According to Department of Labor statistics, the oil and gas industries added 75,000 workers in the same period.

On November 2, 2012, *Forbes* reported that Obama's green jobs were costing "taxpayers big bucks."[57] The publication noted that total wind, solar, and nuclear subsidies under Obama, after four years in office, had produced a mere 252,000 jobs. *Forbes* concluded the following: "The bottom line is that green-energy jobs cost taxpayers, on average, 15 times more than oil, gas, and coal jobs. Wind-backed jobs cost 25 times more. With improvements in hydraulic fracturing and horizontal drilling, this cost gap may actually grow in the coming years, rather than decline as renewables advocates often assume." The mainstream media in 2012, at the end of Obama's first term in office, was willing to report honestly on the complete failure of Obama's wind and solar energy adventures. The Obama administration had failed to create the millions of green energy jobs promised, despite billions of dollars in federal taxpayer spending. But some ten years later, with Biden in the White House, the mainstream media censored any adverse green energy reporting. By 2021, the mainstream media lavished praise on Rep. Alexandria Ocasio-Cortez's Green New Deal, believing once again that green energy would create millions of new jobs.

Former World Bank adviser and global warming advocate Gordon Hughes issued a disappointing 2011 report written for the Global Warming Policy Foundation in Great Britain.[58] Hughes admitted the left's hyping of green jobs stimulating the economy turned out to be a myth. Hughes, a professor of economics who currently volunteers to lecture part-time at the University of Edinburgh, initially believed green energy claims. In his report, entitled "The Myth of Green Jobs," Hughes concluded that there are "no sound economic arguments to support an assertion that green energy policies will increase the total level of employment in the medium or longer term when we hold macroeconomic conditions constant."[59]

Hughes conceded that more people might be employed in manufacturing wind turbines and constructing wind farms. Still, he noted, "this neglects the diversion of investment from the rest of the economy."[60] He stressed that generating green energy is highly capital intensive in that "generating electricity from renewable energy sources will involve a capital cost that is 9-10 times the amount required to meet the same demand by relying upon conventional power plants."[61] Financing green energy projects by taxing carbon use diverts money from other forms of business investment such that the immediate impact will be approximately neutral. He commented that in almost every country globally, some argue that green energy policies will promote innovation and the development of new industries. But in the end, "the numbers do not add up."[62]

Rep. Alexandria Ocasio-Cortez: "The World Is Going to End in 12 Years!"

On February 7, 2019, Rep. Alexandria Ocasio-Cortez (D-NY) and Sen. Ed Markey (D-MA) introduced a resolution calling for a Green New Deal into the House and Senate. Among the Democrat 2020 presidential contenders, the following signed on to the measure as cosponsors: Senators Bernie Sanders (I-VT), Kirsten Gillibrand (D-NY), Elizabeth Warren (D-MA), and Amy Klobuchar (D-MN). House Speaker Nancy Pelosi (D-CA) refused to bring the Green New Deal to the House floor for a vote. On March 25, 2019, the resolution failed to advance in the Senate, with most Democrats voting "present." The issue, however, has moved to the front and center of the Democrat socialist agenda for the 2020 presidential campaign.

Yet, the Democrats running for president in 2020 realized the political necessity of endorsing the Green New Deal or risk losing the support of the Democratic Party's ecological/environmental radicals who had embraced the radical left's socialist critical race theory as enthusiastically. In the Democratic 2020 presidential debates, serious candidates who wanted to advance in the primaries realized the need to articulate the following type of campaign rhetoric:

> There's a lot of people now that are blowing back on the Green New Deal. They're like, "Oh, it's impractical! Oh, it's too expensive! Oh, it's all of this!" If we used to govern our dreams that way, we would never have gone to the moon. We need to be bold again in America. We need to have dreams that other people say are impossible. When the planet has been in peril in the past, who came forward to save Earth from the scourge of Nazi and totalitarian regimes? We came forward.
> —**Sen. Cory Booker** (D-NJ), speech in Mason City, Iowa, February 8, 2019[63]

> Here's the truth: climate change is real, and it is an existential threat to our country, our planet, and our future. With each passing day, the imminent threat of climate change grows—and we see it in everything from more instances of extreme weather to rapidly melting glaciers. According to a harrowing report from the U.N. Intergovernmental Panel on Climate Change, we have a shrinking window to take drastic action to cut carbon emissions and make meaningful change to save our planet. The Green New Deal is a bold plan to drastically shift our country to 100% clean and renewable energy. We will repair our country's crumbling infrastructure, upgrade buildings across the nation, and dramatically cut greenhouse gas emissions.
> —**Sen. Kamala Harris** (D-CA), "Green New Deal," February 8, 2019[64]

So, when I talk about taking on the fossil fuel industry, what I am also talking about is a just transition. All right? We can create, and what the Green New Deal is about, it's a bold idea. We can create millions of good paying jobs. We can rebuild communities in rural America that have been devastated. So, we are not anti-worker. We are going to provide, make sure that those workers have a transition. New jobs, healthcare and education.

—Sen. Bernie Sanders (I-VT), Democratic debate, second round in Detroit, Michigan, on night one, July 30, 2019[65]

On January 19, 2019, at the Women's Unity Rally at Foley Square in New York City, Rep. Ocasio-Cortez declared that young Americans fear "the world is going to end in 12 years if we don't address climate change." Ocasio-Cortez called the fight to mitigate the effects of climate change her generation's "World War II."[66] Ocasio-Cortez appears to have arrived at this twelve-year estimate from the 2018 IPCC report that concluded the world must reach a net zero emissions (NZE) goal by 2050. As noted earlier, to achieve this goal, we had to reduce carbon dioxide emissions dramatically by 2030. But while the IPCC predicted dire climate changes would occur, they never said the "world was going to end" if we did not meet the NZE carbon-emission-reduction target by 2030.

On January 21, 2019, in an interview at the Riverside Church in Harlem, New York City, at a Martin Luther King Jr. Day event, Ocasio-Cortez expanded her remarks. "I think that the part of it that is generational is that millennials and Gen-Z and all these folks that come after us are looking up and we're like, 'The world is going to end in 12 years if we don't address climate change,' and your biggest issue is—your biggest issue is, 'How are we going to pay for it?' And, like, this is the war; this is our World War II."[67] Ocasio-Cortez, in her hysteria, is on the record insisting the world is going to end and we are all going to die unless we comply fully with her "transformative" Green New Deal.

Even Patrick Moore, the cofounder and former president of Greenpeace, says that climate change is "a scam" that has been "taking over science with superstition and a kind of toxic combination of religion and political ideology."[68] John Coleman, the late cofounder of The Weather Channel, echoed these sentiments. In 2024, Coleman slammed the six-hundred-page Federal Assessment of Climate Change issued by an Obama administration team of more than three hundred specialists guided by a sixty-member federal advisory committee. "When temperature data could no longer be bent to support global warming, they switched to climate change and now blame every weather and climate event on CO_2 despite the hard, cold fact that the 'radiative forcing' theory they built their claims on has totally failed to verify," Coleman said in his blog on May 7, 2014. "The current bad science is all based on a theory that the increase in the amount of carbon dioxide in the atmosphere from the exhaust of the burning of fossil fuels leads to a dramatic increase in 'the greenhouse effect' causing temperatures to skyrocket uncontrollably. This theory has failed to verify and is obviously wrong." In the 2014 blog, Coleman attacked "the politically funded and agenda driven scientists" who he claimed had built their careers on this theory while living well on the $2.6 billion a year the federal government pays funding global warming research. Coleman ended the blog by characterizing the federal climate assessment report as "a destructive episode of bad science gone berserk."[69] In an October 2014 open letter in which he attacked the United Nations Intergovernmental Panel on Climate Change, Coleman wrote: "The ocean is not rising significantly. The polar ice is increasing, not melting away. Polar Bears are increasing in number. Heat waves have actually diminished, not increased. There is not an uptick in the number or strength of storms (in fact storms are diminishing). I have studied this topic seriously for years. It has become a political and environment agenda item, but the science is not valid."[70]

How Do We Pay for the Green New Deal?

Suppose we detail the elements in the Green New Deal introduced by Representative Ocasio-Cortez and Senator Markey into Congress. In that case, we begin to understand the extent to which the plan requires a fundamental restructuring of the U.S. economy. The resolution called for the federal government to create an ambitious Green New Deal program aimed at accomplishing the following goals through a ten-year national mobilization:

- achieve net-zero greenhouse emissions;
- meet 100 percent of the power demand in the United States through clean renewable and zero-emission energy sources;
- repair and upgrade the infrastructure in the United States, including eliminating pollution and greenhouse gas emissions as much as technologically feasible;
- build or upgrade to energy-efficient, distributed, and "smart" power grids, providing affordable access to electricity;
- through electrification, upgrade all existing buildings in the United States and build new structures to achieve maximum

energy efficiency, water efficiency, safety, affordability, comfort, and durability;
- overhaul transportation systems in the United States to eliminate pollution and greenhouse gas emissions from the transportation sector as much as is technologically feasible, through investment in (i) zero-emission vehicle infrastructure and manufacturing; (ii) clean, affordable, and accessible public transportation; and (iii) high-speed rail;
- spur massive growth in clean manufacturing in the United States by removing pollution and greenhouse gas emissions from manufacturing and industry as much as is technologically feasible, including by expanding renewable energy manufacturing and investing in existing manufacturing and industry; and
- work collaboratively with farmers and ranchers in the United States to remove pollution and greenhouse gas emissions from the agricultural sector as technologically feasible.[71]

Ocasio-Cortez and Markey also included as an integral part of their Green New Deal objectives a laundry list of social justice initiatives not related to climate policy.[72] Reporting on the resolution introduced into Congress, the *Washington Post* listed the following included social justice provisions:

- guarantee a job with a family-sustaining wage, adequate family and medical leave pay, paid vacations, and retirement security to all people of the United States;
- provide all people of the United States with (i) high-quality health care; (ii) affordable, safe, and adequate housing; (iii) economic security, and (iv) access to clean water, clean air, healthy and affordable food, and nature; and
- Provide resources, training, and high-quality education, including higher education, to all people of the United States.[73]

An overview of the Green New Deal released by Representative Ocasio-Cortez's office summarized that the plan seeks a "massive transformation of our society" that could "rid the country of fossil fuels" and "create millions of family supporting-wage [i.e., union] jobs." Proponents touted the Green New Deal as a broad economic plan that would create an environmentally sound country with benefits for everyone, even for those who do not want to work.[74]

Ocasio-Cortez's congressional office attached FAQs to the document. The FAQs made clear the Green New Deal calls for a "full transition off fossil fuels and zero greenhouse gases," lamenting the ten-year goal may be too fast to "be able to fully get rid of farting cows and airplanes." Her Green New Deal proposal aimed to outlaw all automobiles and other vehicles that run on gasoline. The plan sought to close manufacturing plants and industries that burn coal or other hydrocarbon fuels. The plan called for rigorous inspection of all buildings in the United States to identify the structures of that ecosystem either through the use of hydrocarbon fuels, the failure to include adequate insulation, or the inability to limit the greenhouse gases they emit into the atmosphere.[75]

Douglas Holtz-Eakin, a former head of the Congressional Budget Office, and a team from the American Action Forum estimated that the transition to a power sector with zero greenhouse gas emissions in ten years would require a capital investment of $5.4 trillion by 2020. In addition, the annual operation, maintenance, and capital-recovery costs would be $387 billion. Constructing enough high-speed rail to make air travel unnecessary would cost between $1.1 and $2.5 trillion. In conclusion, the American Action Forum study estimated it would take some $93 trillion to implement fully the Green New Deal, including all the social justice programs packaged within the plan's renewable energy proposals.[76] To appreciate the magnitude of this goal, understand the GDP of the United States in 2018 was approximately $20.5 trillion.[77]

As is typical in all these Democrat socialist government-funded plans, the taxpayer would pay the bill, with users (exempted from paying fees) expected to get the benefits. Representative Ocasio-Cortez has suggested imposing a typical "tax the rich" wealth tax of 70 percent on those earning more than $10 million a year to pay for her Green New Deal.[78] Assuming that it would be sufficient to pay the enormous costs estimated of the cornucopia of energy benefits and social justice proposals in the plan without tanking economic growth, the question remains: If Ocasio-Cortez's enthusiasm for solar and wind power was sufficiently powerful and cost-efficient to replace hydrocarbon fuels, why did the Obama administration fail?

Obama Sold U.S. Oil Rights to the Chinese

While President Obama was promoting green energy in the United States, the Obama administration was busy selling China unprecedented rights to U.S. oil reserves. The first significant intrusion of China in the U.S. oil and natural gas market traced back to the Obama administration's decision in October 2009 to allow state-owned Chinese energy giant CNOOC to purchase a multi-million dollar stake in 600,000 acres of South Texas oil and gas fields. By allowing China to have equity interests in U.S. oil and natural gas production, the Obama administration reversed a 2005 Bush administration policy. That decision blocked China based on national security concerns from an $18.4 billion deal in which China planned to purchase California-based Unocal Corp.

China's two giant state-owned oil companies acquiring oil and natural gas interests in the USA are the following:

- China National Offshore Oil Corporation (CNOOC): a Chinese company that is 100 percent owned by the government of the People's Republic of China; and
- Sinopec Group: a Chinese company whose largest shareholder is an investment company owned by the government of the People's Republic of China.

On March 6, 2012, the *Wall Street Journal* compiled the following state-by-state list of the $17 billion in oil and natural gas equity interests CNOOC and Sinopec acquired in the United States during Obama's first term:

- **Colorado:** CNOOC gained a one-third stake in 800,000 acres in northeast Colorado and southwest Wyoming in a $1.27 billion pact with Chesapeake Energy Corporation.
- **Louisiana:** Sinopec has a one-third interest in 265,000 acres in the Tuscaloosa Marine Shale after a broader $2.5 billion deal with Devon Energy.
- **Michigan:** Sinopec gained a one-third interest in 350,000 acres in a more significant $2.5 billion deal with Devon Energy.
- **Ohio:** Sinopec acquired a one-third interest in Devon Energy's 235,000 Utica Shale acres in a more significant $2.5 billion deal.
- **Oklahoma:** Sinopec has a one-third interest in 215,000 acres in a broader $2.5 billion deal with Devon Energy.
- **Texas:** CNOOC acquired a one-third interest in Chesapeake Energy's 600,000 acres in the Eagle Ford Shale in a $2.16 billion deal.
- **Wyoming:** CNOOC also has a one-third stake southeast Wyoming after the $1.27 billion pact with Chesapeake Energy, referenced above. Sinopec gained a one-third interest in Devon Energy's 320,000 acres as part of a more significant $2.5 billion deal.[79]

Fu Chengyu had served as chairman of both CNOOC and Sinopec. Also, on March 6, 2012, in a separate story, the *Wall Street Journal* described that Fu Chengyu developed China's strategy of buying oil interests in the United States. The scheme involved China taking a low-profile approach by implementing the following plan: "Seek minority states, play a passive role, and, in a nod to U.S. regulators, keep Chinese personnel at arm's length from advanced U.S. technology."[80] The United States and Europe are committing a form of energy suicide by moving from low-cost but highly efficient hydrocarbon fuels to less powerful and more expensive wind and solar energy. For years, China has pursued the opposite strategy by buying, warehousing, and stockpiling oil from global markets.[81]

China appears not to have bought seriously into any aspect of the Democrats' global warming/climate change hysteria. Not only is China importing a record amount of oil, but China also continues to lead the world in both air pollution and greenhouse gas emissions that include both CO_2 and methane emissions. Over the past three decades, China's CO_2 emissions have more than tripled. In 2019, China emitted over one-quarter of the total global greenhouse gas emissions. That year was the first time China's greenhouse gas emissions exceeded the greenhouse gas emissions of all other developed countries combined.[82] Air pollution in China is a severe problem. Scientific studies have shown that anthropogenic aerosol emissions and changes in cloud cover in China have seriously reduced solar radiation. Observational radiation data from 119 locations across China show that the photovoltaic (PV) potential to generate electricity decreased, on average, between 11 to 15 percent between 1960 and 2015.[83]

Although China has led the world in manufacturing solar panels, the growth of wind and solar energy in China has slowed since 2018. As noted, the Chinese government's funding for renewable energy has faltered. Therefore, upgrades to the transmission infrastructure for solar and wind are lagging. In 2018, wind accounted only for 5.2 percent of China's energy, and solar for 2.5 percent, even at their height. In September 2019, Michael Standaert, a freelance journalist based in South China and known for covering environment, energy, and climate change policy, wrote an article published by the Yale School of Environment. "With its renewable energy growth slowing and its fossil fuel use rising, analysts fear that China's emissions may not level off by 2030, the target set in the Paris Climate Agreement, which would be a significant setback for efforts to slow global warming," he warned. "Renewable energy proponents are now seeking to avert a continued slowdown in China's alternative energy sector and spark new green energy growth."[84] On June 11, 2021, China announced the government was stopping all subsidies from the government budget for new wind or solar power stations.[85] During the Trump administration, the Chinese government was concerned about a trade war with the United States. However, now under the Biden administration, China has resumed its push to become the world's largest economy. China has opted for more efficient, lower-costing hydrocarbon fuels to achieve this goal as fast as possible. In so doing, China moved away from renewable solar and wind energies primarily because they are less reliable and more costly, mainly because of increased transmission infrastructure expenses.

China Blames Global Warming on Overpopulation

The Malthusian overpopulation fears resurface amidst the global hysteria over global warming and climate change caused by anthropogenic CO_2 emissions. China has embraced the argument that having fewer people on Earth would be positive for reducing greenhouse gas emissions. At the 2009 United Nations Climate Change Conference in Copenhagen, Denmark, Zhao Baige, China's vice minister of National Population and Family Planning Commission, said the following: "Dealing with climate change is not simply an issue of CO_2 emission reduction, but a comprehensive challenge involving political, economic, social, cultural and ecological issues, and the population concern fits right into the picture." Although China's family planning policy has received substantial criticism over the past three decades, Zhao claimed that China's population program had made an outstanding historic contribution to the well-being of society. He argued that China's population control measures have resulted in 400 million fewer births, translating into eighteen million fewer tons of carbon dioxide emissions a year.[86]

Thomas Wire of the London School of Economics has advanced similar research. In August 2009, Wire published a technical report entitled "Fewer Emitters, Lower Emissions, Less Cost. Reducing Future Carbon Emissions by Investing in Family Planning."[87] Wire's research was motivated by a UN effort to show family planning is justified to reduce CO_2 emissions by a cost versus benefits analysis. Wire argued that every seven dollars spent on family planning would reduce carbon emissions by one ton. Spending $7 on essential family planning to reduce carbon emissions by more than one ton is cost-effective, Wire argued. By comparison, reducing carbon emissions by reforestation would cost $13, $24 to use wind technology,

$51 for solar power, $93 for introducing hybrid cars, and $131 for electric vehicles. Or, as the *Economist* summed up Thomas Wire's research, we are encouraged to conclude: "A world with fewer people would emit less greenhouse gas."[88]

"Woke" Economics, the Modern Monetary Theory, and Global Warming/Climate Change

"Woke" political sensibilities and critical race theory have dominated the politically correct narratives of the Biden administration. With this "woke" consciousness, a new economic theory termed the Modern Monetary Theory (MMT) now dominates the monetary policies of the Federal Reserve and the fiscal policies of the U.S. Department of the Treasury. MMT embraces the idea that the United States now has a fiat currency. The U.S. Treasury can simply manufacture new money by adjusting accounting entry computer blips on federal registers. Printing more money, MMT enthusiasts explain, is the simple answer to how we can pay for a Green New Deal. Even if the Green New Deal costs trillions, we can simply "print" electronically whatever amount is needed.

When President Richard Nixon took the United States totally off the gold standard in 1971, the birth of MMT was inevitable. As explained by what has become the classic textbook of MMT economics, the theory maintains a sovereign state that creates and operates in its currency can always "create" more fiat currency to pay whatever obligations the sovereign state government creates.

William Mitchell, L. Randall Wray, and Martin Watts, three of the leading proponents of MMT, in their college textbook, *Macroeconomics*, explain the core principles of MMT as follows:

> The most important conclusion reached by MMT is that the issuer of a currency faces no financial constraints. Put simply, a country that issues its own currency can never run out and can never become insolvent in its own currency. It can make all payments as they come due. For this reason, it makes no sense to compare a sovereign government's finances with those of a household or a firm.[89]

So, sovereign nations create fiat currency when the government no longer ties money creation to hard assets such as gold or silver. From there, the MMT logic flows to conclude that as long as a sovereign nation issues its currency, the federal government can simply hypothecate (i.e., create out of thin air) whatever money is needed to pay government debts. So too, there are no limits to printing money, such as would be imposed on a currency tied to gold and silver. Under MMT, a sovereign government no longer needs to be worried about budget deficits. A sovereign nation operating under fiat currency like the United States can simply create the money required to pay for whatever the government considers worth funding. MMT allows the Federal Reserve and the U.S. Treasury to support the Green New Deal's ambitious welfare programs, including government-paid health care, guaranteed government employment, and guaranteed minimum annual incomes. Under MMT, no limits remain in fiscal policy to limit government spending. The government does not have to balance expenditures with tax revenues any longer. Even if spending exceeds tax revenues by billions or trillions of dollars, the U.S. government does not need to worry. By operating under MMT, the U.S. can simply print more money to pay any government debts, regardless of the magnitude of the debt.

Under MMT, the government's only reason to impose taxes is to take currency from households and firms to reduce aggregate demand in the economy as a whole, as a measure needed to prevent hyperinflation. In other words, the government only needs to impose taxes as an inflation-control measure. The goal is to avoid the type of hyperinflation experienced in the German Weimar Republic in the 1920s and '30s. The Biden administration understands the need to tax under MMT. But under MMT, government spending comes first, followed by the government raising taxes.

After getting through new trillion-dollar government spending programs in its first year, the Biden administration started calling for significant tax increases. The Biden administration asked Congress to increase taxes for capital gains, raise the marginal rates on income taxes paid by top income earners, and increase taxes on corporations. The magnitude of the tax increases needed to prevent hyperinflation run the risk of depressing economic activity. The concern is that massive tax increases could slow or eliminate economic growth. With slowed economic growth, the economy could go into a recession that significantly raises unemployment and underemployment in the United States. While trillion-dollar deficit government spending may be needed to fund ambitious programs like the Green New Deal, the corresponding increases in taxation required to prevent hyperinflation are likely to be counterproductive to robust activity in the private economy. However, like Obama, Biden has promised that his green energy policies will result in economic growth and new green jobs in the hundreds of thousands.

Conclusion

To date, the practical realities of the U.S. economic experience since 1971 give us serious reasons to reject the enthusiasm of MMT that the government can print with abandon and without potentially disastrous financial consequences. Inflation in the United States from 1970, the last year of the gold-backed U.S. dollar, to date has been 561 percent, as measured by the Consumer Price Index (CPI). In the same time frame, the increase in the U.S. price of gold (1970 to date) has been 4,314 percent. The U.S. national debt doubled under President Obama and has continued rising to a level approaching $30 trillion, which has exceeded U.S. GDP since 2018.

If the Obama administration had succeeded with green energy, we would feel optimistic about the repeat green energy policies of the Biden administration. Given the failure of the Obama-era green energy initiatives, the Biden administration appears to be a redo of expensive "Solyndra Syndrome" green energy failures. What seems to be the case from the Obama years is that even with massive government financial support backed by regulation and law, solar and wind power fail to be economically viable ventures.[90] What reason is there to imagine that the experience under Biden will be any different? Or, we can ask more directly: If wind and solar power worked as promised and were a reliably robust and economically sound energy alternative to hydrocarbon fuels, why would we need a junior, first-term member of Congress, a professed socialist representing New York's Fourteenth Congressional District (parts of the Bronx and Queens) to propose an FDR-like, government-mandated, sweeping Green New Deal to shove hydrocarbon fuels into the dust bin of world energy history?

PART II

The Science of Energy, Global Warming, and Climate Change

CHAPTER 5

Sun Heats Earth

CO_2–a Trace Element, the Weather Thermometer, the Importance of Clouds, the Maunder Minimum, the Little Ice Age, and the Chilling Stars

The climate is always changing.
—Henrik Svensmark and Nigel Calder, *The Chilling Stars*, 2007[1]

There is a widespread, quasi-religious assumption that nature was in a delicate state of balance before it was upset by the activities of humans. As part of this belief system, the CO_2 content of the atmosphere that existed before we started altering it is assumed to be t hat which the Earth "prefers."
—Roy W. Spencer, *The Bad Science and Bad Policy of Obama's Global Warming Agenda*, 2010[2]

To achieve its remarkable projection of future temperatures, the report [IPCC First Climate Assessment Report 1990] had to argue that the global warming of the twentieth century was largely due to carbon dioxide and other greenhouse gases. The role of the Sun had to be minimized.
—Nigel Calder, *The Manic Sun: Weather Theories Confounded*, 1997[3]

PH.D. METEOROLOGIST ROY W. SPENCER, formerly a senior scientist for climate studies at NASA's Marshall Space Flight Center, and currently a principal research scientist at the University of Alabama in Huntsville, charges the

IPCC with expressing the amount of CO_2 released into the atmosphere in a manner intentionally designed to alarm a scientifically naïve public. The IPCC typically reports the amount of CO_2 in the atmosphere in units of tons, claiming that total global emissions are running in the range of thirty billion tons per year. Yet, the IPCC fails to specify that the total weight of Earth's atmosphere is approximately five quadrillion tons. A quadrillion is the number "1," followed by fifteen zeroes. Expressed in other terms, one trillion equals 1,000 billion, and one quadrillion equals 1,000 trillion. So, one quadrillion amounts to one million billion, a number that, if expressed with zeroes, would be written as 1,000,000 billion. Appreciating this, the total amount of annual global CO_2 emissions amounts to thirty billion divided by 1,000,000 billion, for a result of .00003 percent, read as "three hundred-thousandth of 1 percent."

Like Julian Simon, whose Ph.D. in business administration made him suspect to the Ph.D. economists who dominated the study of natural resource economics, Roy Spencer has a Ph.D. in meteorology, the advanced academic degree associated with weather casting. As a Ph.D. meteorologist, Spencer brings to the study of global warming an understanding of how climate functions, giving him a unique perspective on Earth's ongoing ability to cool itself. Spencer shares another important characteristic with Simon in that both were optimists about the human condition on Earth. While Simon questioned why so many people believe false bad news about the environment, natural resources, and the population, Spencer asks why so many people believe such bad news about hydrocarbon emissions of CO_2. "As most of us have learned in school, atmospheric carbon dioxide is just as necessary for life on Earth as oxygen," Spencer has written. "Without CO_2 there would be no photosynthesis, and therefore no plants, and no animals, and no people either." Yet Spencer noted that with the "supposed threat of global warming," former Vice President Al Gore managed to get the Supreme Court to claim carbon dioxide as a pollutant. Spencer also observed that Gore had referred to our emissions of CO_2 as equivalent to treating the atmosphere as an "open sewer."[4] Spencer echoes Julian Simon when he correctly insists that Earth is much more resilient than most scientists claim. "You might say that rather than 'hot, flat, and crowded,' I believe the Earth to be cool, round, and spacious."[5]

Carbon Dioxide in the Atmosphere: "A Trace Element"

In his 2008 book, *Climate Confusion*, Spencer examined the upward trend in atmospheric carbon dioxide concentration as measured at the Mauna

Loa Observatory in Hawaii. He noted that Mauna Loa was chosen as a monitoring site "because it is relatively isolated from any major urban areas, which tend to have elevated concentrations of carbon dioxide." He also noted that the rise in CO_2 since 1958, as recorded by the Mauna Loa Observatory, looks dramatic when the units of concentration are measured in parts per million (ppm). When the CO_2 concentration is expressed in molecules, the dramatic nature of the picture evaporates. "The current concentration of about 380 ppm means that for every million molecules of air, 380 of them are carbon dioxide," he wrote. "Or alternatively, for every 100,000 molecules of air, 38 of them are carbon dioxide." Spencer concluded: "This small fraction reveals why carbon dioxide is called one of the atmosphere's 'trace gases.' There simply isn't very much."[6]

He stressed that the rise of CO_2 in the atmosphere shown on the Mauna Loa Observatory graph from 2000–2005 demonstrated that the atmosphere had added one molecule of CO_2 to every 100,000 molecules of air "every five years or so."[7] Spencer pointed out as a greenhouse gas, CO_2 "is believed to be causing a surface 'warming tendency' because it makes Earth's natural greenhouse effect a little stronger—the 'radiative blanket' is slightly denser." Because CO_2 absorbs infrared energy, the global warming theory asserts more infrared energy is now coming into Earth (from the sun) than is escaping Earth (back into outer space), such that Earth's atmosphere "must heat up as a result." Spencer pointed out scientists say warming tendency because in all complex systems like Earth's atmosphere, "a single change can be expected to cause other responses" that act to dampen the warming trend, "counteracting it with other offsetting tendencies." He also pointed out that should we reach a doubling of the preindustrial atmospheric carbon dioxide concentration, projected for later in this, the twenty-first century, "we will have enhanced the Earth's natural greenhouse effect by about 1 percent."[8]

Weather: A Self-Regulating Thermometer

Spencer correctly stresses that Earth's climate system is "possibly the most complex physical system we know."[9] Water vapor accounts for between 70 and 90 percent of all greenhouse gases, combined with clouds that have a significant greenhouse effect even though clouds are not a gas but consist of water droplets and ice crystals. The amount of infrared radiation absorbed by Earth on average is very close to the amount of sunlight absorbed by Earth since the infrared energy emitted from outer space to Earth is virtually zero. Earth warms during the day when the sun shines and cools at night when infrared energy escapes into outer space because the sun is not shining. Spencer noted that "the combination of solar heating and infrared energy transfers are continuously trying to make Earth's surface unbearably hot and the upper atmosphere unbelievably cold."[10] The purpose of what we call weather is "to move heat from where there is more, to where there is less." Spencer stressed that these flows of heat demonstrate one of the most fundamental laws in science, the second law of thermodynamics. In simple terms, the second law of thermodynamics "just states that energy tends to flow from where there is more to where there is less."[11]

Spencer sums up his conclusion as follows:

> Every gust of wind that blows, every cloud that forms, every drop of rain that falls, all happen as part of processes which continuously move excess heat from either the surface to higher in the atmosphere, or from low latitudes (tropical regions) to higher latitudes (polar regions).[12]

The action of water vapor that cools and warms the planet's surface demonstrates the complexity of Earth's climate. "When the surface water is evaporated to form vapor, it removes heat from the surface," Spencer explains. "After that, the vapor then helps warm the surface through the greenhouse effect. Of course, both of these effects are happening at the same time, continuously."[13]

As noted earlier, Spencer argues Earth's temperature operates by the fundamental principles of thermodynamics that include the following principle: the temperature of an object will increase if the rate of heat gain exceeds the rate of heat loss by the object. The reverse is also true: the temperature of an object will decrease if the rate of heat loss exceeds the rate of heat gain by the object. Thus, Spencer correctly notes, a triggering event (known as a *forcing factor*) causes an increase or a decrease in temperature. Turning on a burner on the stove is a *forcing factor*. The heat from the burner causes the water to heat because the burner transfers heat to the water. The water temperature will rise until it reaches a specific temperature that scientists call a *steady state* or *equilibrium*.

In addition to triggering events, the energy transfer depends on what is called *positive and negative feedback*. *Positive feedback* is anything that adds to the heat transferred into the pot. A lid on the pot, for instance, is *positive feedback* that causes the water to get hotter. *Negative feedback* is anything that takes away from the heat transferred into the pot. Taking the lid off the pot, for instance, is negative feedback because, without the lid, water vapor evaporates as the water's temperature rises.[14] The sun is the *forcing factor* and the infrared radiation is a *feedback factor*. Applying these principles to temperature change in climate, Spencer explained the impact of negative feedbacks as follows:

> *Global warming* will occur if the amount of sunlight absorbed by the Earth is increased (e.g., from less low cloud cover) or if the amount of infrared radiation lost to space is decreased (e.g., from more greenhouse gases, more water vapor, and more high cloud cover).
>
> *Global cooling* will occur if the amount of sunlight absorbed by the Earth is decreased (e.g., from more low cloud cover), or if the amount of infrared radiation lost to space is increased (e.g., from less greenhouse gases, less water vapor, and less high cloud cover).[15]

For the sake of argument, Spencer accepted that humanity's CO_2 emissions had increased the atmospheric CO_2 concentration by some 25 percent since the Industrial Revolution. He even conceded the 2009 IPCC estimate that anthropogenic CO_2 had increased Earth's CO_2 concentration by as much as 40 percent. "This [i.e., the IPCC 2009 estimate] has caused an estimated 1.6 watts per square meter of extra energy to be trapped, out of the estimated 235 to 240 watts per square meter that the earth, on average, emits to outer space on a continuous basis," Spencer calculated. His conclusion: "I find it amazing that the scientific community's purported near-certainty that global warming is manmade rests on a forcing mechanism—a radiative imbalance—that is *too small to measure*."[16]

Two energy flows determine the energy balance of Earth: the rate at which Earth absorbs solar energy (a *forcing factor*) and the rate at which Earth loses infrared energy to outer space (*a feedback factor*).[17] Clouds are the most significant and most uncertain feedback factor in the atmosphere. "I cannot overstate the importance of the uncertainty over cloud feedbacks," Spencer wrote. "At the least theoretically, clouds could either save us from global warming, or cook us."[18] He pointed out that the two sides of the role clouds play in global warming involve forcing and feedback, i.e., cause and effect. "Forcing (cause) would be the clouds causing a temperature change," he distinguished. "Feedback (effect) would be causation flowing in the opposite direction, with temperature causing a cloud change. This effect then feeds back upon the original temperature change, making it larger or smaller."[19] Spencer wondered why IPCC researchers assumed that warmer temperatures caused a decrease in the cloud cover rather than the decrease in cloud cover causing the warmer temperatures. Working with a computational physicist, Spencer investigated the cause-versus-effect issue with a simple climate model. They found out that "clouds causing a temperature change could give the illusion of positive feedback even when we specified negative feedback in the climate model."[20]

Spencer concluded that because IPCC researchers were not careful about distinguishing cause and effect when observing cloud and temperature variations, they fooled themselves into believing the climate system is more sensitive than it is. In other words, IPCC climate models erred by predicting that failure to decarbonize would produce significant increases in global warming because they built clouds as positive feedbacks into their climate models.

Prominent geologist Dr. Robert Giegengack, the former chair of the Department of Earth and Environmental Sciences at the University of Pennsylvania, has lamented "the enormity of the hubris that leads us to believe that we can 'control' climate by controlling anthropogenic emission of CO_2." Giegengack put into perspective the political nature of global warming alarmists. In 2019, Giegengack explained the following:

> If anthropogenic CO_2 is contributing to climate warming now under way, nothing we are doing, or contemplating doing, can have any measurable effect on that warming. Global Warming/Climate Change has evolved into a semi-religious campaign advanced by well-intentioned people who feel, deep in their hearts, that they are 'saving the planet.' It beggars the imagination to assert that the natural factors that drove the warming trend from 18,000 years ago to about 300 years ago (with some unexplained temperature reversals) abruptly stopped operating at the end of the Little Ice Age to accommodate our political need to attribute climate variability to human industrial activity. Today's climate is close to the coolest it has been in 540,000,000 years, and the atmospheric concentration of CO_2 is close to the lowest it has been. Climate models are instructive, but they lead to scenarios, not predictions. They can be manipulated to yield desired outputs.[21]

Dr. Giegengack has done field research on six continents, conducted peer-reviewed studies on the geological archives of climate, and spent much of his academic career doing fieldwork on the history of climate. He has authored some 200 peer-reviewed papers.

Yet, Phil Jones, the former director of the Climatic Research Unit (CRU) and a former professor in the School of Environmental Sciences at the University of East Anglia in Norwich, England, disagrees. Jones laments that climate scientists like Giegengack still think the sun plays a more essential role in Earth's climate than does anthropogenic CO_2. After attending a 2011 conference on solar variability and climate held in Tenerife, the largest of Spain's Canary Islands off West Africa, he said the following:

> Many in the solar-terrestrial physics community seem totally convinced that solar output changes can explain most of the observed changes we are seeing.[22]

Jones lamented that solar-terrestrial physicists are "so set in their ways." In the next chapter, we will examine the Climategate scandal in which Jones played a central role.

The PDO and Climate Change: A Natural Cycle

The Pacific Decadal Oscillation (PDO) is a regional shift in weather patterns over the northern regions of the Pacific Ocean that shifts between positive warming phases and negative cooling phases approximately every thirty years. Spencer has argued that the PDO is one of nature's alternative ways to move heat around Earth. "The most fundamental function of both the oceanic and the atmospheric circulations is to transport heat around the globe, from regions where excess solar heating occurs to regions where there is less solar heating," Spencer has explained.[23] He observed that the IPCC's 2007 report concluded the following: "Most of the observed increase in global-average temperatures since the mid-20th Century is very likely due to the observed increase in anthropogenic greenhouse gas concentrations."[24] The IPCC is confident that global warming in the past fifty years has been a human cause. But since there was a slight global cooling from the 1940s to the late 1970s, the IPCC refers primarily to global warming over thirty years.

Yet, as Spencer had noted in print, both in his 2010 book *The Great Global Warming Blunder* and published articles in peer-reviewed climate journals,[25] that thirty-year interval was the same time the PDO was in its "positive" or warming phase.[26] The last time the PDO changed phase was in 1977. The 1977 PDO phase change ended the cooling trend that started in the 1940s. "Contrary to what you may have heard in news reports, the recent warming in the Arctic is probably not unprecedented," Spencer wrote. "It was just as warm in the late 1930s and early 1940s when the PDO was also in its positive, warm phase. There were newspaper reports of disappearing sea ice and changing wildlife patterns back then, too. Most of the all-time high-temperature records in the United States were set in the 1930s. The Northwest Passage was navigated without an icebreaker between 1940 and 1942, yet satellite observations of it opening up again in 2007 were claimed to have recorded an unprecedented event."[27]

What drew Spencer's attention to the PDO was proof that natural climate variability can cause temperature changes on Earth by changing global average cloudiness. "One of the primary mechanisms the Earth has for cooling itself is the production of clouds, which reflects some of the solar energy that reaches the Earth back to outer space," he commented. "Because the average effect of clouds on the Earth's climate is to cool it, any natural change in global average cloudiness can also be expected to cause global warming or global cooling."[28] Spencer's breakthrough came from studying satellite data. The PDO is not a temperature index but an index of how weather patterns organize over the North Pacific Ocean. Spencer hypothesized the PDO might cause a slight fluctuation in cloud cover resulting from those circulation changes.[29] He and his research colleagues at the University of Alabama in Huntsville found confirming satellite data. "Satellite observations of radiative forcing of the Earth from 2000 through 2008 suggest that the Pacific Decadal Oscillation causes natural cloud variations of a magnitude that a simple climate model indicated would be sufficient to explain most [up to 75 percent of the long-term temperature trend] of the temperature variations during the twentieth century."[30] This satellite data supported Spencer's original claim that a mere 1 percent change in naturally occurring processes can cause global warming or cooling, without any reference to human-created CO_2.

Spencer made it clear he wrote *The Great Global Warming Blunder* for the following reason:

> My main purpose in this book is not to claim that the PDO necessarily constitutes the largest single mechanism of climate change—although that is a possibility. Instead, my aim is to demonstrate that the "scientific consensus" that global warming is caused by humans is little more than a statement of faith by the IPCC. There is evidence of natural climate change all around us if scientists would just take off their blinders.[31]

Spencer solved the "cause and effect" problem, concluding the PDO-induced cloud changes caused the temperature changes because the temperature response came *after* the forcing (i.e., *after* cloud activity either increased or decreased), not before.[32] Spencer commented that his academic training in meteorology gave him an insight into natural weather processes that many IPCC scientists lack, even though they try to act like meteorologists. "We meteorologists understand that the processes controlling clouds, 'nature's sunshade,' are myriad and complex," he wrote. Spencer also noted that some IPCC scientists had chided television meteorologists for second-guessing them on climate change. The IPCC scientists argued that "meteorologists deal with weather, not climate, and therefore should not question the judgment of climate experts when it comes to global warming." But Spencer turned this argument around: "I contend that climate variability cannot be understood without first understanding the complexities of weather. After all, climate is average weather, and if you don't understand what controls variations in weather then you won't be able to understand all the potential sources of climate change."[33]

Spencer summarized his first conclusion as follows:

> The first conclusion is that recent satellite measurements of the Earth reveal the climate system to be relatively insensitive to warming influences, such as humanity's greenhouse gas emissions. This insensitivity is the result of more clouds forming in response to warming, thereby reflecting more sunlight back to outer space and reducing that warming.[34]

He pointed out that this process, known as negative feedback, is analogous to opening your car window or putting a sunshade over the windshield as the sun begins to heat the interior. He added: "An insensitive climate system does not particularly care how much we drive SUVs or how much coal we burn for electricity." A corollary to this first conclusion is that IPCC climate models are flawed for considering cloud changes to be positive feedback. The result is that the IPCC mathematical models of the climate produce too much warming in response to humanity's greenhouse gas emissions. He further commented: "Without the high climate sensitivity of the models, anthropogenic global warming becomes little more than a minor academic curiosity."[35]

Spencer explained his second major conclusion as follows:

> The second major conclusion of this book is closely connected to the first. If the carbon dioxide we produce is not nearly enough to cause significant warming in a climate system dominated by negative feedback, then what caused the warming we have experienced over the last fifty years or more? New satellite measurements indicate that most of the global average temperature variability we have experienced in the last 100 years could have been caused by a natural fluctuation in cloud cover resulting from the Pacific Decadal Oscillation (PDO).[36]

Spencer stressed his primary conclusion that PDO causes cloud changes would explain most of the significant variations in global average temperature since 1900, including 75 percent of the warming trend.[37] Spencer hypothesized that the PDO might cause a slight fluctuation in cloud cover resulting from changing weather patterns over the North Pacific Ocean. In this case, the PDO would constitute the forcing. The reduction in cloud cover would function as negative feedback allowing Earth's weather thermostat to restore an energy balance concerning the heat absorbed from the sun.[38] Spencer summed up his major conclusion with the following statement: "The climate system itself can cause its own climate change, supporting the widespread public opinion that global warming might simply be part of a natural cycle."[39]

The importance of Spencer's work is that as a trained Ph.D. meteorologist, he permits us to understand the function and the complexity of weather on Earth. Weather on Earth serves as a mechanism for moving heat back into outer space and shifting heat from warmer locations on Earth to cooler areas. Spencer reminds us that clouds can act as heating mechanisms (*forcing effect*), for instance, when low-hanging clouds are sparser, permitting more rays from the sun to heat Earth. Or clouds can serve as cooling mechanisms (*feedback mechanisms*) when low-hanging clouds allow more heat to escape from Earth to outer space as infrared energy. Spencer also points out that it is essential to note whether the buildup or thinning of clouds precedes or follows changes in global temperature. If the cloud activity follows the change in global temperature, the cloud activity is an effect of the temperature change, not a cause of the temperature change.

The same analysis will hold for CO_2. Does the buildup of CO_2 in the atmosphere correlate in geological time with temperature changes on Earth? If the quantity of CO_2 in the atmosphere does not correlate statistically with changes in Earth's temperature, CO_2 is not an essential mechanism for causing Earth's temperature to change. Similarly, suppose the quantity of CO_2 in the atmosphere changes after a temperature change occurs on Earth. In that case, the amount of CO_2 is a function of the temperature change, not a cause of the temperature change. Weather patterns like the Pacific Decadal Oscillation affect global temperature changes by focusing on the amount of infrared energy retained or released by Earth given changes in cloud cover (here acting as a *feedback mechanism*). Other ocean patterns that affect Earth's temperature include La Niña natural climate cycles (i.e., cooler-than-average sea currents in the Pacific Ocean) or El Niño climate cycles (i.e., warmer sea currents in the Pacific Ocean).

Effect of the Sun on Earth Temperatures

Virtually all the attention in the public debate over global warming "has been focused on the contribution man might be making to shaping the climate by producing gases [i.e., CO_2 in particular] which make it harder for heat from the earth to escape back into space." Russian scientists have forced us to acknowledge that "nothing like enough attention had been paid to the source of all that heat in the first place: the giant radiant ball of fire in the heavens without which no life could exist, and which is far and away the most powerful determinant of all the variations in climate on earth."[40]

In a career that has stretched over decades, Habibullo Abdussamatov, the head of space research at St. Petersburg's Pulkovo Astronomical Observatory in Russia, has argued that total sun irradiance (TSI) is the primary factor responsible for causing climate variations on Earth, not CO_2. By examining historical records of sun activity, Abdussamatov has argued that an active sun characterized by vigorous sunspot activity is the primary factor responsible for Earth's temperature increases. Correspondingly, he has argued that an inactive sun characterized by minimal sunspot activity is the primary actor responsible for Earth's temperature decreases. Abdussamatov's research has led him to conclude that the sun is the primary cause of global warming and global cooling.

Abdussamatov has concluded that changes in the increases and decreases in TSI, characterized by the number of different manifestations of its sunspot-forming activity, are the primary determinant of global temperature on Earth. In an article published in *Earth Sciences* in 2020, he stated his conclusions as follows: "The Sun, being the main source of energy for the Earth, controls the climate system, and even the smallest long-term changes in TSI can have serious consequences for the climate."[41] In the same article, Abdussamatov examined changes in Earth's temperature and the atmospheric concentration of CO_2 over the past 420,000 years as determined by ice core data drilled from a depth of over 3,768 meters near the Vostok Station, Antarctica, during the glacial/interglacial cycles. "A rise in concentrations of greenhouse gases has begun every time after warming begins and ended after the warming was replaced by cooling," he concluded.

In that paper, he wrote the following:

> It is worth emphasizing that the temperature starts to decrease, after reaching its highest values in the glacial/interglacial cycles, despite the fact

that the concentration of greenhouse gases continues to grow. The peaks of the carbon dioxide concentration have never preceded the warming, but on the contrary always took place 800±400 years after it, being its consequence, i.e., they have always been a natural consequence of the temperature increase caused by long-term growth of the incoming annual solar energy.[42]

Abdussamatov concluded that CO_2 could not cause global warming because the Vostok ice core study proved that higher levels of CO_2 concentration occur after warming, not before.

Abdussamatov's analysis that the "Sun heats Earth" as the primary cause of global warming and global cooling supports Edward Maunder's work at the Greenwich Royal Observatory documenting the decline in sunspot activity between 1615-1715. Maunder found that in one thirty-year period, only fifty sunspots were observed and documented, instead of the usual 40,000 to 50,000 typically recorded.[43] The sun's inactive period, known today as the Maunder Minimum, coincided with the Little Ice Age that lasted from approximately 1300 to 1850. Abdussamatov's independent research correlated the onset and the passing of the Little Ice Age to what he found was a bicentennial cyclical increase/decrease of TSI as the sun rotated between periods of active sunspot activity and minimal sunspot activity.[44]

A Coming New Little Ice Age?

Abdussamatov has also predicted that from 1990, the decreased solar activity as the sun enters a new minimum will lead Earth to a new Little Ice Age. This prediction flies directly in the face of CO_2 alarmists who have bet everything on their presumption global warming is irreversible as long as humans continue to burn hydrocarbon fuels to propel capitalist societies, with the U.S. being their chief culprit. In a 2015 paper published in *Thermal Science*, Abdussamatov predicted the following:

> The portion of the solar energy absorbed by the Earth is decreasing. Decrease in the portion of total solar irradiance absorbed by the Earth since 1990 remains uncompensated by the Earth's radiation into space at the previous high level over a time interval determined by the thermal inertia of the Ocean. A long-term negative deviation of the Earth's average annual energy balance from the equilibrium state is dictating corresponding variations in its energy state. As a result, the Earth will have a negative average energy balance also in the future. This will lead to the

beginning of the decreasing in the Earth's temperature and of the epoch of the Little Ice Age after the maximum phase of the 24th solar cycle approximately since the end of 2014.[45]

Again, in this paper, Abdussamatov dismissed the importance of CO_2 to temperature changes on Earth, writing as follows: "Negligible effect of the human-induced carbon dioxide emission on the atmosphere has insignificant consequences."[46]

Global warming true believers strongly object, arguing we entered Solar Cycle 25 in 2019. Cycle 25 would be the twenty-fifth solar cycle switching between solar maximums and solar minimums recorded since 1755 when extensive recording of sunspot activity began. On September 15, 2020, NASA and the National Oceanic and Atmospheric Administration (NOAA) announced their analysis and prediction that Cycle 24 ended with a solar minimum in December 2019.[47] If the sun were entering a more active Cycle 25, it would be in line with the predictions of global warming true believers that Earth is entering a critical phase of global warming. The point was that a more active sun would produce a warmer Earth. Global warming true believers were comfortable because if Earth continued to warm, the IPCC could blame the warming on steadily increasing anthropogenic CO_2 emissions.

Interestingly, Abdussamatov and other similar Earth climate scientists began to gain traction as media attention focused on the argument that Earth was entering a new Little Ice Age. Global warming true believers stepped up efforts to push out a counterargument with dubious scientific evidence. They remained locked into the argument that the sun is not responsible for global warming. In time-series analysis, trends are too complex to discern immediately, given that many year-to-year variations appear to be trend-reversing but end up just being typical deviations that were not statistically significant. The statistical problem of picking the correct sequence of years to discern actual patterns is inherent to time-series analyses. In September 2020, NOAA published data predicting sunspot numbers that made clear NOAA expected the sun was entering another "full-blown" grand solar minimum in Cycle 25 from mid-2025 to 2031, with the result that Solar Cycle 25 would be yet another historically weak cycle similar to the cooling trend experienced in Cycle 24, a similarly weak solar cycle.[48] Yet, some two weeks later, on September 15, 2020, a NOAA press release ignored available sunspot data that confirmed sunspot numbers were likely to drop off sharply between 2025 and 2030. The truth was that NOAA had already published predictive data suggesting the sun was entering a continuing solar minimum trend that would last until at least the 2040s. But to admit the sun was entering a Cycle 25 solar minimum that could last until 2040 upset IPCC climate scientists who understood a solar minimum most likely would predict a cooler Earth.[49]

Solar Cycle 25 Continues Weak Sun Pattern of Cycle 24

On August 4, 2020, Valentina Zharkova, a mathematician and astrophysicist at Northumbria University in Newcastle upon Tyne, U.K., published an editorial supporting the argument that with Cycle 25, the sun had entered a new minimum phase. Zharkova acknowledged that the new minimum would significantly reduce solar activity and the sun's magnetic field. She predicted these developments could lead to a noticeable decrease in terrestrial temperature comparable to the Maunder Minimum. "Currently, the Sun has completed solar cycle 24—the weakest cycle of the past 100+ years—and in 2020, has started cycle 25," Zharkova wrote. "During the periods of low solar activity, such as the modern grand solar minimum, the Sun will often be devoid of sunspots. This is what is observed now at the start of this minimum, because in 2020 the Sun has seen, in total, 115 spotless days (or 78%), meaning 2020 is on track to surpass the space-age record of 281 spotless days (or 77%) observed in 2019." Zharkova also demonstrated Cycle 24 leading into Cycle 25 showed a significant reduction of the sun's magnetic field. She also correctly predicted that reducing the sun's magnetic field would increase galactic and extra-galactic cosmic rays. This increase in cosmic rays hitting Earth would lead to the formation of clouds in the terrestrial atmosphere. Increased cloud formation on Earth would have a forcing effect that would lower Earth's temperatures consistent with a Maunder-like minimum. In coming to this last conclusion, Zharkova pointed to the research of Danish scientist Henrik Svensmark, the subject of the next section of this chapter. In conclusion, Zharkova predicted this new solar minimum that began in 2020 would last until 2053, with a reduction in the average terrestrial temperature by up to 1°C (compared to an average terrestrial temperature reduction of 1.4°C during the Little Ice Age) that will extend through Solar Cycles 25–27 in the decades 2031–2043.[50]

As sunspot data began to emerge in 2021 confirming Zharkova's hypothesis, Dr. Scott McIntosh, the deputy director at the National Center for Atmospheric Research (NCAR), a private research center funded by the National Science Foundation, went on the record. McIntosh claimed that in June 2021, Solar Cycle 24 had not yet ended. McIntosh claimed a "termination event" caused when oppositely charged bands of magnitude collide at the equator of the sun was imminent. McIntosh predicted this future termination event that had not yet happened would mark the beginning of Solar Cycle 25, which he believed would be one of the sun's most active cycles ever observed.[51] McIntosh's statement gave the appearance of a Hail Mary pass thrown into the endzone in the last seconds of an otherwise losing football game. McIntosh appeared to be hoping for a dramatic "termination event" that would jar the sun out of the observed Cycle 25 minimum. His panic demonstrated just how threatened climate change true believers were that a new Little Ice Age would expose their anthropogenic CO_2 global warming theories as fraudulent.

Still, NASA persists in touting the politically correct global warming narrative. In analyzing global surface temperatures and total sun irradiance measured on Earth (watts/meter2) since 1880, NASA continues to claim the amount of solar energy received by Earth has followed the sun's natural eleven-year cycle of small ups and downs with no net increase since the 1950s. Note: every eleven years, the magnetic field of the sun flips such that the north and south poles switch places. The first cycle, Solar Cycle 1, began in 1755 when solar scientists on Earth started to track sunspots. Solar Cycle 24 lasted eleven years from December 2008 to 2019. We are currently in Solar Cycle 25.

As noted earlier in this chapter, NASA continues to insist it is "extremely unlikely" that the sun has caused the observed global warming trend over the past half century. "The Sun can influence Earth's climate, but it isn't responsible for the warming trend we've seen over recent decades," NASA argues. "The Sun is a giver of life; it helps keep the planet warm enough for us to survive. We know subtle changes in Earth's orbit around the Sun are responsible for the comings and goings of the ice ages [i.e., a reference to Milankovitch Cycles, a topic we will cover subsequently in chapter 8]. But the warming we've seen in recent decades is too rapid to be linked to changes in Earth's orbit and too large to be caused by solar activity." NASA argued that while global cooling has accelerated in the stratosphere, global warming is still occurring because of a continued buildup of anthropogenic CO_2 near the surface of the earth. This explanation does not require the sun to get "hotter."[52] NASA's argument is a concession that Solar Cycle 24 demonstrated minimum sunspot activity and that Cycle 25 is not starting any stronger. NASA climate scientists have joined forces with global warming true believers. The scientists and global warming true believers have rolled the dice betting that Earth's temperature will continue to warm over the next few decades.

Yet, the National Centers for Environmental Information of the NOAA acknowledges, on the organization's website, that for the past 120,000 to 150,000 years, since the end of the most recent glacial period, Earth has been in an interglacial period called the Holocene.[53] The NOAA website continues to note that glacial-interglacial periods have waxed and waned throughout the Quaternary Period (the last 2.6 million years). The NOAA website further points out that glacial-interglacial cycles have a frequency of about 100,000 years since the middle Quaternary. This pattern would suggest Earth is about to end an interglacial warming period with the possibility of resuming a new Little Ice Age.

Cosmic Rays, Solar Activity, and the Climate

In 1911 and 1912, Austrian physicist Victor Hess made a series of ascents in a balloon to take measurements of the radiation in the atmosphere. He was trying to determine the source of ionizing radiation that registered on an electroscope. On April 7, 1912, Hess made an ascent of 5,300 meters during a near-total eclipse of the sun. He measured the ionization rate in the atmosphere and found it increased to approximately three times the ionization at sea level. When the ionization of the atmosphere did not decrease during the eclipse, he reasoned the source of the radiation could not be the sun. Instead, Hess claimed the radiation came from a natural source of high-energy particles coming from outer space that he called "cosmic rays." In 1936, Hess shared a Nobel Prize in Physics for his discovery.[54]

Subsequent research has established that the high-energy particles Hess called "cosmic rays" consist of pervasive elements in the universe. NASA research shows roughly 90 percent of cosmic ray nuclei are hydrogen (protons) and 9 percent are helium (alpha particles). Hydrogen and helium are the most abundant elements in the universe and the origin point for stars, galaxies, and other large structures in space. The remaining 1 percent involve other elements, all the way up to uranium.[55] These cosmic rays from outer space collide with particles in the upper atmosphere, creating more particles, mainly *pions*. These pions are produced "when protons get stuck in a magnetic field inside the shockwave of the supernova and crash into each other."[56] The charged pions tend to decay quickly, emitting particles called *muons*. Muons are one of the fundamental subatomic particles, similar to electrons but weighing more than 207 times as much.[57] Just to be clear, recall that ionization is the process by which an atom gains a negative or a positive charge by gaining or losing electrons, or when one atom or molecule combines with another atom or molecule that already has a charge. The resulting charged particle (atom or molecule) is called an ion: positively charged ions are called *cations*, while negatively charged atoms are called *anions*. The lowest energy cosmic rays arrive at Earth from the sun in what is known as the solar wind. Sources of the highest energy gamma rays in our galaxy, the Milky Way, come from the remnants of exploding stars, supernovae, like the famous Crab Nebula. Other sources of ultra-high gamma rays come from other galaxies, where supermassive black holes may also drive the acceleration that sends ultra-high-energy gamma rays into space, some of which hit Earth.[58]

The key that is pertinent to our discussion here is this: How do cosmic rays affect the climate on Earth? In 1991, Danish scientists Eigil Friis-Christensen and Knud Lassen of the Danish Meteorological Institute published an important paper in *Science*. They examined a statistical correlation between sunspot activity and the increase or decrease of temperatures in the Northern Hemisphere during the twentieth century.[59] Friis-Christensen and Lassen realized the impact their findings had on the anthropogenic CO_2 theory of global warming. "Although the Northern Hemisphere temperature record includes a significant net increase during the last 130 years, which could partly be caused by the increased greenhouse effect [i.e., by a nearly exponential increase in the concentration of CO_2 in the atmosphere], the temperature record does show a considerable departure from this long-term trend from 1940 to 1970 [i.e., when Earth reflected cooling temperatures despite continuing rises in the concentration of CO_2 in the atmosphere]. During these years the temperature decreased, simultaneously with a decrease in solar activity as indicated by the variation of the solar cycle length."

In the next paragraph, Friis-Christensen and Lassen make their most crucial conclusory point. They commented that the high correlation between a decrease in solar activity and a decrease in temperatures observed in the Northern Hemisphere from 1940 to 1970 "could reduce the importance of measured greenhouse gases relative to the direct influence of solar variability." In their following two sentences, the authors speculated why the greenhouse effect appeared inoperative to explain the global cooling measured in the Northern Hemisphere from 1940 to 1970. They wrote: "This result would not necessarily indicate that an increased greenhouse effect does not exist—it could just mean that other effects may be counteracting the greenhouse effect. In particular, it has been debated whether increased cloudiness due to increased global pollution could have a cooling influence on the climate, similar to the effects due to volcano eruptions."[60]

The breakthrough in understanding that the solar effect was not due to pollution but to cosmic rays came in 1995 when Friis-Christensen teamed up with Danish physicist Henrik Svensmark, a professor in the Division of Solar System Physics at the Danish National Space Institute (DTU Space). Svensmark studied data compiled by the NASA Goddard Institute for Space Studies (NASA GISS) at Columbia University. This data on global cloud cover had been collected by the International Satellite Cloud Climatology Project (ISCCP) established by the World Climate Research Program (WCRP) in 1982. The ISCCP collected satellite readings worldwide recording changes in cloud cover between 1983 and 1990. Svensmark was intrigued that the NASA cloud data showed a strong correlation between the extent of cloud formation and the relative intensity of cosmic rays. Here was the link Friis-Christensen had been looking for to explain why solar activity appeared to have more of an impact upon Earth's temperatures than could be explained by the greenhouse gas effect.[61]

The Chilling Stars

In 1996, Svensmark and Friis-Christensen published a paper in the *Journal of Atmospheric and Solar-Terrestrial Physics* reporting on their search for a physical mechanism that would explain the observed relationship that correlated cosmic ray activity with the formation of clouds on Earth.[62] The two scientists reported that an observed variation of 3–4 percent of the global cloud cover varied enormously with cosmic ray flux. Increased cosmic ray activity led to increased cloud cover, while reduced cosmic ray activity led to reduced cloud cover. The two scientists also reported the earth's magnetic field increased when the active sun produced active sunspots. The earth's magnetic field decreased when the sun entered a minimum with decreased sunspot activity. The earth's magnetic field acted as a shield blocking the cosmic rays incoming from outer space. They concluded that a sufficiently robust physical mechanism existed to explain how and why cosmic ray activity incoming from outer space, as impacted by solar magnetism, could increase or decrease Earth's cloud cover, which would increase or decrease Earth's surface temperatures.

To determine this mechanism and overcome the political resistance to publish their findings, Svensmark launched on a ten-year quest. This quest involved research in theoretical physics and the collection of reliable Earth data on cosmic rays, the sun's magnetism, cloud creation on Earth, and surface temperature recordings. Svensmark struggled to find the minimum funding needed to conduct scientifically rigorous experimental laboratory tests as a final step. He never underestimated the challenge he faced. "The discovery seemed crazy at first," Svensmark and his coauthor Nigel Calder explained in the 2007 book they published in the U.K., entitled *The Chilling Stars: A Cosmic View of Climate Change*. "Who would think that the ordinary clouds that decorate the sky take their orders from exploded stars far off in space? Or that the climate obeys the swarms of atomic particles that rain down on us from the Milky Way?" Yet, Svensmark was confident he was right. In that book, Svensmark and Calder continued: "But a recent experiment reveals how the trick is done, and thereby alters much of what scientists believed they knew about the weather, the climate, and the long history of life on Earth."[63]

In a lecture entitled "The Connection between Cosmic Rays, Clouds, and Climate," at a climate conference in Munich, Germany, held November 23–24, 2018, Svensmark reported on his decade-long quest. He had found the physical mechanism that explained the physics and the chemistry behind their theory that explained the previous correlation researchers like Habibullo Abdussamatov had found between sun activity and Earth temperatures. In a final chart to that lecture, Svensmark explained his conclusions as follows:

- Cosmic rays, high-energy particles raining down from exploded stars, knock electrons out of air molecules. This produces ions, that is, positive and negative molecules in the atmosphere.
- The ions help the formation clusters of mainly sulphuric acid and water molecules to form and become stable against evaporation. This process is called nucleation and results in small clusters (aerosols). These small aerosols need to grow nearly a million times in mass to have an effect on clouds.
- The second role of ions is that they accelerate the growth of small aerosols into cloud condensation nuclei – seeds on which liquid water droplets form to make clouds. The more ions, the more aerosols become cloud condensation nuclei.

From these three points, Svensmark drew the following implications:

- When the sun is lazy, magnetically speaking, there are more cosmic rays and more low clouds, and the earth's temperature is cooler.

- When the sun is active, fewer cosmic rays reach the earth, and with fewer low clouds, the world warms up.
- The sun became unusually active during the twentieth century and, as a result, we experienced "global warming" effects on Earth.
- Cooling and warming of around 2°C have occurred repeatedly over the past 10,000 years, as the sun's activity and the cosmic ray influx have varied.
- Over millions of years, much more significant variations of up to 10°C occur as the sun and Earth, traveling through the galaxy, visit regions with more or fewer exploding stars.[64] We will cover this point more extensively in the next section of this chapter.

On August 25, 2011, CERN (the Conseil Européen pour la Recherche Nucléaire, or in English, the European Council for Nuclear Research, located in Geneva, Switzerland) reported on the CLOUD experiment (Clouds Leaving Outdoor Droplets). This experiment tested the hypothesis proposed by Svensmark and his colleagues in 1997 that cosmic rays play a significant role in the formation of low-lying clouds on Earth. The CERN press release announcing the results of the CLOUD experiment reported findings consistent with Svensmark's theory. The press release specified the following:

> The CLOUD results show that trace vapors assumed until now to account for aerosol formation in the lower atmosphere can explain only a tiny fraction of the observed atmospheric aerosol production. The results also show that ionization from cosmic rays significantly enhances aerosol formation. Precise measurements such as these are important in achieving a quantitative understanding of cloud formation and will contribute to a better assessment of the effect of clouds in climate models.[65]

The results of CERN's CLOUD experiment strongly supported Svensmark's argument. The CERN press release emphasized that results from the CLOUD experiment showed that sulfuric acid and water vapor a few kilometers up in the atmosphere could rapidly form clusters. Cosmic rays enhance the formation rate by up to tenfold or more. "Atmospheric aerosols play an important role in the climate," the press release further specified. "Aerosols reflect sunlight and produce cloud droplets. Additional aerosols would therefore brighten clouds and extend their lifetime."[66] Other cosmic rays entering the lower atmosphere when the sun was less active could create clouds in sufficient quantity to block out sunlight, thus constituting a forcing effect on global cooling. The day before the press release, on August 24, 2011, scientists at CERN published in the journal *Nature* their preliminary research results showing a connection between cosmic rays and the aerosol nucleation required to produce clouds, making public the CLOUD experiment results that supported Svensmark's theory.[67]

We also note here that Svensmark's cosmic ray theory supports the contention that a less active sun in Cycles 24 and 25 could predict the coming of a new ice age. This new ice age could occur even if humans keep burning hydrocarbon fuels that emit more CO_2 into the atmosphere. The challenge Svensmark's research represents to climate change true believers is obvious. In their 2007 book *Scared to Death*, British authors Christopher Booker and Richard North noted that in 1992, when a Danish delegation had suggested to the IPCC the influence of the sun on climate as a topic worthy of future research, the IPCC summarily rejected the proposal. In 1996, when the IPCC's overall chairman Professor Bert Bolin was asked to comment on Friis-Christensen and Svensmark's findings, he angrily dismissed them as "scientifically extremely naïve and irresponsible."[68]

The Milky Way Galaxy's Spiral Arms, Ice-Age Epochs, and the Cosmic Ray Connection

Nir J. Shaviv, a scientist at the Racah Institute of Physics at the Hebrew University of Jerusalem, advanced Svensmark's theory by publishing a paper in *New Astronomy* in 2003.[69] Shaviv argued that variability in the cosmic ray flux (CRF) hitting Earth may also be affected by Earth's position in the Milky Way Galaxy. The Milky Way is a spiral galaxy consisting of four major spiral arms in which are clustered most of the stars constituting the Milky Way. Earth is in a smaller arm (or "spur") known as the Orion Arm, positioned between the Sagittarius and Perseus Arms.[70] Note: the spiral arms of our galaxy are named after the constellations where the largest part of the spiral's stars cluster.

The earth travels with the solar system around the Milky Way Galaxy, taking approximately 225–250 million years (i.e., one "cosmic year") to complete one journey around the galaxy's center.[71] On the journey orbiting the Milky Way Galaxy, the earth passes through one of the galaxy's four spiral arms once every approximately one hundred million years, taking some ten million years to go through each spiral arm. Thus, in each cosmic year, the sun and Earth can be expected to transit no more than two of the galaxy's spiral arms.[72] Shaviv estimated that the sun passes through a spiral arm once every 135 million years. When Earth transits a spiral arm along with our solar system, there is a higher chance Earth will pass "nearby" an exploding star (i.e., a supernova). Recall that supernovae generate the higher number of cosmic rays that Svensmark's cosmic ray theory predicts will cause a greater increase of low-altitude clouds whose reflective power functions as a forcing effect for creating a period of global cooling on Earth.[73] Thus, Shaviv's theory was that when Earth passes through the galactic spiral arms, the cosmic ray flux hitting Earth increases. The result is that the average low-altitude cloud cover (LACC) on Earth will increase, with the average global temperatures on Earth reducing as a consequence. Shaviv hypothesized that the passing of Earth through the galaxy's spiral arms is an essential factor determining the reoccurrence of ice age epochs on the planet. Analyzing Earth's ice age periods back one billion years, Shaviv found a strong correlation between Earth's transition of the galaxy's spiral arms and the amount of cosmic ray flux that resulted. Cosmic ray flux increased when Earth passed through a spiral arm, resulting in a growth of the low-altitude global cloud cover and reduced temperature that facilitated the occurrences of the ice ages.

In a paper published in 2003, Shaviv collaborated with Canadian geochemist Ján Veizer, who had created a geochemical reconstruction of Earth's temperature over the past half a billion years. The two scientists concluded that as much as 75 percent of the variance in Phanerozoic geological history (i.e., the past 545 million years) could be attributed to CRF variations likely due to solar system passages through the spiral arms of the galaxy.[74] Shaviv and Veizer also concluded that cosmic ray flux, not CO_2, was the dominant factor affecting climate variability on multi-million-year time scales.[75] As we will see in the next section, CO_2 levels in geological time have been some ten times higher than today, during the end-Ordovician glaciation over 440 million years ago.[76] Given the much higher CO_2 concentrations in geological time, extending through several different glacial eras, Shaviv and Veizer concluded that CO_2 "is not likely to be the principal climate driver."[77] Veizer's independent study measuring calcium and magnesium isotopes in fossilized seashells over the 500 million years in which Earth's sea creatures have made seashells had provided evidence convincing him Earth had experienced a major "warming-cooling cycle every 135 million years, a time period that coincides with no earthly phenomenon."[78] Veizer's work with Shaviv was inspired once Shaviv explained to Veizer that "[the] cosmic rays strik[e] the Earth cycle up and down over 135 million years as our solar system pass through one of the bright arms of the Milky Way."[79]

Legitimate scientific debate yet rages over the Milky Way Galaxy structure and physics, especially given discoveries resulting from the Gaia space observatory satellite launched by the European Space Agency in 2013.[80] Yet, the confirming scientific research conducted by CERN in their noted CLOUD experiment makes refuting Svensmark's key point nearly impossible. We have additional experimental confirmation from CERN's research that cosmic rays affect low-altitude cloud formation on Earth. Thus, among Svensmark and Shaviv's more interesting corollary conclusions is that periodic glaciation epochs on Earth are related to the star formation rate in the Milky Way Galaxy.

CO_2 Concentrations on Earth in Geological Time

Some 4.5 billion years ago, Earth's atmosphere was primarily composed of carbon dioxide, with a CO_2 concentration (as measured today) of approximately one million parts per million (i.e., 1,000,000 ppm) compared to about 420 ppm today. Some 500 million years ago, the CO_2 concentration fell to around 7,500 ppm, about eighteen times today's levels. Between twenty-five million and nine million years ago, the CO_2 atmospheric concentration appeared to have varied between 180 and 290 ppm. With the inception of the Industrial Revolution, the air's CO_2 content increased to above 400 ppm registered today.[81]

Daniel H. Rothman, a professor at the Department of Earth, Atmospheric, and Planetary Sciences at MIT, published an important paper in the *Proceedings of the National Academy of Sciences* in 2002.[82] His research established that over most of the geologic record of the past 500 million years, Earth's CO_2 concentration fluctuated between values two to four times greater than those of today. However, over the past 175 million years, the data shows a long-term decline in the air's CO_2 content. Again, we encounter an inherent problem with time-series analysis. What is the proper period to identify the actual trends the data reflects? Since the inception of the industrial age, CO_2 levels in the atmosphere have risen. Yet, the trend if we look back over the past 175 million years is different. If we look back over the past 175 million years, the CO_2 levels in the atmosphere have continued to drop, including through today.

Still, IPCC adherents want to blame global warming and climate change disasters on human beings burning increasing amounts of hydrocarbon fuels since the dawn of the industrial age. The argument demands that IPCC adherents can establish CO_2 levels today are at historically high levels. From the evidence just presented, this argument fails when we examine the scientific evidence of atmospheric CO_2 concentrations over the history of geological time. As we have just noted, the scientific evidence is not clear the industrial age burning of hydrocarbon fuels has caused a rise in atmospheric CO_2 levels. Looking over the past 175 million years, CO_2 levels today have continued to drop, such that the recent rise since the industrial age does not change the trend curve when the time-series analysis extends back millions of years. In other words, to make their argument work, IPCC adherents fall into a classic trap of time-series statistics by choosing a time period for their analysis that is nonrepresentative of the data as a whole. Clearly, with the historical record of CO_2 in Earth's atmosphere going back 4.5 billion years, the trend curve of CO_2 concentrations dropping over the past 175 million years receives further confirmation.

The IPCC argument also falls victim to the classic logical fallacy known in Latin as *post hoc, ergo propter hoc*, or in English, "after this, therefore because of this." In other words, IPCC adherents claim a rise in CO_2 atmospheric concentration in Earth's atmosphere since the industrial age (which has not happened with the time-series analysis stretches back millions of years) has a causal link with today's historically dangerous global warming, resulting in catastrophic climate change. So, IPCC adherents must erase both the evidence of high atmospheric CO_2 levels over the past millions of years and the many interglacial warming periods that occurred over the past millions of years. But finally, even if it were true that CO_2 levels today are at geologically historical highs and also true that we are currently going through global warming at unprecedented highs over the past millions of years, IPCC adherents fall into the *post hoc, propter hoc* logical error. In other words, the IPCC argument also demands IPCC adherents must establish that the increased burning of hydrocarbon fuels is the only relevant factor on Earth that has changed since the beginning of the industrial age. If changes in Earth's temperature are correlated with other environmental factors,[83] e.g., the sun entering a grand solar minimum or Earth passing through an arm of the Milky Way, the argument fails because the causal link between high concentrations of CO_2 and global warming needs independent proof. Thus, even if global CO_2 concentrations were today at historically unprecedented highs and even if Earth were experiencing today a historically unprecedented period of global warming, the IPCC argument fails because IPCC adherents have assumed rather than proven anthropogenic burning of hydrocarbon fuel caused the global warming, when the cause could have been a grand solar minimum in our current Sun Cycle 25 or the earth passing through an arm of the Milky Way.

Rothman's conclusion was correct because he used scientific evidence of CO_2 atmospheric concentrations over the past millions of years to prove the point of his conclusion, namely, that CO_2 "does not exert dominant control on Earth's climate."

Many scientific studies published in peer-review journals strongly suggest that rises in CO_2 lag by periods of hundreds of years of warming in geologic history. The clear implication would be that "temperature is the independent variable that appears to induce changes in CO_2,"[84] not the other way around, as the IPCC argues dogmatically. In 2003, Nicolas Caillon, a scientist at the French Atomic Energy Commission, and his colleagues published a paper in *Science* magazine that is important to the discussion here. Caillon and his research team studied air bubbles from some 240,000 years ago trapped in the Vostok ice core in Antarctica. They found that the increase in CO_2 concentrations lagged Antarctic deglacial warming by an estimated 800 to 1,000 years.[85] The study concluded: "This confirms that CO_2 is not the forcing that initially drives the climatic system during a deglaciation. Rather, deglaciation is probably initiated by some isolation forcing, which influences first the temperature change in Antarctica (and possibly in part of the Southern Hemisphere) and then the CO_2."[86] The study attributed the subsequent increase in CO_2 to warming-induced CO_2 out-gassing from the Southern Hemisphere oceans.

Disinformation: "Proof" Solar Activity Does Not Cause Warming

In an article published in *Forbes* in 2019, entitled "Why Solar Activity and Cosmic Rays Can't Explain Global Warming," senior science contributor Marshall Shepherd spearheaded the disinformation campaign. From his first two sentences, Shepherd exuded disdain to the point of disgust for climate deniers who failed to see the wisdom of his counterattack. "As a climate scientist, I hear my share of myths about what is causing climate change or why it is a 'hoax.'" Shepherd wrote, kicking off his article. "I call them 'zombie theories' because they just will not die." Next, Shepard comments that "sun and its variability" is one of the arguments that "will not die." He continued: "I am pretty sure I've had to spray 'climate science repellent' on that nagging 'mosquito' numerous times. This week I heard a variation of this myth involving cosmic rays. Here is a science-based debunking of the solar-cosmic ray myth."

The magic bullet, Shepherd argued, was NASA data on total solar irradiance (TSI) tracked by satellites since 1978 that showed global temperatures rising 1°C in the twentieth century, even though TSI stayed relatively stable [showing no rise] during the same period. Shepherd wrote: "To anyone that has studied climatology and not astrophysics [i.e., Svensmark, Shaviv, and Veizer], this is a 'clear as a bell' signal that warming is related to greenhouse gases in the troposphere rather than the sun getting 'hotter' or other hypotheses [i.e., cosmic rays affecting cloud formation on Earth]."[87]

Methodological Difficulty of Reconstructing Historical TSI Data

In 1999, S.K. Solanki and M. Fligge, astronomers from the Institute of Astronomy, ETH-Zentrum in Zurich, Switzerland, published an article in *Geophysical Research Letters* entitled "A Reconstruction of Total Solar Irradiance Since 1700." This article explains the limitations of relying upon TSI data.[88] Solanki and Fligge wrote: "Precise measurements of the irradiance have only been made since 1978, whereas longer time series are required to establish a possible relationship with climate." In attempting to reconstruct sun irradiance data going back nearly three hundred years, Solanki and Fligge had to rely upon historical sunspot records that were admittedly sparse. "This paucity necessitates a simplification of the modelling process," Solanki and Fligge admitted, acknowledging the obvious, namely, that the "main uncertainty, however, lies in the reconstruction of quiet-sun irradiance variations." In the scope of geological time, data extending thirty years is insufficient to establish a reliable trend curve in time-series data.

Statisticians experienced with the "noise" in most observational time-series data understand the analytic problem fully. Advanced time-series methodology requires applying complex mathematical formulas known as Fourier analysis. Fourier analysis can reveal patterns hidden in the type of noisy data that is typical in the geoscience measurements, including measurements of TSI.[89] Geoscience time-series data is noisy; for instance, temperature measurements over time scatter above and below a mean number, making pattern determinations difficult. Fourier analysis transforms ordinary numbers into common logarithms in a base 10. The logarithmic meta-numbers can smooth out data to make patterns more discernable. The point is that discerning actual patterns in geoscience time-series analysis is complicated by the scattering "up and down" of the various data points taken at over hundreds, thousands, or millions of years.

As noted above, another problem of time-series data is that a trend that appears in one period may disappear if another period is chosen for the analysis. Selecting the right period for the research is an inherent problem with mutual fund reporting in the financial services industries. A mutual fund may be ranked number one in "Year 1" through "Year 10." But that same mutual fund may rank low if the period for analysis is "Year 1" through "Year 15." Much of the data generated in geosciences, including climate science, are time-series data, and these problems of statistical analysis make determining cause and effect problematic.

We just saw this problem in action when we realized that for most of the geologic record of the past 500 million years, Earth's CO_2 concentration fluctuated between values two to four times greater than those of today. However, over the past 175 million years, the data shows a long-term decline in the air's CO_2 content. Which is the actual pattern: the much higher CO_2 concentration over the past 500 million years or the reduction in CO_2 content over the past 175 million years? The data of these two time periods justifies the conclusion we drew earlier: rather than being a driver of Earth's climate or temperature, the atmospheric concentration of CO_2 is a dependent variable with no apparent correlation to Earth's climate or temperature through geological time. Understanding the time frame problem inherent in time-series analysis highlights the fraudulent nature of the IPCC time-series analyses that base much of their anthropogenic CO_2 argument by cherry-picking warmer global temperatures in the 1990s while ignoring the cooling period of 1940 through 1970.

Climate change true believers like Shepherd do not want to admit that TSI data with any chance of being accurate at most extends back to the satellite era, starting around 1978. A period as short as from 1978 to today is hardly sufficient for historically valid time-series analysis. Data that extends no farther back in time than 1978 yields, at most, some three decades of data on solar irradiation that may turn out to be an aberration. In other words, data from 1978 may mislead us because the period between 1978 and today may not be representative of a different or more extended period in geological time. The point is that data from 1978 to today might constitute nothing more than a deviation from a more meaningful mean that correlation or regression equation analysis might establish if we had data from a more extended period that more truly captures the nature of the phenomenon under observation. Put simply, thirty years of satellite-measured TSI data is, at most, interesting but by no means definitive or conclusive, especially when we realize that the proper frame of analysis demands we have reliable, scientific time-series data stretching back over the many millions of years of relevant geologic time.

In contrast, Svensmark, Shaviv, and Veizer tested their theories with reliable data precisely measured over millions of years. Svensmark and Shaviv, for instance, have validated in their scientific studies the historical variance in geologic time by examining traces of chemicals like argon-36 found in iron meteorites. The isotope argon-36 is well known to be helpful in the various established methods scientists employ to measure the quantity of galactic cosmic rays found in iron meteorites. Why? An isotope is an element that contains equal numbers of protons and electrons but different numbers of neutrons in the nuclei. For instance, carbon-12 and carbon-14 have six protons and six electrons, but carbon-12 is a carbon atom with six protons and six neutrons while carbon-14 is a carbon atom with six protons and eight neutrons. Hence, an isotope differs in atomic mass but not in chemical properties.

While nearly all the argon in Earth's atmosphere is argon-40, the isotope argon-36 is more common in the extraterrestrial universe because argon-36 is produced by stellar nucleosynthesis in supernovae. When argon-36 is found in an iron meteorite, the argon-36 measures the galactic cosmic ray flux the meteorite experienced on its journey to Earth. By testing for argon-36 (and other isotopes known to be created by stellar events), Svensmark, Shaviv, and Veizer were able to establish that the cosmic ray exposure of the iron meteorites clustered around troughs and peaks in the geological record provide scientific evidence tracing the beginning and end of ice ages going back one billion years in geological time, thereby establishing a relationship between CRF and Earth's climate. True believers wanting to argue TSI measurements prove the sun does not affect Earth's temperature have no similar time-sensitive, valid, and reliable measure for total sun irradiance in geological history.

Veizer noted that cosmic rays generate cosmogenic nuclides, such as beryllium-10, carbon-14, and chlorine-36, that serve as indirect proxies for solar activity and can be measured, for instance, in ancient sediments, trees, and shells. He noted that other proxies, including hydrogen and oxygen isotopes, can reflect past temperatures, since carbon isotopes can measure CO_2, or boron isotopes can measure for the acidity of ancient oceans, etc. Veizer noted that comparing temperature records from geological and instrumental archives with the trends for these isotope proxies should enable scientists to determine which of the two alternatives—CO_2 or the sun and cosmic rays—were responsible for climate variability. The scientific rigor of the historical analysis of proxy isotopes as a measure of geoscience variables is lacking in the TSI measures available today.

Yet, climate change true believers are undisturbed that they lack equally rigorous scientific analysis of TSI data over geological history. Undeterred, climate change true believers continue to insist upon a proposition contrary to common sense, namely, the argument that the sun doesn't play an essential role in explaining global warming.

The level, almost nonvarying time-series data emerging since 1978 from TSI measurements has bothered critically thinking scientists like Blanca Mendoza of the Institute of Geophysics in Mexico. In 2005, she published an article in *Advances in Space Research* entitled "Total Solar Irradiance and Climate." In that article, she argued the TSI variations reconstructed for past geological time do not account for the temperature changes we know have occurred.[90] Here is her concluding paragraph:

> Some current challenging questions on TSI variations and climate are the following: Are TSI variations from cycle to cycle mainly represented by sunspot and facular [i.e., granular structures on the sun hotter or cooler than the surrounding photosphere outer shell] changes? Does TSI variations always parallel the solar activity cycle? Is there a long-term component of the TSI? Closely related to the former, is the TSI output of the quiet Sun constant? If there is not a long-term trend of TSI variations, then we need amplifying mechanisms of TSI to account for the good correlations found between TSI and climate.[91]

A critical study reviewing the scientific literature on solar variability and its possible effects on Earth's climate was published in 2006 by three scientists: J. Beer and his colleague M. Vonmoos, both from the Swiss Federal Institute of Environmental Science and Technology in Dübendorf, Switzerland, together with R. Muscheler of the NASA/Goddard Space Flight Center in Greenbelt, Maryland. Published in *Space Science Reviews* and entitled "Solar Variability Over the Past Several Millennia," the article began with the following sentence: "The Sun is the most important energy source for the Earth." The three scientists asked this question: What role does the Sun play in climate change?[92]

Beer, Vonmoos, and Muscheler correctly noted that direct observations of solar irradiance via satellites above Earth's atmosphere have only been made since 1978, while observations of sunspots have been made and recorded for approximately four centuries. Isotopes with cosmogenic radionuclides, such as beryllium-10 or carbon-14, are available proxy measures of the sun's radiance over geological time because the isotopes are preserved in ice cores, in sedimentary rock layers, and in tree rings. These proxy isotopes provide the only reliable method to infer solar irradiance variability on a millennial time scale. Beer and his colleagues concluded the following:

> In order to establish a quantitative relationship between solar variability and solar forcing it is necessary to extend the records of solar variability much further back in time [i.e., more than the TSI satellite recordings that only began in 1978] and to identify the physical processes linking solar activity and total and spectral solar irradiance.[93]

Beer and his colleagues stressed that radiometers operating outside the atmosphere on satellites had dispelled the long-standing scientific belief TSI is a solar constant. While the changes in TSI recorded by satellites are small, the authors felt TSI variations, even though small, have an important impact on Earth's temperatures. The authors noted that "the observed changes of the TSI over an 11-year cycle are very small (0.1%), corresponding to an average temperature change of 1.5 Kelvin of the photosphere [i.e., the Sun's outer shell] and, on Earth, to a global forcing change of $0.25 Wm^{-2}$ [i.e., 0.25 Watts per square meter] (averaging over the globe and taking into account the albedo [Earth's reflective capability] of 30%)." They concluded: "This led many people to conclude that, even if the solar constant is not constant, the changes are too small to be climatically relevant without invoking some strong amplification mechanisms."

Yet, Beer and his colleagues disagreed: "This conclusion seems premature, first because there is no doubt that there are positive feedback mechanisms in the climate system." They explained, "A cooling for example, leads to growing ice sheets which increases the albedo and thus the cooling." The authors insisted that the feedback mechanisms that influence glacial-interglacial cycles are related to even small changes in the TSI. "General circulation models show that a change of the TSI by 0.1 percent over decades to centuries [as detected by satellite data since 1978] is not negligible." They added that the change of 0.1 percent in the TSI associates with far more significant changes in the solar system's U.V. (ultraviolet) part.

This analysis means that even if TSI changes are small, the earth's climate system is complex, with feedback systems that can be positive and negative, a point we saw developed by Roy Spencer at the beginning of the chapter. With this analysis, Beer and his colleagues raised a scientifically severe challenge to the contention that the sun plays no role in global warming. TSI readings have varied little in recent years while global temperatures have increased. Beer and his colleagues have explained that even small changes in the TSI can produce significant shifts in Earth's temperature. The team also attacked true believers on TSI measures by insisting on the following: "There are no physical reasons why the emission of radiation from the Sun should not show larger fluctuations on longer time scales up to 3×10^7 years [30 million years]. Other stars, although not exactly of the same type as the Sun, show considerably larger fluctuations." Here they remind us that the solar system is not stable over millennia. There is no reason in the physical record over geological time to assume anything different.

Beer and his colleagues pointed out that the only solution would be to conduct on TSI over geological time using the same type of work with isotopes that Svensmark, Shaviv, and Veizer have done with their work on cosmic rays and their impact on Earth's climate. They elaborated the point as follows:

> The first step, the extension of solar variability, can be achieved by using cosmogenic radionuclides such as beryllium-10 in ice cores. After removing the effect of the changing geomagnetic field on the beryllium-10 production rate, a 9,000-year long record of solar modulation was obtained. Comparison with paleoclimatic data provides strong evidence for a causal relationship between solar variability and climate change.[94]

We cannot directly measure sun activity accurately going back millions of years because we lack TSI data recorded over geological time. Yet, Beer and his colleagues noted that cosmogenic radionuclides such as carbon-14 and chlorine-36 offer the unique opportunity to extend the reconstruction of solar activity back into geological time. They explained the point as follows: "Cosmogenic radionuclides are produced in the Earth's atmosphere by the interaction of galactic cosmic rays (GCRs) with nitrogen, oxygen, and argon. Before reaching Earth, GCR have to cross the heliosphere where they are subject to solar induced modulation effects." The three concluded by commenting that the solar activity on the energy spectrum of galactic cosmic rays is described by the so-called solar modulation function designated by φ, the Greek capital letter phi. Beer and his colleagues were confident this methodology would establish that sun irradiation impacts Earth's temperature and climate because "the φ-record is characterized by high and persistent variability throughout the Holocene."[95]

Conclusion

This analysis of this chapter places the sun as a primary cause of heating or cooling Earth. Put simply, when the sun is less active, Earth's temperatures cool; when the sun is more active, Earth's temperatures warm. The analysis of this chapter also clarifies that water vapor—as reflected in ocean effects such as the PDO, La Niña, El Niño—and clouds are far more critical as heat/cooling *feedback factors* than carbon dioxide. This chapter should also clarify that water vapor (e.g., the PDO, La Niña, and El Niño) is the most abundant greenhouse gas in the atmosphere, acting as a head-generating or repressing *feedback factor*. The scientific record of CO_2 in geological time reveals that more significant percentages of CO_2 have been in Earth's atmosphere millions of years before human beings walked the face of the earth. If the geological record had established that increases in CO_2 always triggered global warming, those who demonize anthropogenic CO_2 would have a strong case. But the historical record has evidence of the opposite being the case. Temperature increases precede increases in CO_2, a relationship we will continue to explore in more depth in subsequent chapters.

Finally, we have argued that the sun is at center stage in causing Earth to change temperatures. We have aimed in this chapter to refute climate change true believers who insist the sun is nothing more than a yellow globe in the sky that plays no significant role in determining Earth's temperature.

CHAPTER 6

Climategate

True Believers Falsify Data, the Hockey Stick, the East Anglia University Hack, and the Built-in Failure of Climate Change Computer Models

The data set of proxies of past climate used by Mann, Bradley, and Hughes [authors of the original publication of the "Hockey Stick" appearing in Science, *April 1998] for the estimation of temperatures from 1400 to 1980 contains collation errors, unjustifiable truncation or extrapolation of source data, obsolete data, geographical location errors, incorrect calculation of principal components and other quality control errors. We detail these errors and defects.*

—**Stephen McIntyre** and **Ross McKitrick**, *Energy & Environment*, 2003[1]

Quite apart from what the Hockey Stick tells us about the positioning of the IPCC in the global warming debate, the panel's need for a sales tool also suggests something important about the overall case for manmade global warming. None of the corruption and bias and flouting of rules we have seen in the course of this story would have been necessary if there is, as we are led to believe, a watertight case that mankind is having a potentially catastrophic effect on the climate. What the Hockey Stick affair suggests is that the case for global warming, far from being settled is actually weak and unconvincing.

—A.W. Montford, *The Hockey Stick Illusion*, 2010[2]

> *The "hockey stick" concept of global climate change is now widely considered totally invalid and an embarrassment to the IPCC.*
> —**Don Easterbrook**, *Evidence-Based Climate Science*, 2011[3]

> *For more than two decades I was in the crosshairs of climate change deniers, fossil fuel industry groups and those advocating for them—conservative politicians and media outlets. This was part of a larger effort to discredit the science of climate change that is arguably the most well-funded, most organized PR campaign in history. Now we finally have reached the point where it is not credible to deny climate change because people can see it playing out in front of their eyes.*
> —**Michael E. Mann**, *The Guardian*, 2021[4]

WHEN HE PUBLISHED HIS FIRST infamous "Hockey Stick" article in 1998, Michael E. Mann was in his early thirties. He was a relatively unknown recent adjunct/research assistant professor in the Department of Geosciences at the University of Massachusetts. He received his Ph.D. from the Department of Geology and Geophysics at Yale University.[5] The article, published in *Nature* on April 23, 1998, bore the academic-sounding title, "Global-scale temperature patterns and climate forcing over the past six centuries."[6] His two secondary coauthors were Raymond S. Bradley, a colleague in the Department of Geosciences at the University of Massachusetts, and Malcolm K. Hughes, from the Laboratory of Tree Ring Research at the University of Arizona.

The "Medieval Warm Period" Disappears

Four years earlier, in 1994, Hughes had coauthored an article in *Climatic Change* entitled "Was There a 'Medieval Warm Period, and If So, When and Where?'"[7] In this article, Hughes utilized tree ring analysis to argue that the Medieval Warm Period and the subsequent Little Ice Age were regional phenomena, limited largely to Europe but not experienced in other parts. In the article, Hughes concluded: "The generalized behavior of the global climate of the last millennium as a Medieval Warm Period followed by a Little Ice Age, each one or more centuries long and global in extent, is no longer supported by the available evidence."[8]

One of the leading critics of Mann's work is accountant Andrew William (known as "A. W.") Montford, a British writer and editor/owner of the Bishop Hill blog (Bishop-Hill.net). In his 2010 book entitled *The Hockey Stick Illusion*, Montfort commented on the importance of Hughes's 1994 paper as follows: "On its own, these findings might look interesting but otherwise unremarkable. But in the context of the temperature history of the last thousand years, their impact on the climate debate was potentially explosive."[9] As Montford noted, after Hughes's 1994 paper, the Medieval Warm Period became "less warm."[10]

What Hughes accomplished in his paper was to set the stage for Mann's Hockey Stick argument, namely that the only rise in global temperatures in the last thousand years was since the Industrial Revolution. Montford continued his comments on Hughes's paper by noting Hughes was attempting to flatten out the temperature rises during the Medieval era and minimize the temperature drops during the Little Ice Age. Montford commented that by flattening out the temperature changes, Hughes was preparing the argument that current temperatures are dramatically higher than anything seen in previous times. Dramatically higher temperatures in recent years would be "powerful evidence" that anthropogenic global warming was having a "serious and deleterious effect on the world's climate."[11] Montford argued what Hughes was attempting to accomplish was "to overturn a well-embodied paradigm."[12] Montford summed up the impact of the Hughes paper as follows: "The flatter the representation of the medieval period in the temperature reconstructions, the scarier were the conclusions."[13]

The "Hockey Stick" Is Born

Mann appears to have designed his 1998 *Nature* paper to advance Hughes's 1994 argument. The highlight of the 1998 paper was a graph of Northern Hemisphere temperatures from 1400 to 1980 that looked like a hockey stick. A long handle over 500 years showed Earth's temperatures were relatively flat from 1400 until the start of the twentieth century when temperatures shot up in an almost straight line. But from the twentieth century on, the temperatures went dramatically up, thus forming the "blade" of the "hockey stick." Mann's method was to search the archives to find some 112 temperature indicators that he could put into a mathematical model as temperature proxies in the effort to reconstruct historical Earth temperatures back to the year 1400.

As Montford pointed out, the Hockey Stick graph proved a highly effective promotional tool. Here is how Montford described the Hockey Stick graph in Mann's 1998 paper:

The key graphic in the paper was a reconstruction of Northern Hemisphere temperature for the full length of the record from 1400 right through to 1980. The picture presented was crystal clear. From the very beginning of the series the temperature line meandered gently, first a little warmer, then a little cooler, never varying more than half a degree or so from peak to trough. This was the 500-year long handle of the Hockey Stick, a sort of steady state that had apparently reigned, unchanging, throughout most of recorded history. Then suddenly, the blade of the stick appeared at the start of the twentieth century, shooting upwards in an almost straight line. It was a startling change and it was this that made the Hockey Stick such an effective promotional tool, although to watching scientists, the remarkable thing about the Hockey Stick was not what was happening in the twentieth century portion—that temperatures were rising was clear from the instrumental record—but the long flat handle. The Medieval Warm period had completely vanished. Even the previously acknowledged "regional effect" now left no trace in the record. The conclusions were stark: current temperatures were unprecedented.[14]

As Montford correctly noted climate change true believers embraced the long flat handle erasing the Medieval Warm Period and the Little Ice Age, such that even the "regional effect" Hughes had previously recognized was now gone. "The conclusions were stark," Montford noted, "current temperatures were unprecedented." Mann instantly achieved international celebrity status. The mainstream media worldwide proclaimed that Mann's tree ring data proved three recent years—1990, 1995, and 1997—were the warmest years since 1400. The Hockey Stick graph portrayed the twentieth century as significantly warmer than the five centuries that preceded it, with the global climate change in the twentieth century assumed to be linked to the burning of fossil fuels.[15] The importance of Hughes's paper is reflected in the fact that he joined Mann's team, becoming one of Mann's two coauthors on Mann's 1998 Hockey Stick paper.

The following year, 1999, Mann, Bradley, and Hughes followed up their first paper with a second entitled "Northern Hemisphere Temperatures During the Past Millennium: Inferences, Uncertainties, and Limitations," published in *Geophysical Research Letters*.[16] The second analysis, intended to make the first paper more authoritative, examined Northern Hemisphere temperatures back 1,000 years. The conclusion that the twentieth century was the warmest of the millennium reinforced the impact of blaming global warming on the human use of hydrocarbon fuels. "School books told children that the Hockey Stick meant the world had to change," Montford wrote, commenting on how influential Mann's Hockey Stick analysis had become. "Politicians told voters that only they could save people from the threat it demonstrated. Insurers, newspapers and magazines, pamphlets and websites were all in thrall to its message; the Hockey Stick swept all before it."[17]

The IPCC Embraces Michael Mann

Before Michael Mann published his Hockey Stick graph, the IPCC appeared not to have realized the extent to which the Medieval Warm Period and the Little Ice Age destroyed the IPCC narrative that CO_2 was the only significant driver of Earth's temperature and climate. In 1990, the IPCC published a graph in its first *Climate Change: Scientific Assessment*—a climate assessment the organization would never have published had Mann published his graph a decade earlier. In retrospect, the graph the IPCC published in 1990 repudiated the data Mann's Hockey Stick displayed. In the IPCC 1990 graph, they appear to have reported the truth about the Medieval Warm Period and the Little Ice Age. The graph showed accurately that Earth's temperatures during the Medieval Warm Period were higher than today from the late tenth to early thirteenth centuries (about AD 950-1250), characterizing the period as the "Medieval Climatic Optimum."[18] The same 1990 IPCC graph presented the Little Ice Age as a period "which resulted in extensive glacial advances in almost all alpine regions of the world between 150 and 450 years ago so that glaciers were more extensive 100–200 years ago than now nearly everywhere."[19] The executive summary of the climate assessment report also acknowledged both the Medieval Warm Period and the Little Ice Age, noting the following:

> Over the last two million years, glacial-interglacial cycles have occurred on a time scale of 100,000 years, with large changes in ice volume and sea level. During this time, average global surface temperatures appear to have varied by about 5–7°C. Since the end of the last ice age, about 10,000 BP, globally average surface temperatures have fluctuated over a range of up to 2°C on time scales of centuries or more. Such fluctuations include the Holocene Optimum around 5,000–6,000 years ago, the shorter Medieval Warm Period around 1000 AD (which may not have been global) and the Little Ice Age which ended only in the middle to late nineteenth century. Details are often poorly known because palaeo-climactic data are frequently sparse.[20]

THE TRUTH ABOUT ENERGY

By the time Mann published his Hockey Stick graph in 1998, the IPCC had come to realize the 1990 climate assessment admissions threatened the IPCC's central contention that anthropogenic CO_2 production is the sole cause of global warming. As climate skeptic John Daly pointed out on his blog, the 1990 IPCC account of climatic history contains two severe difficulties for the present global warming theory:

1. If the Medieval Warm Period was warmer than today, with no greenhouse gas contribution, what would be so unusual about modern times being warm also?
2. If the variable sun caused both the Medieval Warm Period and the Little Ice Age, would not the stronger solar activity of the 20th century account for most, if not all, of the claimed 20th century warmth?[21]

Dr. David Deming, a professor at the College of Earth and Energy at the University of Oklahoma, explained this background in a testimony he gave in 2006 to a full committee hearing on climate change and the media held by the U.S. Senate Committee on Environment and Public Works. Deming explained that he was a geologist and a geophysicist who specialized in temperature and heat flow. He related that in 1995, he published a paper in the academic journal *Science*. But between 1990 and 1998–1999, when Mann started publishing his graph, the global warming community and the IPCC began scrambling to find a way to erase the Medieval Warm Period and the Little Ice Age any way they could. Here is what Deming testified happened next:

> In that study, I reviewed how borehole temperature data recorded a warming of about one degree Celsius in North America over the last 100 to 150 years. The week the article appeared, I was contacted by a reporter for National Public Radio. He offered to interview me, but only if I would state that the warming was due to human activity. When I refused to do so, he hung up on me.[22]

Deming continued to relate another eye-opening experience he had following the 1995 publication of his paper in *Science*:

> I had another interesting experience around the time my paper in *Science* was published. I received an astonishing email from a major researcher in the area of climate change. He said, "We have to get rid of the Medieval Warm Period."

Deming explained to the Senate committee why the global warming community was so desperate over this issue:

> The Medieval Warm Period (MWP) was a time of unusually warm weather that began around 1000 AD and persisted until a cold period known as the "Little Ice Age" took hold in the 14th century. Warmer climate brought a remarkable flowering of prosperity, knowledge, and art to Europe during the High Middle Ages.
>
> The existence of the MWP had been recognized in the scientific literature for decades. But now it was a major embarrassment to those maintaining that the 20th century warming was truly anomalous. It had to be "gotten rid of."
>
> In 1769, Joseph Priestly warned that scientists overly attached to a favorite hypothesis would not hesitate to "warp the whole course of nature." In 1999, Michael Mann and his colleagues published a reconstruction of past temperature in which the MWP simply vanished. This unique estimate became known as the "hockey stick," because of the shape of the temperature graph.[23]

Deming told the Senate committee he was astounded at the enthusiasm with which the scientific community and the media accepted Mann's Hockey Stick graph uncritically. He testified on this point as follows:

> Normally in science, when you have a novel result that appears to overturn previous work, you have to demonstrate why the earlier work was wrong. But the work of Mann and his colleagues was initially accepted uncritically, even though it contradicted the results of more than 100 previous studies. Other researchers have since reaffirmed that the Medieval Warm Period was both warm and global in its extent.
>
> There is an overwhelming bias today in the media regarding the issue of global warming. In the past two years, this bias has bloomed into an irrational hysteria. Every natural disaster that occurs is now linked with global warming, no matter how tenuous or impossible the connection. As a result, the public has become vastly misinformed on this and other environmental issues.[24]

By erasing the Medieval Warm Period and the Little Ice Age after Mann started publishing his two Hockey Stick papers in 1998–1999, the IPCC conveniently could make these analytic problems simply go away.

But with the publication of the IPCC's *Third Assessment Report* in 2001, the IPCC changed course. In 2001, the IPCC enthusiastically embraced his Hockey Stick graph. Following Mann, the organization erased the Medieval Warm Period and minimized the Little Ice Age while declaring that 1998 and 1999 were the warmest years ever in global history. The IPCC included the Hockey Stick in the "Summary for Policymakers" on page three of the *Third Assessment Report*'s volume entitled *Climate Change 2001: The Scientific Basis*. The policymaker summary argued, "the rate and duration of warming of the 20th century have been much greater than in any of the previous nine centuries." The summary contended it was likely "that the 1990s was the warmest decade and 1998 the warmest year in the instrumental record, since 1861."[25] Mann's two published papers were cited extensively throughout the report (some eighty footnote mentions). It listed Mann as a contributing or lead author for several of the report's chapters.

The National Oceanic and Atmospheric Administration (NOAA) quickly picked up on Mann's work, making the Hockey Stick officially accepted by the U.S. climate bureaucracy, declaring 1998 the warmest year in recorded history. NOAA's Global Climate Report for 1998 reported the most significant temperature anomaly occurred that year, "making it the warmest year since widespread instrument records began in the late Nineteenth Century." The NOAA report claimed "the second warmest year was 1997, and seven of the ten warmest years" in history "were in the 1990s."[26]

Mann Twists Tree Ring Data

The data Mann used to construct his Hockey Stick draft relied heavily on a previous study of high-altitude pine trees in Great Basin National Park in Nevada and from the Sierra Nevada Mountains in California. This study was created by Donald A. Graybill, a research physicist from the Laboratory of Tree-Ring Research, University of Arizona, Tucson, and Sherwood B. Idso, from the U.S. Water Conservation Laboratory, Agricultural Research Service, in Phoenix, Arizona, who published the paper in *Global Biochemical Cycles* in 1993.[27]

Ironically, Mann relied on data collected by Idso who had since gone on to become the president of the Center for the Study of Carbon Dioxide and Global Change. That center, a 501(c)(3) nonprofit organization, rejected the IPCC premise that anthropogenic CO_2 causes climate change. Graybill and Idso's study presumed temperatures had warmed but, as we shall next see, Graybill and Idso, in the conclusion of their published study on high-altitude pine trees that Mann relied upon for source data, were not convinced CO_2 was the temperature driving factor causing global temperatures to rise.[28]

"Our research supports the hypothesis that atmospheric CO_2 fertilization of natural tree growth has been occurring from at least the mid- to late-19th century," Graybill and Idso concluded.[29] Yet, the abstract to their paper made clear that Graybill and Idso did not have the data to support the conclusion that anthropogenic CO_2 was the cause. The abstract makes this point as follows:

> The growth-promoting effects of the historical increase in the air's CO_2 content are not yet evident in tree-ring records where yearly biomass additions are apportioned among all plant parts. When almost all new biomass goes into cambial enlargement, however, a growth increase of 60% or more is observed over the past two centuries. As a result, calibration of tree-ring records of this nature with instrumental climate records may not be feasible because of such growth changes. However, climate signals prior to about the mid-19th century may yet be discovered by calibrating such tree-ring series with independently derived proxy climate records for this time.[30]

Please also note the following: the Graybill-Idso study did not yield a database that could be extrapolated as a measure of global temperatures because the data was drawn from one type of tree, conifers, growing in high-altitude, specifically, in the northwestern regional section of the United States, an area with distinctive mountainous terrain and temperature characteristics.[31] Graybill and Idso's "1993 study of one group of trees in one untypical corner of the USA seemed a remarkably flimsy basis on which to base an estimate of global temperatures going back 1,000 years," Christopher Booker and Richard North pointed out in their 2007 book *Scared to Death*.[32]

Another irony is that the Graybill-Idso study confirmed CO_2 plays a negative feedback function in that when the conifers take CO_2 out of the atmosphere to absorb the CO_2 as fertilizer, this reduces the temperature-forcing effect of the atmospheric CO_2. The CO_2 absorbed by the conifers is no longer in the atmosphere with the potential to function as a temperature-increasing greenhouse gas.

Hockey Stick Data Analysis Deconstructed

Two mathematically adept Canadians not explicitly trained in climatology conducted the data analysis that undermined the validity of Mann's methodology. Stephen McIntyre, a retired mining company executive living in Toronto, teamed up with Ross McKitrick, an economics professor in

Ontario at the University of Guelph. Although Mann was resistant and uncooperative, McIntyre and McKitrick managed to get ahold of the dataset Mann used in his original paper to construct the Hockey Stick analysis. The two men held no punches launching a damaging critique of Mann's methodology in the quotation cited at the start of this chapter. In short, McIntyre and McKitrick found Mann's database to be so riddled with methodological problems that it was not possible to audit Mann's database. Yet, by going back to academically authenticated temperature data and collating it correctly, McIntyre and McKitrick demonstrated that Mann had manipulated his data to create the illusion modern warming was unprecedented in geological history. McIntyre and McKitrick replicated Mann's statistical analysis on a corrected database prepared with substantially improved quality control. McIntyre and McKitrick found "that their own method [i.e., the methodology Mann et al. used], carefully applied to their own intended source data, yielded a Northern Hemisphere temperature index in which the late 20th century is unexceptional compared to the preceding centuries, displaying neither unusually high mean values nor variability." McIntyre and McKitrick's analysis of Mann's data with the errors corrected did show the Medieval Warm Period with temperatures registered higher than today.[33]

Mann's initial response to McIntyre and McKitrick's critique was that the two were not using the data he used to draw his conclusions. A tug-of-war followed, precipitated by Mann's refusal to hand over his data and his computer analysis algorithms. Finally, when McIntyre and McKitrick got ahold of Mann's data, they found his programs used hundreds of lines of code written in Fortran. They found this suspicious because Fortran is an early high-level mainframe computer programming language that is today largely obsolete. They argued professionally trained statisticians could have quickly written Mann's statistical procedures in a few lines of more reliable code. Particularly damaging to Mann's methodology, McIntyre and McKitrick found the Fortran code contained an unusual data transformation that had never been reported in print. McIntyre and McKitrick also found that when they subjected Mann's data to the rigors of modern statistical analysis, R^2 coefficients [a regression analysis measure of explained variance/total variance] would have demonstrated Mann's conclusions fell well short of statistical significance. They concluded the results Mann obtained were spurious.[34]

In layman's terms, the point was that by applying modern data management and statistical testing techniques to Mann's database, McIntyre and McKitrick could not replicate the Hockey Stick results Mann had published. Even more damaging, when McIntyre and McKitrick analyzed meaningless random datasets of nothing but "red noise," Mann's algorithms nearly always produced a hockey stick–shaped diagram.[35] "Had the IPCC done the kind of rigorous review that they boast of, they would have discovered there was an error in a routine calculation step (principal component analysis) that falsely identifies a hockey stick shape as the dominant pattern in the data," McKitrick argued in a 2005 memorandum he published with the House of Lords in the U.K. Parliament. "The flawed computer program can even pull out spurious hockey stick shapes from lists of trendless random numbers."[36] In a paper presented at an international conference on managing climate change that the Australian Asia Pacific Economic Cooperation Study Centre at Monash University hosted in April 2005 in Canberra, Australia, McKitrick explained the problem with Mann's analysis as follows: "In 10,000 repetitions on groups of red noise, we found a conventional PC [principal component] algorithm almost never yielded a hockey stick-shaped PC1 [principal component #1, the dominant data pattern], but the Mann algorithm yielded a pronounced hockey stick-shaped PC1 over 99% of the time."[37] Even the telephone book would have come out like a hockey stick, commented the authors Christopher Booker and Richard North in their book, *Scared to Death*.[38]

As a cautionary methodological note, please consider another problem with Mann's reliance on tree ring analysis to determine Earth temperatures historically. Tree ring data is inherently suspect of being not conclusive. Measures of tree ring growth are notoriously unreliable proxies for estimating Earth temperature because tree rings form in one short period of the tree's annual growth. Hence, tree ring growth measures are not necessarily reflective of temperatures for the year.[39]

Climategate

On November 17, 2009, an unidentified hacker began publishing thousands of emails and other documents taken from a server at the Climatic Research Unit (CRU) at the University of East Anglia (UEA) in Norwich, England. The CRU was a source of much of the data supporting the anthropogenic CO_2 theory, responsible for co-compiling the HadCRUT global temperature series, in conjunction with the U.K. Met Office Hadley Center. The hacked emails created a firestorm for the IPCC. Phil Jones, then the director of the CRU and a professor in the School of Environmental Sciences at UEA, had exposed that Mann introduced an obvious bias into his analysis. Mann appeared determined to tag CO_2 as the sole culprit for global warming, even if coming to that conclusion required falsifying or otherwise altering data. Jones, on November 16, 1999, sent one of the most damaging emails to Mann and the other two authors of the Hockey Stick papers published in 1998:

> I've just completed Mike's *Nature* trick of adding the real temps to each series for the last 20 years (i.e., from 1981 onwards) and from 1961 for Keith's to hide the decline.

"Mike" is Michael Mann, the lead author on the *Nature* paper. "Keith" is Keith Briffa, the deputy director of the CRU and a lead author of an article that showed a decline in the growth of tree rings reflecting a decline in temperatures starting in 1961, despite the buildup of CO_2 emissions in those years. The problem the CRU government scientists were trying to solve was known as a "divergence problem," in that several tree ring studies, including Briffa's, showed this decline since 1961 in tree ring growth that implied something else other than CO_2 was causing the temperature decline. Mann and his colleagues were concerned Briffa's published conclusions would undermine the public's faith in paleo-estimates of global temperature derived from tree ring growth analysis.[40] Mann's solution to the problem was to eliminate the embarrassing data by replacing Briffa's tree ring measures with thermometer records of global temperature to remove the decline revealed by the tree ring analysis, thereby hiding the decline.[41]

The hacked emails revealed a disturbing pattern. The IPCC scientists had falsified data in published studies purporting to prove anthropogenic CO_2 was causing global warming. The hacked emails also showed the IPCC scientists were willing to falsify data and lie to intimidate critics. Dr. Kelvin Kemm, formerly a scientist at South Africa's Atomic Energy Corporation, summarized Climategate as "the release of thousands of emails to and from client scientists who had been (and still are) collaborating and colluding to create a manmade climate crisis that exists in their minds and computer models."[42] In 2008, Jones wrote to Mann, asking him to delete any emails he might have had with Briffa, as well as asking Mann to see if he could get Eugene Wahl and Caspar Ammann, two U.S. researchers supportive of Mann's analysis, to do the same. The email made clear that researchers associated with the CRU wanted to erase damaging emails to hide them from Freedom of Information Act requests. This 2008 email led the U.K.'s Information Commissioner's Office to conclude the university was "'acting so as to prevent intentionally the disclosure of requested information,' and thus requests were 'not dealt with as they should have been under the legislation.'"[43]

Other leaked emails showed Jones and Mann conspiring to block the publication of climate skeptic scientific articles, as well as threatening to fire any editor who relied on peer review to publish views questioning the anthropogenic CO_2 theory. In August 2007, Mann sent an email suggesting he wanted to destroy McIntyre's reputation to discredit his criticism of Mann's Hockey Stick analysis. Other emails showed Mann refusing to let anyone examine his statistical analysis. Specifically, Mann was resistant to allowing anyone to challenge his R^2 regression numbers that were critical to determining the degree of correlation (i.e., a statistical measure assessing the strength of independent variables in the outcome of dependent variables). Mann insisted his statistical analyses were his private property, despite public grants subsidizing his research. In an apparent admission of an even more suspicious crime, Jones initially suggested he lost critical data before admitting he had destroyed the data in question.[44]

"I have been talking [with] folks in the States about finding an investigative journalist to investigate and expose him [i.e., Steve McIntyre]." The timing of the released hacked emails, coming one month before the Copenhagen Climate Change Conference scheduled for December 2009, undermined the moral high ground of leading climate change advocates like Jones and Mann. It also destroyed the summit's plan to legally bind the Western countries to the extension of the Kyoto Protocol, a commitment to the year 2050 for developing renewable energies [45]

On November 1, 2019, South African nuclear physicist Kemm published a ten-year retrospective on Climategate. He summarized the scandal as follows:

> Then on the morning of 17 November 2009 a Pandora's box of embarrassing CRU information exploded onto the world scene. A computer hacker penetrated the university's computer system [at the Climate Research Unit (CRU) of the University of East Anglia] and took 61 Megs of material that showed the CRU had been manipulating scientific information to make global warming appear to be the fault of mankind and industrial CO2. Among many other scandals, the shocking leaked emails showed then-CRU-director Prof. Phil Jones boasting of using statistical "tricks" to remove evidence of observed declines in global temperatures.
>
> In another email, he advocated deleting data rather than providing it to scientists who did not share his view and might criticize his analysis

Non-alarmist scientists had to invoke British freedom of information laws to get the information. Jones was later suspended, and former British Chancellor Lord Lawson called for a Government enquiry into the embarrassing exposé.

The affair became known as "Climategate," and a group of American University students even posted a YouTube song, "Hide the Decline," mocking the CRU and climate modeler Dr. Michael Mann, whose use of the phrase "hide the decline" in the temperatures had been found in the hacked emails.[46]

In his 2018 book *The Politically Incorrect Guide to Climate Change*, Marc Morano pointed out the Climategate scandal "revealed that the UN IPCC was simply a lobbying organization portraying itself as a science panel." Morano noted that if the IPCC failed to find CO_2 as the global warming culprit, the IPCC "would no longer have a reason to continue studying it—or to be in charge of offering 'solutions.'"[47] Rex Murphy of the Canadian Broadcasting Corporation stressed that Climategate "'pulls back the curtain on a scene of pettiness, turf protection, manipulation, defiance of freedom of information, lost or destroyed data and attempts to blacklist critics or skeptics of the global warming cause.' Murphy added, 'Science has gone to bed with advocacy and both have had a very good time.'"[48] Phillip Scott, emeritus professor of biogeography at the University of London School of Oriental and African Studies, summed up succinctly Mann's statistical manipulations. "Any scientist ought to know that you just can't mix and match proxy and actual data," Scott explained to the *Daily Mail* in London. "They're apples and oranges. Yet that's exactly what he did."[49]

Despite the data flaws of the original Hockey Stick analysis and the revelations of fraudulent climate science by Jones, Mann, and their colleagues, the disinformation campaign launched by climate change true believers continue to insist Mann's analysis was correct. "Most researchers would agree that while the original hockey stick can—and has—been improved in a number of ways, it was not far off the mark," wrote Michael Le Page, a reporter for the *New Scientist*, in a 2007 article. "Most later temperature reconstructions fall within the error bars of the original hockey stick. Some show far more variability leading up to the 20th century than the hockey stick, but none suggest it has been warmer at any time in the past 1,000 years than in the last 20th century."[50] Coming out of the Hockey Stick controversy and the Climategate disclosures, the disinformation campaign launched by the anthropogenic CO_2 community has decided to hold the fort on the argument that Earth's temperature over the past 1,000 years has never been warmer than the temperatures recorded in 1998–1999.

Evidence for the Medieval Warm Period and the Little Ice Age

Amid the controversy, climate change true believers decided to shift their tactics to claim that the Medieval Warm Period was a regional phenomenon, not one experienced by Earth as a whole. In a 2021 article, Frédérik Saltré and Corey J.A. Bradshaw of Flinders University in Adelaide, Australia, who are supported by the government-funded Australian Research Council, wrote an article entitled "Climate explained: what was the Medieval Warm Period?"[51] To introduce their argument, Saltré and Bradshaw explained why they felt this piece was necessary to write. "We are living in a world that is getting warmer year by year, threatening our environment and way of life," they stated as a fact a premise they presumed the reader would accept without proof. "But what if these climate conditions were not exceptional?" they continued. "What if it had already happened in the past when human influences were not part of the picture?" That was the argument Saltré and Bradshaw intended to attack. Saltré and Bradshaw's goal was to discredit the proposition these climate conditions had occurred previously, before the onset of the Industrial Revolution and the increased human burning of hydrocarbon rules. They calculated that a reinterpretation of the Medieval Warm Period was necessary to achieve that purpose. "The often-mentioned Medieval warm period seems to fit the bill," Saltré and Bradshaw continued. "This evokes the idea that if natural global warming and all its effects occurred in the past without humans causing them, then perhaps we are not responsible for this one. And it does not matter because if we survived one in the past, then we can surely survive one now." From there, Saltré and Bradshaw argued it "it's just not that simple."

Beginning their rebuttal, the two authors agreed that during the Medieval era "some regions experienced temperatures exceeding those recorded during the period between 1960 and 1990." This was their main point of refutation, namely the argument that "the Medieval warm period was by and large a regional event." In contrast, they argued, today we are experiencing "a global increase in atmospheric gases such as carbon dioxide." Consider the following sentence as typical of their argument: "While the northern hemisphere, South America, China and Australasia, and even New Zealand, recorded temperatures of 0.3-1.0°C higher than those of 1960-1990 between the early ninth and late 14th centuries, in other areas such as the eastern tropical Pacific Ocean, it was much cooler than today." In order to diminish the importance of a solar maximum as the likely cause of the Medieval Warm Period, Saltré and Bradshaw explained, "an increase in solar radiation and decrease in volcanic eruptions created a La Niña-like event that changed the usual patterns [with] [s]tronger trade winds pushing more warm water toward Asia." In other words, this complex warming pattern and not the sun by itself, Saltré and Bradshaw maintained, created many different weather patterns, including wetter conditions in Australasia, droughts in the southern U.S. and Central America, and heavy rains accompanied by flooding in the Pacific Northwest and Canada. As demonstrated by the two authors, the point of disinformation campaigns is to substitute an alternative explanation. Here, in dealing with the Medieval Warm Period, Saltré and Bradshaw's goal was to diminish the importance of the sun in causing a global Medieval Warm Period without worrying about whether scientific evidence supported the alternative explanation.

The attempt to disappear the Medieval Warm Period and the Little Ice Age flies in the face of several thousand scientific publications over decades that establish the global impact of both. Don Easterbrook, emeritus professor of geology at Western Washington University, made the point in his 2011 book *Evidence-Based Climate Science: Data Opposing CO_2 Emissions as the Primary Source of Global Warming*, "Oxygen isotope studies in Greenland, Ireland, Germany, Switzerland, Tibet, China, New Zealand, and elsewhere, plus tree-ring data from many sites around the world, all confirm the existence of a global MWP [Medieval Warm Period]." The well-respected science publisher Elsevier in the Netherlands published Eastbrook's book. "Evidence that the MWP was a global event is so widespread that one wonders why Mann et al. ignored it." Easterbrook further commented: "Thus, it came as quite a surprise when Mann et al., on the basis of a single tree-ring study, concluded that neither the MWP nor the Little Ice Age actually happened, and that assertion became the official position of the 2001 Intergovernmental Panel on Climate Change (IPCC)." Again, Easterbrook expressed his frustration, writing, "The Mann et al. 'hockey stick' temperature curve was so at odds with thousands of published papers, one can only wonder how a single tree-ring study could purport to prevail over such a huge amount of data."[52]

In 2003, Willie Soon at the Harvard & Smithsonian Center for Astrophysics and Sallie Baliunas at the Mount Wilson Observatory published in *Climate Research* an article that surveyed climatic and environmental changes over the past 1,000 years. "Climate proxy research provides an aggregate, broad perspective on questions regarding the reality of Little Ice Age, Medieval Warm Period and the 20th century surface thermometer global warming," Soon and Baliunas noted in conclusion to their twenty-page article. "The picture emerges from many localities that both the Little Ice Age and Medieval Warm epoch are widespread and near-synchronous phenomena," they continued. "Overall, the 20th century does not contain the warmest anomaly of the past millennium in most of the proxy records, which have been sampled world-wide. Past researchers implied that unusual 20th century warming means a global human impact. However, the proxies show that the 20th century is not unusually warm or extreme."[53] As further confirmation, the Center for the Study of Carbon Dioxide and Global Change has provided a compilation of published scientific studies providing evidence of the Medieval Warm Period in Africa, Antarctica, Asia, Australia, New Zealand, Europe, North America, Northern Hemisphere, and South America.[54]

Though the climate change true believers try as hard as they can to attempt to explain away the Medieval Warm Period and the Little Ice Age, the published documentation extending over many decades is extensive. Even Mann was at a loss to explain why the evidence for both is solid in Europe and the Northern Hemisphere. In a 2002 encyclopedia, Mann speculated that the Medieval Climatic Optimum was a regional phenomenon. In his encyclopedia entry Mann wrote the following: "Increased northward heat transport by an accelerated Atlantic thermohaline ocean circulation during Medieval times may have warmed the North Atlantic and neighboring regions, causing the warmest temperatures to be evident in Europe and lands neighboring the North Atlantic."[55] He further asserted that the Little Ice Age was a European phenomenon. "A variety of factors thus may have contributed to both the moderate warmth of the Northern Hemisphere and the more sizable and distinct North Atlantic/European warming during the early centuries of the second millennium," he concluded.[56] Yet, statements like "may have warmed" and "may have contributed" are distinctly short of rigorous scientific proof. But such qualified assertions are perfectly acceptable in a climate change court where true believers accept disinformation as gospel truth.

Evidence That CO_2 Lags Temperature Change

Despite the determination of the IPCC to blame CO_2 for global warming, the historical record is that CO_2 lags temperature warming, not the reverse. We have touched on this point before, but now let's dig even further into the scientific evidence. For example, multiple published scientific studies of CO_2 in air bubbles uncovered in the Antarctic Vostok core from over the past 420,000 years have found that as the climate cooled into an ice age, the decrease in atmospheric CO_2 lagged temperature by several thousand years.[57] Other studies of Antarctic ice cores showed that CO_2 concentrations increased by eighty to one hundred parts per million by volume some 600 ± 400 after global warming caused the last three glacial terminations—again proving CO_2 changes in geological time lagged temperature changes, not the other way around.[58]

Global temperatures cooled in the mid-1940s, when CO_2 emissions were increasing—the exact opposite of what the IPCC insists had to happen. "Global temperatures began to cool in the mid-1940s at the point when CO_2 emissions began to soar," Easterbrook noted in *Evidence-Based Climate Science*. "Many of the world's glaciers advanced during this time, and recovered a good deal of the ice lost during the 1915-45 warm period. Although CO_2 emissions soared during this interval, the climate cooled, just the opposite of what should have happened if CO_2 causes global warming."[59] A study of atmospheric CO_2 from January 1980 to December 2001 again found CO_2 changes lag temperature changes. Correlation analysis of CO_2 and temperature showed CO_2 lagging global surface temperature by eleven to twelve months, lagging global surface air temperature by nine and a half to ten months, and lagging global lower troposphere temperature by nine months.[60]

Once again, we see that the IPCC confuses cause and effect. For instance, a Great Climate Shift occurred in 1977 when the global cooling since the mid-1940s ended. Yet even though global warming resumed in 1977, that year was not proof a rise in CO_2 caused the shift from cooling to warming. "The abruptness of the shift in Pacific sea-surface temperatures and corresponding change from global cooling to global warming in 1977 is highly significant and strongly suggests a cause-and-effect relationship. The rise of atmospheric CO_2, which accelerated after 1945, shows no sudden change that could account for the 'Great Climate Shift,'" noted Easterbrook.[61]

Michael Mann Plays the Victim Card

From the opening paragraph of the prologue to his 2012 book *The Hockey Stick and the Climate Wars*, Michael Mann let the reader know how badly he had been treated by climate deniers, i.e., anyone who refuses to take his anthropogenic CO_2 doctrines as gospel. Mann led off with the following saga:

Climategate

On the morning of November 17, 2009, I awoke to learn that my private e-mail correspondence with fellow scientists had been hacked from a climate research center at the University of East Anglia in the United Kingdom and selectively posted on the Internet for all to see. Words and phrases had been cherry-picked from the thousands of e-mail messages, removed from their original context, and strung together in ways designed to malign me, my colleagues, and climate research itself. Sound bites intended to imply impropriety on our part were quickly disseminated over the Internet.[62]

In the following sentence, Mann blamed what he considered a defamatory attack on "a coordinated public relations campaign" organized by "groups affiliated with the fossil fuel industry." Yet before the first paragraph of the book's prologue ends, Mann played the victim, claiming his Hockey Stick analysis withstood ill-spirited and money-motivated attacks. "Though our work was subsequently vindicated time and again, the whole episode was a humiliating one—unlike anything I'd ever imagined happening," he wrote. "I had known that climate change critics were willing to do just about anything to try and discredit climate scientists like myself. But I was horrified by what they now had stooped to."[63]

Nowhere in the book did Mann accept responsibility for the emails he composed, received, and exchanged. The self-righteous moral outrage Mann expressed to open the book drips of self-pity. Mann's assertion that he is a climate scientist above reproach, demeaned by thieves who stole his emails, substituted for any honest acceptance of responsibility. His disgrace traced back to words and thoughts he authored himself. He then shared these thoughts and words in emails with colleagues. What Mann resented was that this group of like-minded IPCC scientists got exposed. Truthfully, Mann and his colleagues were acting like a conspiracy of climate activists who were determined to force geological history into their Procrustean bed. Whatever it took, including falsifying data, lying, and intimidating critics, Mann and his colleagues were set on a course to blame human beings and the Industrial Revolution for emitting CO_2, the chemical compound we exhale in breathing. If Mann was so concerned that anyone might read or publish his emails, why did he write them in the first place?

Yet, while Mann is hypersensitive about anyone who dares criticize his scientific methodology or challenge his hypothetical conclusions, he readily bares knuckles when attacking those who dare disagree with him. Mann vehemently condemns Marc Morano, to whom this book is dedicated. Why? Because Morano uses his website ClimateDepot.com to challenge the secular ideology of anthropogenic CO_2 true believers. In his chapter on climate deniers, Mann wrote a subsection devoted to attacking Morano that he titled "Swiftboating Comes to Climate Change."[64] Mann immediately reveals his leftist political orientation by characterizing "swiftboating" as "the art of the smear campaign." Having coauthored with John O'Neill in the 2004 book *Unfit for Command: Swift Boat Veterans Speak Out Against John Kerry*, I can attest that Morano played a constructive role in that campaign. Morano assisted the Swift Boat Veterans to get out the truth about John Kerry's purple heart frauds during his military service in Vietnam and the anti-U.S. posture Kerry played after leaving Vietnam in serving as a lead spokesperson for the Vietnam Veterans Against the War. Predictably, Mann comes to the defense of Kerry, denying any merit to the Swift Boat campaign that opposed Kerry's presidential run in 2004. "That attack had taken one of Kerry's greatest strengths—he had been awarded three Purple Hearts for his service in Vietnam, while his opponent George W. Bush, had avoided active duty—and, through a perversion of revisionist history, turned it instead into a perceived weakness," Mann insisted.

Next, Mann alleged Morano "became the pit bull of the climate change denial movement." In 2006, Morano took a job with the Senate Committee on Environment and Public Works, which Senator James Inhofe chaired. Mann now insisted Morano switched careers to become a hit man for Inhofe. Mann insisted that while working for Inhofe, Morano launched swiftboat-like attacks as before, shifting his target from Kerry to climate scientists like Mann. "Undaunted after his position with Inhofe was terminated in 2009, Morano headed back through the revolving door, this time hired by a Scaife- and ExxonMobil-funded entity known as the Committee for a Constructive Tomorrow (CFACT) to run a new Web site called ClimateDepot.com," Mann wrote, implying Morano's views on climate issues were money motivated. Throughout his 2021 book, Mann demonstrates his obsession with Morano by returning to attack him by name six more times. Mann was also wrong to claim ExxonMobil had funded Morano or ClimateDepot.com. ExxonMobil supported the Paris Climate Accords and funds research to support the IPCC.

The double standard in Mann's moral stance is apparent. But Mann typically portrays Morano's criticisms as cheap, unfair, mean-spirited, politically biased, or money motivated. Yet, when Mann attacks Morano, Mann assumes we will all agree that climate deniers are paid stooges for fossil fuel, big oil multinational conglomerates and the right-wing-monied political hacks who do big oil's bidding. Mann spares no civility dressing down his critics. Mann reduced all who dare criticize his climate assertions to those who, in the 1970s, accepted millions from tobacco giants for lending their "scientific credibility to advocacy efforts aimed at downplaying the health threats posed by the smoking of tobacco."[65]

Mann was particularly dismissive of Edward Wegman, a professor of statistics at George Mason University. Wegman wrote a particularly damaging report on Climategate and Mann's Hockey Stick analysis for a Senate committee. In what became known as the "Wegman Report," the conclusions were that there is no evidence that Mann had any significant interactions with mainstream statisticians as he developed his Hockey Stick graph. The Wegman Report concluded: "Overall, our committee believes that Mann's assessments that the decade of the 1990s was the hottest decade of the millennium and that 1998 was the hottest year of the millennium cannot be supported by his analysis."[66] Mann capsulized his attack on the Wegman Report as follows:

> In summary, then, the supposed independent review by Wegman et al. turned out to be a partisan hatchet job from the start. Wegman was handpicked by Republican Party operatives working for Joe Barton [Republican member of the House of Representatives from Texas, 1985–2019, and the former chairman of the House Committee on Energy and Commerce who commissioned the Wegman Report]. Barton and his staff rejected the National Academy of Science's offer of an impartial review so they could manufacture a report whose content they could control. Wegman had accepted a Faustian bargain when he agreed to author the report.[67]

Those who have any remaining doubt that Mann thinks anthropogenic CO_2 will create more climate disasters than even Al Gore should examine his 2008 book *Dire* Predictions: *Understanding Global Warming*, subtitled *The Illustrated Guide to the Findings of the IPCC*.[68] The slick book contains lavish illustrations and full-color photographs showing every kind of human catastrophe imaginable, ranging from floods, droughts, famine, species extinction, and pollution, to polar meltdown, all caused by anthropogenic CO_2. Page after page, we see heartbreaking photos of people in distress from what the book alleges were catastrophes caused by anthropogenic CO_2 emissions. On one spread, Mann hung the emotion-throbbing question in the air: "What is the value of the life of a starving child in Bangladesh as measured in cheap barrels of oil?" Mann's final judgment is clear: climate deniers have made a pact with the Devil, while Mann stands on the side of angels trying desperately against the odds to save the planet from capitalism and its burning of hydrocarbon fuels.[69]

Mann Embraces the Green New Deal

In his 2021 book, *The New Climate War: The Fight to Take Back Our Planet*,[70] Michael Mann was open about the extent to which he agrees the neo-Marxist vast agenda of social justice issues that form the heart of the Green New Deal legislative agenda are central to his global warming and climate change concerns. In his 2021 book Mann announced his verdict that the climate change true believers have secured the argument in worldwide public opinion that the Green New Deal must be implemented in all its various parts. He declared that Earth "has now warmed into the danger zone," such that we are in a climate "war" in which the "enemy" is the "forces of denial and delay" employing public relations tactics funded by big oil conglomerates "to stymie climate action."[71] Mann described the forces of denial and delay as "the fossil fuel companies, right-wing plutocrats, and oil-funded governments that continue to profit from our dependence on fossil fuels."[72] All who opposed Mann's increasingly political agenda were dismissed as desperate fools fighting hopelessly against the established truth that anthropogenic CO_2 will destroy Earth in a blaze of hydrocarbon-fueled, corporate greed–driven burning of oil, natural gas, and coal. "Outright denial of the physical evidence of climate change simply isn't credible anymore," Mann insisted. "So they have shifted to a softer form of denialism while keeping the oil flowing and fossil fuels burning, engaging in a multipronged offensive based on deception, distraction, and delay. This is the *new climate war*, and the planet is losing."[73] He claimed that "the enemy has masterfully executed a deception campaign—inspired by those of the gun lobby, the tobacco industry, and beverage companies."[74]

In chapter 2, "The Climate Wars," Mann declared that Climategate was "the opening skirmish" in the new climate war. Here, Mann insisted, the "forces of denial and inaction all but conceded that they could no longer make a credible, good-faith case against the basic scientific evidence." Hence, he insisted those who dare criticize his science have simply adopted the strategy of lying. "That's what Climategate was all about," Mann pronounced. "Prevarication has become so normalized in the era of Trump (who lies so often that journalists have a hard time keeping up with the count) that climate-change deniers have felt emboldened to dissemble with abandon." In the following sentence, Mann insisted that

Climategate

"a majority of the public" now accepted "the reality of climate change," such that climate deniers targeted "a shrinking majority of people who are motivated by ideology and tribal political identity over fact—a subset of the 'conservative base.'"[75] Chapter after chapter dripped with this ideological intolerance, dismissing those with legitimate scientific questions, while characterizing as ignorant throwback trolls motivated by atavistic politics and greed. Mann stopped short of demanding criminalizing climate deniers and burning their books. Consider this paragraph-long polemic demonizing his critics that Mann wrote to end his second chapter:

> The forces of inaction—that is, fossil fuel interests and those who do their bidding—have a single goal—inaction. We might therefore call them *inactivists*. They come in various forms. The most hard-core contingent—the deniers—are, as we have seen, in the process of going extinct (though there is still a remnant population of them). They are being replaced by other breeds of *deceivers* and *dissemblers*, namely, *downplayers*, *deflectors*, *dividers*, *delayers*, and *doomers*—willing participants in a multipronged strategy seeking to deflect blame, divide the public, delay action by promoting "alternative" solutions that don't actually solve the problem, or insist that we simply accept our fate—it's too late to do anything about it anyway, so we might as well keep the oil flowing. The climate wars have thus not ended. They have simply evolved into a new climate war.[76]

One wonders if the science were indeed on Mann's side would he have needed to be so prosaic with his invective? Mann's demonizing of anyone who dared criticize him had intensified, evidencing the transformation that climate change true believers believe they could accomplish, making their new enemy, anthropogenic CO_2, into today's secular form of Satan. With his 2021 book, Mann had elevated himself into an evangelist position, judging his critics as "evil," and opposing his secular religion that believes we can save the planet if we just kneel at the chapel Mann sanctified. When does the inquisition start, with critics such as Marc Morano placed on the rack, tortured until he confesses his climate sins?

The logical result in Mann's now apparent embrace of neo-Marxism was his decision to champion New York Congresswoman Alexandria Ocasio-Cortez. He described her as "a strong, smart, bold, young, powerful Latina," who proposed the Green New Deal as "a nod to President Franklin Roosevelt's New Deal, a major government initiative of the 1930s that used massive government stimulus spending in an effort to lift the United States out of the Great Depression."[77] Yet, Mann cautioned about endorsing the effort by Democrat Senator Ed Markey to add into the Green New Deal legislative proposals a government program to provide workers with guaranteed government jobs at family-sustaining wages. Mann parted with Markey not because he disagreed with these neo-Marxist social objectives, but because Mann was afraid Markey's championing of social causes gave "raw meat" to critics like "Murdoch's Fox News presenters when it comes to misrepresentation and bad-faith arguments."[78] His embrace of neo-Marxist social justice ideology was evident in Mann's insistence a "leaked document that emerged in June 2020 revealed that fossil fuel companies, including Chevron, were behind a PR campaign aimed at exploiting the spring 2020 Black Lives Matter protests to sow racial division within the climate movement."[79]

In this book, Mann openly endorsed the neo-Marxist political movement to oust Donald Trump from the White House. He proclaimed he was optimistic on the political side because the "2018 midterm elections in the United States resulted in a historic swing toward Democrats." He gushed with praise over the arrival on the political scene of "rock star" newcomers like Alexandria Ocasio-Cortez. He also claimed some Republicans "seemingly aware of the shift in public perception" have switched grounds, no longer "seeking to challenge the basic scientific evidence behind human-caused climate change."[80] Even in widely publicized current events, Mann saw a parallel between critical race theory and the global warming/climate change movement. "Triggered by the horrific killing, captured on video of a forty-six-year-old black man, George Floyd, by Minneapolis police, a similar tipping point on attitudes toward racial justice seems to have taken place in early summer 2020," Mann declared. "It is not unreasonable to speculate that we might be close to a tipping point on climate as well."[81]

Mann rejected on purely tactical grounds Naomi Klein's call to overthrow free-market capitalism through mass resistance. He also dismissed her insistence that climate change is inseparable from other social problems, including income inequality, corporate surveillance, misogyny, and white supremacy.[82] Yet, he embraced globalism when it came to tackling climate issues. Mann compared his mission to the global governmental effort launched to fight the COVID-19 pandemic. Mann concluded his book by insisting there is "no path of escape from climate-change catastrophe that doesn't involve policies at societal decarbonization." He argued policies to implement societal decarbonation "requires intergovernmental agreements, like those fostered by the United Nations Framework on Climate Change [UNFCCC] that bring the countries of the world to the table to agree on critical targets." He conceded that the 2015 Paris Accords "did not solve the problem, but it put us on the right path, a path toward limiting warming below dangerous levels." To make the point, he quoted *The Matrix* that "there's a difference between knowing the path and walking the path." He ended with a rousing admonition: "So we must build on the initial progress in future agreements if we are to avert catastrophic planetary warming."[83]

IPCC Issues Code Red for Human-Caused Global Heating

On August 9, 2021, the IPCC released a Working Group report entitled *Climate Change 2021: The Physical Science Basis*, as part of the IPCC's Sixth Assessment Report.[84] The report concluded that human-induced climate change was widespread, rapid, and intensifying, with some trends now irreversible for centuries to millennia ahead. A United Nations press release[85] stressed that human-induced climate change was already affecting many weather and climate extremes in every region across the whole of Earth's climate system, with scientists observing changes in the atmosphere, oceans, ice flows, and on land. The press release argued that there was still time to limit climate change. The newly released IPCC report suggested that substantial and sustained reductions in CO_2 and other greenhouse gas emissions could stabilize global temperatures in twenty to thirty years.

The press release quoted United Nations Secretary-General António Guterres as saying the IPCC's report was nothing less than "a code red for humanity." He stressed that "the alarm bells are deafening, and the evidence is irrefutable."[86] Guterres noted the internationally agreed threshold of 1.5°C above preindustrial levels of global heating was perilously close. "We are at an imminent risk of hitting 1.5 degrees in the near term," he said. "The only way to prevent exceeding this threshold is by urgently stepping up our efforts and pursuing the most ambitious path." The UN chief said solutions were clear. "Inclusive and green economies, prosperity, cleaner air and better health are possible for all, if we respond to this crisis with solidarity and courage," he emphasized.[87] He added that ahead of the COP26 climate conference scheduled for Glasgow in November 2021, "all nations—especially the advanced G20 economies—needed to join the net zero emissions coalition." He asked the G20 countries to reinforce their promises to slow down and reverse global warming through credible, concrete Nationally Determined Contributions (NDCs).[88] Before he worked for the U.N., Guterres served as the president of Socialist International (SI), described as a global network of socialist parties seeking to establish democratic socialism around the world. Guterres became the leader of Portugal's Socialist Party in 1992 and then the country's prime minister when the socialists won a major electoral victory in 1995.[89]

Almost immediately, the world press began churning out warnings like the *Politico* headline on the day of the IPCC report release: "'Get scared': World's scientists say disastrous climate change is here."[90] The *Politico* report stressed that the "long-feared era of disastrous climate change has arrived." The article cited examples of "extreme rainfall in Germany and China, brutal droughts in the western U.S., a record cyclone in the Philippines, and compound events like the wildfires and heat waves from the Pacific Northwest to Siberia to Greece and Turkey." The *Politico* article stressed the nations of the world had no choice now but to take drastic, immediate actions to implement zero net emissions policies. "The IPCC report underscores the overwhelming urgency of this moment," U.S. climate envoy John Kerry said, as reported by *Politico*.[91] Kerry insisted the upcoming UN climate summit in Glasgow needed to be a turning point.

Marc Morano responded immediately to the IPCC Sixth Assessment Working Group report: "The United Nations is a self-interested lobbying organization," he said, pointing out that the IPCC was just using science to drum up fear for a political agenda.[92] In a televised Fox News interview, Morano explained that the IPCC was a bureaucratic science institution. He charged that the UN expected Democratic Party politicians in the United States to latch on to this UN report while claiming every weather event around the globe is another example of a human-induced CO_2 disaster. Instead of debating science, he charged that the UN was determined to scare people into supporting a UN climate treaty and the Green New Deal. "It's like the mafia coming into your neighborhood and saying, 'I'd hate something bad to happen to you with that climate—if you pay us, we can protect you from the climate.'" Morano argued that the UN said that if we didn't go Marxist, the climate will kill us. "It's time for us to say hell no, to the UN, to the Green New Deal and expose this corrupt science."[93]

The Complexity of Earth, Weather, and Science

In July 2021, one month before the publication of the above-discussed IPCC Sixth Assessment, a little-noticed but crucial scientific article appeared in *Nature Communications*, reporting on an exciting and potentially important study. The study found that scientists had underestimated the amount of carbon drawn into Earth's interior. Subduction zones (i.e., zones where tectonic plates collide and drive into Earth's interior) lock the carbon away at depth, rather than circulating the carbon to resurface in the form of volcanic emissions.[94] The study was highly reputable, conducted by an international team of scientists from various institutions, ranging from the Department of Earth Sciences at the University of Cambridge, U.K., to the Institut für Mineralogie at WWU Münster, Germany, among many others.

A University of Cambridge press release on the study emphasized that the plate subduction of CO_2 into the earth was an underappreciated methodology the earth uses to remove CO_2 from the atmosphere.[95] "We currently have a relatively good understanding of the surface reservoirs of carbon," the study's lead author Stefan Farsang explained. "But [we] know much less about Earth's interior carbon stores, which cycle carbon over millions of years." While carbon is emitted to the atmosphere as CO_2 in many different ways, the University of Cambridge press release elaborated, "but there is only one path in which it can return to the Earth's interior: via plate subduction." The press release continued:

> Here, surface carbon, for instance in the form of seashells and microorganisms which have locked atmospheric CO_2 into their shells, is channeled into the Earth's interior. Scientists had thought that much of this carbon was then returned to the atmosphere as CO_2 via emissions from volcanoes. But the new study reveals that chemical reactions taking place in rocks swallowed up at subduction zones trap carbon and send it deeper into the Earth's interior—stopping some of it from coming back to Earth's surface.[96]

Studies such as this demonstrate the incredible complexity of Earth's weather and climate systems.

The Built-in Failure of IPCC Climate-Change Computer Models

Let's ask a fundamental question: Why does the IPCC insist global warming causes extreme weather conditions? The IPCC Fourth Assessment Report in 2007 suggested a warmer climate generated more moisture that could lead to increases in some regions of both droughts and floods.[97] This suggestion has morphed into the hypothesis that a warmer climate will cause various extreme weather conditions, including floods, droughts, and heatwaves.[98] Typically, the IPCC assumes and asserts that global warming will produce more extreme weather events without producing a legitimate scientific explanation detailing the climate mechanism whereby global warming causes an increase in extreme weather events, including hurricanes and tornadoes. When warning that a temperature rise of 1.5°C or 2.0°C would be catastrophic, the IPCC makes clear that a temperature rise of this magnitude would lead to "an increase in heavy rainfall events in some regions, particularly in the high altitudes of the Northern Hemisphere, potentially raising the risk of flooding." In other words, the IPCC is concerned not about global warming per se, but because global warming causes increased precipitation, and high levels of precipitation cause extreme weather.[99]

The IPCC argument focusing on rainfall highlights the widely recognized reality that water vapor accounts for the most significant greenhouse gas effect. Water vapor and clouds provide about 80 percent of the greenhouse gas effect, with carbon dioxide, ozone, and other trace gases constituting the rest.[100] Water vapor is a critical temperature forcing factor because water vapor emits and absorbs infrared radiation at many more wavelengths than any other greenhouse gas.[101] The feedback between surface temperature, water vapor, and Earth's radiation balance is known as "water vapor feedback." Since water vapor emits strongly in the infrared part of the spectrum, the radiative flux at both the top of the atmosphere and the earth's surface increases as the amount of water vapor increases.[102] If the IPCC argument is that increased atmospheric CO_2 raises Earth's temperature, producing increased water vapor, and the increased water vapor causes extreme weather, then are we to assume increased rainfall resulting from higher atmospheric CO_2 concentrations also causes earthquakes, tsunamis, volcanoes, and all the other extreme environmental emergencies the IPCC attributes to global warming?[103]

A major problem with this analysis is that the likely result of adding more water vapor to the atmosphere is that more clouds will cause Earth to cool, not to become warmer. In other words, clouds typically play a negative feedback effect, not a positive feedback effect. More atmospheric water vapor leads to more cloud formation. Clouds, in turn, reflect sunlight and reduce the amount of energy that reaches Earth's surface to create heat. As we have observed, water vapor is the most important greenhouse gas, and as such, water vapor—not CO_2—controls Earth's temperature. Suppose global warming produced by anthropogenic CO_2 increases precipitation as the IPCC assumes. In that case, the IPCC is inconsistent not to observe that increased precipitation means increased water vapor in the atmosphere, hence increased cloud formation. The point is that the IPCC has no rigorous scientific explanation capable of explaining the weather mechanisms by which anthropogenic CO_2 is capable of causing all the many different types of extreme environmental disasters that IPCC attributes to global warming.[104]

Here we return to Ph.D. meteorologist Roy W. Spencer, formerly a senior scientist for climate studies at NASA's Marshall Space Flight Center and currently a principal research scientist at the University of Alabama in Huntsville, whose analysis we reviewed at some length in chapter 5. In his 2010 book, *The Great Global Warming Blunder*, Spencer pointed out that the IPCC and the community of climate change scientists did not fully appreciate the natural radiative forcing mechanisms inherent to the complex working of Earth's weather and climate systems. "Again, I emphasize that the problem stems from previous researchers not accounting for natural cloud variations in the climate system," Spencer insisted.[105] This comment preceded a subsection in the book that Spencer devoted to IPCC-inspired mathematical modeling of Earth's weather and climate—a subsection Spencer appropriately titled "Garbage In, Garbage Out." Here Spencer argued, again correctly, that the IPCC is "concerned only with 'external' sources of forcing, such as manmade pollution, volcanoes, or tiny changes in the output of the sun." Similarly, these (and in particular CO_2) were the only sources the IPCC had in mind when the IPCC mentioned "radiative forcing," not cloud formation. "I believe that this neglect of natural cloud fluctuations has been the Achilles' heel of the so-called scientific consensus on global warming," Spencer commented.[106]

Referring to his analysis of the Pacific Decadal Oscillation (that we covered extensively in chapter 5), Spencer wrote a paragraph that sums up the discussion here:

> What I have demonstrated with the Pacific Decadal Oscillation is just scratching the surface of naturally induced climate change. What if other modes of natural climate variability—such as El Niño, La Niña, the Atlantic Multidecadal Oscillation (AMO), the North Atlantic Oscillation (NAO), and the Arctic Oscillation (AO)—also contribute to changes in global average cloudiness? It is entirely reasonable to hypothesize that one or more of these does. And if the global cloud cover changes, global temperatures will change as well. Again, I emphasize: it would take very small changes in global cloud cover to explain all the temperature variability in the last 2,000 years [as Spencer showed earlier in the book in Figure 1]. The IPCC's assumption that such small natural variations in global cloudiness do not occur is, in my view, arbitrary and scientifically irresponsible.[107]

Spencer's point was that the IPCC's narrow focus on CO_2 lead to a net positive feedback bias in the computer models tracked by the IPCC. "It is well known that positive feedback in these models is what causes them to produce so much warming in response to humanity's greenhouse gas emissions, he concluded. "Without the high sensitivity of these models, anthropogenic global warming becomes little more than a minor academic curiosity."[108]

The IPCC's reasoning is circular, Spencer correctly noted. "The IPCC merely ends up concluding what they assumed to begin with," Spencer wrote. "By ignoring natural climate variability, they 'prove' that there is no need for natural climate variability to explain global warming. They can even claim their explanation is self-consistent—but then, that is true of any circular argument, isn't it?"[109] Spencer stressed that the IPCC ignored the strong negative feedback in the real climate system because IPCC scientists had not been careful about inferring causation. "Climate researchers have neglected to account for clouds causing temperature change (forcing) when they tried to determine how temperature caused clouds to change (feedback)," Spencer concluded. "They mixed up cause and effect when analyzing year-to-year variability in clouds and temperature. You might say they were fooled by Mother Nature."[110]

Spencer summed up, concluding the IPCC climate models simply assumed clouds amplified rather than reduced warming in response to increasing atmospheric carbon dioxide concentrations. "Probably as a result of the confusion between cause and effect, climate models have been built to be too sensitive, with clouds erroneously amplifying rather than reducing warming in response to increasing atmospheric carbon dioxide concentrations," he concluded. "The models then predict far too much warming when the small warming influence of more manmade greenhouse gases is increased over time in the models. This ultimately results in predictions of serious to catastrophic levels of warming for the future, which you then hear about through the news media."[111]

Conclusion

The IPCC's Sixth Assessment Report released in August 2021, books like Michael Mann's lavishly illustrated *Dire Predictions*, and feature films like Al Gore's 2006 *An Inconvenient Truth* aim to motivate us by fear. Unless we decarbonize, presentations like these aim at convincing us, especially the youngest among us, that we have no choice but to obey the United Nations' directives on the climate. In the future, if you lack a "green pass" confirming your compliance with the U.N. IPCC-determined "green" lifestyle, you may forfeit certain fundamental human rights that we assume today in an open capitalist system. Suppose you refuse to limit your carbon footprint to comply with U.N. standards. In that case, you may also forfeit admission to public events, you may lose your permission to buy an airplane ticket, and you may no longer have the right to hold a job, own a credit card, or open a bank account.

When confronted by climate change fear tactics, the average person forgets that climate disasters and extreme weather phenomena have been routine throughout history. In ancient Roman history, the eruption of Mount Vesuvius in AD 79 buried ancient Pompeii. The San Francisco earthquake and fire in 1906 destroyed 28,000 buildings, killed more than 3,000 people, and left some 250,000 residents homeless. In 1950, movements in the Eurasian and Indian plates caused an 8.6 magnitude earthquake that killed over 1,500 people in India, Tibet, and China.

We end this chapter with Spencer's question: "What if more carbon dioxide in the atmosphere turns out to be a good thing for life on Earth?"[112]

CHAPTER 7

Cataclysmic Climate Change

Uniformitarianism versus Catastrophism, the Chicxulub Asteroid, the Deccan Volcanism, the Disappearance of the Dinosaurs, and the Ice Age Cause of the Permian-Triassic "Great Dying" Extinction Event

> *Early science was egocentric, and uniformitarian, in that it assumed that things have always been much as we now see them, at least since the earliest beginning too hazed in ignorance to discuss. A century ago, geologists believed that the mass, volume, and diameter of the earth were fixed inheritances, that the axial obliquity to the ecliptic was immutable, that the earth was a dying body dissipating primal heat from a still molten core, that magnetic north was north and south was south, and always had been so, that physical constants had been and would remain constants, and that continents were fixed permanent features which heaved and sagged from time to time against an ebbing and flooding sea. The geologist's task was to describe and understand the details of a planet on which the really big things had happened eons ago as a prologue before his saga opened.*
>
> —S. Warren Carey, *The Expanding Earth*, 1976[1]

> *Climate has always changed. It always has and always will. Sea level has always changed. Ice sheets come and go. Life always changes. Extinctions of life are normal. Planet Earth is dynamic and evolving. Climate changes are cyclical and random. Through the eyes of a geologist, I would be really concerned if there were no change to Earth over time. In the light of large rapid natural changes, just how much do humans really change climate?*
>
> —Ian Plimer, *Heaven and Earth: Global Warming, the Missing Science*, 2009[2]

Cataclysmic Climate Change

> *Our new understanding of why the dinosaurs and so many of their contemporary species became extinct has revealed the earth as a planet not specifically designed for our well-being and one that continues to be the target of comets as well as asteroids. From time to time, life is rudely interrupted by shattering events on a scale we can barely imagine.*
>
> —Gerrit L. Verschuur, *Impact! The Threat of Comets and Asteroids*, 1996[3]

EVEN OTHERWISE SOPHISTICATED SCIENTISTS, CAPTIVATED by the global warming movement, subtly shifted in their climate paradigms to begin looking at Earth's climate as a structurally fixed system operated by immutable laws. The truth is that Earth's climate is a highly complex natural system in which physical laws are mutable and cataclysmic changes that reset the rules are possible at any moment, including today.

This chapter will explore the subtle interplay between the extraordinarily complex system that controls Earth's weather and climate with natural laws and principles and the disruption of this uniformitarianism with sudden, mighty cataclysmic events. We will see those cataclysmic events fundamentally change Earth's geology, weather, and climate in unpredictable ways. This discussion is essential to understanding the underlying assumptions that invalidate the critical premise of global warming/climate change ideology: namely, that anthropogenic CO_2 generated by the burning of hydrocarbon fuels is a threat to human life on planet Earth.

Uniformitarianism versus Catastrophism

A traditional debate among professional geologists has been historically divided between uniformitarianism and catastrophism. Geologists who hold the uniformitarianism view of Earth tend to view historical changes as occurring gradually because of continuous and uniform processes unfolding methodically over millions of years. Geologists having the catastrophism view see Earth changes over geological time due to sudden and violent events. An essential doctrine of uniformity is "the assumption that the same natural laws and processes operating in the present-day scientific observation have always operated in the past." "Catastrophism is the idea that the Earth's features have remained fairly static until dramatic changes were wrought by sudden, short-lived, violent events."[4]

As we will see in this chapter, climate change true believers see Earth as uniformitarian. The logical coherence of their argument that Earth is warmer today than ever falls apart unless Earth's carbon cycle as a law of nature operates today as it did billions of years ago. Suppose its carbon cycle changes its rules of operation just as catastrophic events dramatically changed Earth's geology and ecology. In that case, determining cause and effect with the CO_2 accumulating in the atmosphere today becomes more problematic. Deciding whether the CO_2 accumulating in the atmosphere today is a plus or a minus for the human condition also becomes difficult. As we consider Lawrence Krauss and Donald DePaolo's publications in this next section, both scientists view the carbon cycle in terms of uniformitarianism. Both scientists assume the only catastrophic event operative today is a result of humans burning hydrocarbon fuels. They believe we burn hydrocarbon fuels because a capitalist economic system emphasizes utilizing the most potent forms of energy that are also cheaply available on Earth today. Thus, authors like Krauss and DePaolo assume the argument for using oil, coal, and natural gas will remain compelling unless the capitalist economic system itself is changed.

Finally, as an introduction to this chapter, understand that both Krauss and DePaolo view hydrocarbon fuels as fossil fuels, i.e., nonrenewable resources that are created by chemical laws that change organic material in sedimentary rocks into hydrocarbon fuels. Thus, the organic theory of the origin of oil presumes hydrocarbon fuels are finite resources, limited to the amount of organic material deposited in sedimentary rocks. That is precisely why both Krauss and DePaolo view the burning of hydrocarbon fuels as releasing into the atmosphere extra CO_2 that otherwise would remain safely trapped in Earth's upper-layer sedimentary rock structures.

Krauss and DePaolo hold the conventional belief that oil is an organic fossil fuel. Believing the amount of organic material available to make oil is a limiting factor, they also believe geological eons are required to transform the resident organic material into hydrocarbon fuels. Their logic compels them to conclude the amount of hydrocarbon fuel on Earth today is all the hydrocarbon fuel we will have. This reasoning allows the two men to believe they can estimate reliably just how much CO_2 remains to be expended by burning the hydrocarbon fuels available before we run out. Because they assume Earth will one day run out of oil, they must also construe the anthropogenic CO_2 problem as finite. In other words, Krauss and DePaolo's assumption that hydrocarbon fuels are organic compels them to conclude that when we run out of hydrocarbon fuels, the problem with anthropogenic CO_2 also goes away. The irony of Krauss and DePaolo's position on global warming means they want to compel major changes in capitalism to decarbonize now when they assume from the start the CO_2 global warming problem is only temporary because sooner or later, we will exhaust Earth's hydrocarbon fuel supply.

As we saw in chapter 1, Julian Simon challenged this by insisting that hydrocarbon fuels (even if organic in origin) should not be considered finite. At the same time, Simon insisted that Earth's supply of hydrocarbon fuels was not infinite. Instead, Simon concluded that it was very unlikely humans would ever exhaust Earth's oil or natural gas supply. As proof of this, Simon referenced the unfounded fear in the early years of the Industrial Revolution that we would run out of coal. Simon also viewed human intelligence as our ultimate natural resource, and he was confident we would find new, more powerful, and more affordable fuel alternatives long before there was any real likelihood we would run out of coal, oil, and natural gas. In chapter 9, we will argue the origin of oil is abiogenic, not organic, in origin. But for the discussion here, the critical point is that virtually all climate change true believers agree with Krauss and DePaolo by just assuming that fossil fuels are organic.

Gale L. Pooley, Ph.D., an associate professor at Brigham Young University—Hawaii and a fellow of the Discovery Institute's Center on Wealth and Poverty, teamed with Marian L. Tupy, the editor of HumanProgress.org, to create what they called the Simon Abundance Index. Keying off Julian Simon's confidence in the abundance of Earth's natural resources and his debate with Paul Ehrlich, Pooley and Tupy created the index to measure the continuing availability of natural resources and their relationship with population growth.[5] In their Simon Abundance Index 2020 evaluation, Pooley and Tupy reported Earth, as a whole, was 57.9 percent more abundant in 2019 than it was in the base year of 1980. They summed up their findings as follows:

> Simon's revolutionary insights with regard to the mutually beneficial interaction between population growth and availability of natural resources, which our research confirms, may be counterintuitive, but they are real. The world's resources are finite in the same way that the number of piano keys is finite. The instrument has only 88 notes, but those can be played in an infinite variety of ways. The same applies to our planet. The Earth's atoms may be fixed, but the possible combinations of those atoms are infinite. What matters, then, is not the physical limits of our planet, but human freedom to experiment and reimagine the use of resources we have.[6]

We tend to forget all the unfulfilled, dire predictions the IPCC, Michael Mann, and Al Gore have issued if we refuse to surrender to U.N.-dictated climate accords. Where are the coastal cities in America that rising sea levels were supposed to destroy? What happened to the predicted extinction of those cute polar bears whose current global population is, with 95 percent certainty, between 22,000 and 31,000, all living healthily in nineteen subpopulations distributed around the Arctic?[7] If Al Gore had made bets in Las Vegas on all his frightening predictions, he would not be the millionaire he is today. If Julian Simon were alive today, he most likely would be publishing offers to wager with the likes of Michael Mann and Al Gore, just as he did with Paul Ehrlich.

But regarding the uniformitarianism versus catastrophism debate, a key point of this chapter is that IPCC adherents view human beings as the ultimate catastrophe—the reason we must decarbonize now rather than simply waiting to run out of hydrocarbon fuels. IPCC adherents believe the earth's natural carbon cycle, without human intervention, would have a natural CO_2 management cycle in place that would never cause global warming. After all, Michael Mann's Hockey Stick graph shows no global warming in earth history until the Industrial Revolution when humans began burning increasing quantities of hydrocarbon fuels. IPCC adherents also dismiss the importance of the sun as a temperature forcing factor on Earth. But if the sun didn't warm Earth and there were no human beings to burn hydrocarbon fuels millions of years ago, how did Earth ever get rid of the constantly recurring ice ages in geologic history that made the planet difficult if not impossible for humans to inhabit in the first place?

A major point of this chapter is to develop the argument that IPCC global warming/climate change adherents see human activity in the capitalist economic system as the ultimate climate catastrophe. As we pointed out earlier, John Holdren understood the psychological necessity to repackage Ehrlich's overpopulation concerns as a climate species survival threat. From there, IPCC adherents like Michael Mann found their most important allies were Green New Deal neo-Marxists wanting to destroy the capitalist system.

"Stable Earth" Ideology: Only Humans Destabilize Earth's Natural CO_2 Dispersion Strategy

Consider, for instance, Lawrence M. Krauss, Ph.D., a theoretical physicist and a climate change true believer who has written several books to popularize scientific ideas. In his 2021 book, *The Physics of Climate Change*,

Cataclysmic Climate Change

Krauss argues that anthropogenic CO_2 generated since the Industrial Revolution has led to an accumulation and concentration of CO_2 in the atmosphere that has upset the natural balance. He assumes a CO_2 equilibrium "has existed on Earth over much of the recent history of the planet since atmospheric abundances achieved their current concentration and life emerged in more or less its present form from hundreds of millions of years ago."[8] Krauss argues that humans upset this natural CO_2 balance by burning hydrocarbon fuels.

Krauss argued that in the natural equilibrium, the atmosphere contains about 600 Gt of carbon, or about 2,200 Gt/yr. of CO_2. A gigaton equals one billion tons, i.e., "1" with nine zeroes after it, written as 1^9. So, Krauss maintains that in the natural equilibrium the earth's atmosphere contains 2,200,000,000,000 tons (2.2 trillion tons) of CO_2. He argues that at this equilibrium, another 2,200 Gt of CO_2 resides in the terrestrial biosphere, while about three times that amount (another 6,600 Gt of CO_2) resides in terrestrial soils. The surface levels of the ocean that are in contact with the atmosphere contain another 3,670 Gt of CO_2, and the deep ocean not in contact with the atmosphere stores a much larger reservoir of 135,790 Gt of CO_2. Krauss details the various processes through which CO_2 gets recycled between the atmosphere, the biosphere, soils, and oceans. "Altogether, this cycling keeps the carbon abundance in these reservoirs and the atmosphere roughly constant," Krauss summed up.[9] The point, according to Krauss, is that Earth's natural CO_2 cycling processes maintain a safe quantity of CO_2 in the atmosphere.

From there, Krauss proceeded to make his case demonizing anthropogenic CO_2 by arguing that today, humanity is emitting over thirty-six billion tons of CO_2 into the atmosphere per year. He concluded that human beings are burning enough hydrocarbon fuels to add 1,000 to 4,000 Gt of carbon (equivalent to 3,670 to 14,680 Gt of CO_2) into the atmosphere in the next 150 years—a quantity he estimated equals seven times the total amount of CO_2 in the atmosphere before the advent of modern civilization.[10] This extra 3,670 to 14,680 Gt of CO_2 emitted by human beings burning hydrocarbon fuels over the next 150 years is what Krauss assumed will just sit there in the atmosphere as extra CO_2.

As we pointed out in chapter 5, climate change true believers tend to express CO_2 in the atmosphere in terms of tons, not in percentages of the atmosphere. Yet, like the IPCC, Krauss fails to consider that the total weight of Earth's atmosphere is approximately five quadrillion tons [i.e., "5" followed by fifteen zeroes]. Appreciating this, Krauss has estimated that over the next 150 years, human beings will add up to 14,680 Gt [i.e., by rounding "14,680" to "15,000" we get "15" followed by twelve zeroes] of CO_2 into the atmosphere. So after 150 years, Krauss calculates that anthropogenic CO_2 will add up to a number that still accounts for only approximately 0.003 percent of Earth's atmosphere—a quantity that still leaves CO_2 as a trace chemical compound in the atmosphere. No wonder Krauss did not express his CO_2 calculations as a percentage of the atmosphere.

The source study Krauss draws upon for this analysis is a 2015 article entitled "Sustainable Carbon Emissions: The geologic perspective," published by Donald J. DePaulo, Ph.D., a professor in the Department of Earth and Planetary Science at the University of California, Berkeley, and on the staff of the Earth Sciences Division of Lawrence Berkeley National Laboratory.[11] DePaulo began his argument by asserting that the earth's carbon cycle has been stable over geologic time going back billions of years. To make this point, he wrote the following:

> Unlike the global temperature signal, the changes in the carbon cycle in the last 100 years are not subtle. These changes have been produced almost entirely by burning of fossil fuel, with a smaller (and less problematical in the long term) contribution from destruction of forests. Discussion of the human-induced changes has in some cases been muddied by comparison to the large rates of carbon exchange between atmosphere, biosphere, and oceans. These large exchange fluxes are neither the ones that have changed drastically nor are they particularly significant for understanding what is happening due to burning fossil fuel.[12]

DePaulo continued by asserting the only disruption in the earth's carbon cycle after billions of years of stability was in the last hundred years, when humans began burning an increasing quantity of hydrocarbon fuels to provide energy for the Industrial Revolution. He continued the above quotation as follows:

> The main change, which is the focus of this study, is *the rate that carbon is moved from deep Earth storage—in rocks—to the atmosphere*. This transfer does happen naturally and is responsible for many familiar aspects of Earth, including the fact that the planet has maintained a hospitable climate that has allowed life to flourish for billions of years. However, in the absence of human actions, the transfer is done mainly by volcanoes, and at a small rate. Fossil fuel burning has increased this transfer rate by

at least 40–50 times, which is not something that can be argued about—the change is so huge that no likely level of uncertainty can change the conclusion that virtually all the transfer of deep Earth carbon to the atmosphere is currently a result of fossil fuel burning and cement production. This radical change represents something that has never before been done on earth, even if we look back hundreds of millions of years. It is the magnitude of fossil carbon emissions that is the problem, and this can be understood in terms of relatively simple concepts and bookkeeping.[13]

DePaolo calculated "98% of the movement of carbon out of geologic reservoirs (coal-, oil-, and gas-bearing sedimentary rocks and limestone) into the atmosphere is due to human activities."[14] DePaolo reasoned as follows: "In most projections of the future of carbon emissions, it is assumed that in the worst case we will burn all the accessible combustible carbon (coal, oil, and natural gas). He estimated the total remaining carbon came to 5,000 Gt in carbon reserves. DePaolo commented that once humans exhaust hydrocarbon fuels on Earth, the natural carbon cycle will kick back in, with Earth's CO_2 returning to preindustrial levels. Like Krauss, DePaolo argued it could take 100,000 years to work atmospheric CO_2 back to preindustrial levels where the natural carbon cycle could once again stabilize atmospheric CO_2. He claimed the key point was if we add 5,000 Gt of carbon to the atmosphere in 300 years, "it will take 100-1,000 times as long to get that carbon out of the surface reservoirs and back into geologic storage through natural mechanisms."[15] DePaolo concluded that returning "to a sustainable carbon cycle requires systematic lowering of the carbon emission intensity of energy production over the next century."[16] He stressed the urgency to decarbonize as follows: "Whatever we decide to do may determine the Earth's climate for the next 10,000 to 100,000 years. In terms of climate and carbon cycles, the 21st is the most important century in Earth's history since the end of the last Ice Age."[17]

As we noted earlier, that CO_2 was more abundant as a percentage of the atmosphere in geological time before human beings existed is a problem for climate change true believers like Krauss and DePaolo. The example DePaolo picked to explain the problem away was the period between 18,000 and 12,000 years ago, the end of the last Ice Age, when glaciers were melting in the Northern Hemisphere and some 150 Gt of carbon "was added to the atmosphere by natural processes."[18] He reasoned that the rate of addition, 150 Gt carbon per 6,000 years, equates to a transfer rate of 0.025 Gt carbon per year, acknowledging that this carbon transfer rate "must be considered extreme by geologic standards."[19] DePaolo has assumed CO_2 emitted in the atmosphere remains there to accumulate in a strictly arithmetic manner. In other words, "x" quantity of CO_2 emitted in Period 1 adds to "y" quantity of CO_2 emitted in Period 2, resulting in "x + y" CO_2 in the atmosphere at the end of Period 2. Still, DePaolo insisted "this 'rapid' rate is almost 200 times smaller than the rate that CO_2 is currently accumulating in the atmosphere (4-5 Gt carbon per year, or about half the total emissions)."[20]

Next, DePaolo assumed that 18,000 to 12,000 years ago, the rapid accumulation of CO_2 in that atmosphere was entirely a result of the immutable natural process of the carbon cycle acting normally. "During the 6,000-year deglaciation period, and analogous to the present, there was more carbon being moved into and out of the atmosphere than was accumulating in the atmosphere," he explained. "Overall, it is estimated that there was about 600 Gt carbon transferred from the deep ocean through the atmosphere and into the land biomass during the 6,000-year deglaciation interval."[21] DePaolo concluded his explanation by stressing that these natural carbon cycle processes do not operate properly today. Why? Because humans are burning carbon fuels that dump a catastrophic amount of CO_2 into the atmosphere, thereby overwhelmingly preventing carbon cycle natural laws from operating. In his final sentence on the Ice Age example, DePaolo wrote that rate of the CO_2 accumulation over the last Ice Age, "(600 Gt in 6,000 years) is about 0.1 Gt carbon per year, still 100 times smaller than the modern rate of 10 Gt carbon per year by burning of fossil fuels."[22]

Finally, DePaolo simply dismissed the paleoclimate CO_2 argument in his discussion of the Paleocene-Eocene Thermal Maximum some fifty-five million years ago. DePaolo still insisted that the rate of transfer of CO_2 into the atmosphere was ten to twenty times slower than the present rate. He estimated that currently "we are in the process of doubling atmospheric CO_2 over a period of about 150 years, which is forty times faster and apparently unprecedented in Earth's history." He noted that "climate models" [i.e., mathematical models of the assumed natural carbon cycle] indicate that a doubling of atmospheric CO_2 concentration produces an increase in Earth surface temperature of 2-5°C." So, he concluded that "at 55 million years ago, the doubling of atmospheric CO_2 caused an already warm Earth with no polar ice caps to become even warmer." He argued today is different. "At present, we are starting from a relatively cold Earth with polar ice caps and rapidly headed for an atmosphere with high CO_2

Cataclysmic Climate Change

concentrations while the polar ice caps are still present," he wrote. "If this situation ever happened before on Earth, the last time was probably about 700 million years ago, long before any kind of complex life was present, and the conditions would have prevailed are so unlike the present that it strains the best minds to create models that describe the Earth's climate at that time."[23]

Cause or Effect—Green Earth Blooming

In the previous chapter, we reviewed the scientific evidence that changes in CO_2 atmospheric concentrations lag temperature increases, not the other way around. In other words, temperature increases are what cause CO_2 to increase. As we noted even earlier, in chapter 5, the ideology of climate change blames humans for the burning of hydrocarbon fuels that have caused a perceived catastrophic increase of CO_2 concentrations since the dawn of the industrial age. As we noted in chapter 6, this reasoning falls victim to the classic logical fallacy known in Latin as *post hoc, ergo propter hoc*, or in English, "after this, therefore because of this." In other words, because the CO_2 concentration in Earth's atmosphere had risen since the Industrial Revolution began, climate change true believers concluded that the current rise in CO_2 was necessarily caused by the industrial age triggering a rise in the human consumption of hydrocarbon fuels.

Let's engage in a thought experiment and ask ourselves how the modern rise in Earth's global temperature could cause a rise in atmospheric CO_2. A warmer Earth is a more productive, greener Earth. A warmer Earth is more receptive to life and human activity. In 1900, the global population was approximately 1.6 billion people. The worldwide population of 2021 totals some 7.9 billion people. Estimates are that Earth's population will top eight billion people between 2022 and 2023. Despite the devastating horror of tens of millions killed in two world wars, a massive explosion in human population in the twentieth century occurred, not accidentally, in the global warmth of this interglacial period.

More people breathing generates more atmospheric CO_2. To climate change true believers, more people means more capitalism. These believers insist that the burning of hydrocarbon fuels in our current capitalist economic system is responsible for some 87 percent of increased CO_2 concentrations, with electricity generation alone accountable for 42.5 percent of global CO_2 emissions.[24] So, for those convinced that humans cause global warming, an increase of the worldwide population as large as that experienced in the twentieth century is a frightening prospect. But the point we are making here is that a warmer Earth is a more productive Earth. A warmer Earth is more abundant, both in terms of the number of people on Earth and the number of people it can support.

Interglacial warm periods are not only good for human beings, but warming is also good for life on the planet. A 2018 study conducted by Pekka Kauppi, Vilma Sandström, and Antti Lipponen, Finnish scientists at the Faculty of Biological and Environmental Sciences and the Helsinki Institute of Sustainability Science at the University of Helsinki, found that contrary to popular belief, global reforestation is occurring today. The Finnish scientists positively correlated a rebirth in forests with the United Nations Human Development Index, such that developed nations want solid and healthy forests.[25] "This indicates that forest resources of nations have improved along with progress in human well-being," the Finnish scientists concluded in their peer-reviewed article. "Highly developed countries apply modern agricultural methods on good farmlands and abandon marginal lands, which become available for forest expansion." The authors continued to explain that developed countries invest in sustainable programs of forest management and nature protection. "A large and persistent sink of carbon has been detected in the forests of the World. Terrestrial ecosystems including forests have become increasingly green."[26] As nations become wealthier in the prosperity of warmer Earth, land management practices change profoundly, leading to reforestation, contrary to popular belief fueled by a Malthusian global mainstream media. "In developed economies the forces of technological advance and agricultural intensification have outweighed the impacts of population growth and improving diets," the Finish scientists stressed. "Therefore, we observe a global diffusion of forest transitions."[27] Greener Earth increases carbon sequestration through increased photosynthesis, resulting from more life-supporting land biomass and ocean biosphere.

Carbon sinks on land and in the oceans are estimated to remove nearly half (about 45 percent) of the CO_2 emitted into the atmosphere by human activity in general, including the burning of fossil fuels.[28] In 2000, an international team of earth scientists reported in *Science* magazine that while there is no "natural savior" waiting to assimilate all the anthropogenically produced CO_2 in the last century, "our knowledge is insufficient to describe the interactions between components of the Earth system and the relationship between the carbon cycle and other biogeochemical and climatological processes."[29] The world's space agencies are "actively working to coordinate ambitious plans for an expanded space-based remote sensing capability that supports atmospheric CO_2 measurements, high resolution maps of land surface type and biomass and ocean biological productivity."[30] As we saw in chapter 5, the Gaia space observatory satellite launched by the European Space Agency in 2013 provided new insight into the structure and operation of the Milky Way. As a result of the increased observational powers the Gaia space observatory satellite provided astronomers, we also have an increased understanding of Earth's positioning and movement in the Milky Way galaxy. Similarly, we should see a tremendous advance in our understanding of Earth's carbon cycle over the next few years, given the space observation and measurement of CO_2 currently in the planning stage.

Given the very complex nature of cloud physics, as well as the complications in observing and measuring changes in Earth's cloud cover, there remains "a large gap in understanding the role of clouds in the terrestrial radiation budget, thus representing a major source of uncertainties in climate model predictions." Despite arguments by climate change true believers that the catastrophic amount of anthropogenic CO_2 currently being emitted will tip Earth's energy balance toward more warming, the uncertainties "are determined by the fact that global mean cloud feedback summarizes many individual cloud processes, which are not fully understood." The result is that clouds "induce the largest uncertainty in the estimation of important climate parameters, as shortwave irradiance."[31] As we saw in the last chapter, the IPCC believes global warming will cause extreme weather events because a warmer Earth produces more precipitation, an increase in Earth's cloud cover, and corresponding water vapor in the atmosphere. NASA satellite data indicates that about 67 percent of Earth's surface is typically covered by clouds, with less than 10 percent of the sky over the oceans completely clear of clouds at any time, compared to overland, where 30 percent of the skies are entirely cloud free.[32] NASA's future reports should measure Earth's cloud cover over time, although these current figures establish the extensive nature of Earth's cloud cover today. As a negative feedback mechanism, increased cloud cover resulting from a warmer planet serves as a sunscreen that reflects the sun's irradiance into outer space, thus serving as a cooling mechanism.

Another factor in Earth's global thermometer, as we discussed in chapter 5, was Roy Spencer's analysis of how ocean currents affect global temperature. In August 2021, NOAA's Climate Prediction Center issued a 70 percent likelihood the United States will see a La Niña winter returning this year between November 2021 and January 2022.[33] As Spencer would predict, a La Niña winter will bring colder temperatures, despite what global warming enthusiasts might like to believe.

The main point in this section is not to argue that carbon sequestration through increased photosynthesis on land and in the oceans will clean the atmosphere of human-generated CO_2. Nor do we anticipate that cloud formation will cancel out warming effects of the current interglacial period by performing the expected balancing function, adding negative cooling feedbacks into the near future of Earth's weather. Yet both these arguments should make clear why climate change true believers commit a fundamental error to assume CO_2 emitted into the atmosphere creates additive global warming. The negative feedback mechanisms of carbon sequestration via land biomass and ocean biosphere serve as natural carbon cycle processes to remove CO_2 from the atmosphere. At the same time, increased water vapor and cloud cover also play a cooling function by blocking sun irradiance from reaching Earth's surfaces.

In summary, without the current warmer global temperatures of the warm interglacial period, we would never have experienced such a dramatic expansion of the world's population that we saw in the twentieth century. Along with this has come a corresponding dramatic increase in productive human activity. The interglacial warming gave rise to the Industrial Revolution, increased world population, and dramatically increased human economic advancement. In other words, global warming caused the increased atmospheric CO_2 that we are experiencing today. Our current interglacial warming period has also caused the expansion of plant life both on land and in the oceans. The economic advancement of this warming period has allowed a rebirth of forests as developing countries increase their standard of living. So, while interglacial warming has caused an increase in atmospheric CO_2, the warming has also increased the ability of Earth's ecosystem to manage the CO_2 safely. The warming's growth has also augmented the power of the biomass and oceans to absorb more CO_2 from the atmosphere through photosynthesis. As the interglacial period we are enjoying extends in time, Earth's natural thermometer functions to intensify negative cooling feedback mechanisms through changes in ocean currents and Earth's cloud cover.

In concluding this subsection, we need to ask climate change advocates what should the "right," "correct," or "perfect" temperature of Earth be? While Earth is estimated to be some 4.6 billion years old, Homo sapiens originated in Africa between 400,000 and 250,000 years ago. Hominoids as a proto-human species dates back some four million years. In geological time viewed as a twenty-four-hour clock, Earth's conditions that allowed the appearance of Homo sapiens occurred as the last minute wound down.[34] So, was Earth's climate "wrong" to have so much CO_2 in the atmosphere eons ago? Yet somehow, climate change true believers continue to insist Earth's climate will not be fixed or possibly survive unless we human beings take control of the environment by decarbonizing. For most of the geological time on Earth, human beings did not exist, yet the planet survived.

The Chicxulub Asteroid

In 1978, the state-owned Mexican oil company Petróleos Mexicanos, known as Pemex, hired geophysicists Glen Penfield and Antonio Camargo to conduct an airborne magnetic survey looking for oil in the Gulf of Mexico north of the Yucatán Peninsula. Penfield noted certain anomalies in the offshore magnetic data. Checking his findings against gravity data from the 1940s, Penfield confirmed that his aerial sighting corresponded with the 1940 data showing a sizable concentric set of gravity anomalies that suggested he had found the crater of a massive meteor that had impacted Earth in prehistoric times.[35]

Independent of Penfield's finding, renowned American Nobel Prize physicist Luis Walter Alvarez published a paper entitled "Extraterrestrial Cause for the Cretaceous-Tertiary Extinction" in the June 1980 edition of *Science* magazine. Alverez coauthored the paper with members of his research team, including his son, Walter Alvarez. The latter was then an associate professor in the Department of Geology and Geophysics at the University of California, Berkeley.[36] Luis and Walter Alvarez were reporting on their search at different locations around the globe for some aspects from the platinum group. Platinum-group elements (i.e., platinum, iridium, osmium, and rhodium) are not common in Earth's crust and upper mantle but are found more commonly in meteorites and average solar system material. At that time, scientists were speculating on the cause of the mass extinctions known to have happened in the boundary between the Cretaceous and Tertiary Periods (also known as "C-T," or "K-T" boundary) at the end of the Mesozoic Era approximately sixty-six million years ago. The C-T boundary is also known as the Cretaceous-Paleogene (K-Pg) boundary, designating the boundary between the Cretaceous Period and the Paleogene Epoch (at the beginning of the Tertiary Period in geological time). The C-T extinction is significant because it marked the end of the dinosaurs.

Before the 1980 publication of the Alvarez team's paper, the scientific speculation centered on the hypothesis that a supernova caused mass extinctions at the C-T boundary. However, the Alvarezes considered the probability of a supernova explosion close enough to the sun in the last 100 million years to be extremely low (a 10^{-9} possibility).[37] The Alvarez team found an anomalous iridium concentration at the C-T boundary in Gubbio, Italy. They concluded this high concentration of iridium was "best interpreted as indicating an abnormal influx of extra-terrestrial material."[38] But unresolved was whether the anomalous iridium concentration came from a supernova, or from an asteroid originating in the solar system. Finally, conclusive for the Alvarez team was their mass spectrometry analysis of the iridium isotope ratio ($^{191}Ir/^{193}Ir$) of the samples they found in C-T boundary sedimentary rock at Gubbio, Italy. The mass spectrometry analysis determined that the isotopic ratio ($^{191}Ir/^{193}Ir$) of the boundary iridium at Gubbio did not differ significantly by more than 1.5 percent of the isotopic ratio typically found in iridium on the earth. That discovery led them to conclude the anomalies in the C-T border "is very likely of solar system origin and did not come from a supernova or other source outside the solar system—for example, during passage of the earth through the galactic arms."[39] The Alvarez team concluded with the hypothesis suggesting "that an asteroid struck the earth, formed an impact crater, and some of the dust-sized material ejected from the crater reached the stratosphere and was spread around the globe." They next hypothesized the dust effectively prevented sunlight from reaching the surface for years. Therefore, the loss of sunlight suppressed photosynthesis, and as a result, "most food chains collapsed, and the extinctions resulted."[40]

Coincidently, Penfield and Camargo presented their findings on the Yucatán anomaly at a 1981 meeting of the Society of Exploration Geophysicists the same week as when researchers studying the C-T boundary had convened in Snowbird, Utah, to debate the Alvarezes. Aware of the Alvarezes' publication in *Science*, Penfield and Camargo summarized their conclusions as follows:

> We would like to note the proximity of this feature in time to the hypothetical Cretaceous-Tertiary boundary event responsible for the emplacement of iridium-enriched clays on a global scale and invite investigation of this feature in light of the meteorite impact-climate alteration hypothesis for the late Cretaceous extinctions.[41]

Cataclysmic Climate Change

David B. Weinreb, an expert in the study of catastrophic events causing mass extinctions, commented on this fortuitous timing coincidence as follows:

> Amazingly, only several months after the Alvarez team announced the elevated iridium in the Gubbio clay, Penfield and Camargo had discovered the smoking gun: a crater of appropriate age, large enough to have been produced by impact with a 10-kilometer wide object. And, more amazingly, nobody seemed to have noticed.

It turned out that petroleum geologists working for the Mexican oil giant Petróleos Mexicanos had discovered the circular Chicxulub (a Mayan word meaning "red devil") structure in the 1950s. In search of oil reservoirs, Petróleos Mexicanos researchers drilled into the structure and collected rock cores. In 1980, immediately after the publication of the Alvarez paper in *Science*, Penfield wrote to Walter Alvarez to inform him of the existence of this structure. He never received a response.[42]

Alan Hildebrand, a scientist with the Geological Survey of Canada, was among the first to rediscover the Penfield and Camargo study and connect it to the 1980 paper the Alvarez team had published in *Science* magazine.[43] In reflecting on this history, please note the importance of isotopic analysis in the geologic research of prehistoric events in order to date cataclysmic events related to climate change. We saw this earlier, in chapter 5, when commenting on the importance of isotopic evidence in determining solar activity.

Scientific studies after 1980 have substantiated that a meteor impact did occur at the C-T boundary, hitting Earth at the Yucatán. The Chicxulub crater clay layer found at the C-T boundary contains tiny glass spheres formed from the flash freezing of molten rock, wood ash, and shocked quartz (grains of quartz subjected to intense pressure). Geologists have concluded that only a severe meteorite impact could explain these features. "The glass spheres formed when melt sprayed into the air from the impact site, the iridium came from fragments of the colliding object, and the shocked quartz grains were produced and scattered by the force of the impact," a popular college geology textbook summarizes. "The wood ash resulted when the forests were set ablaze, conceivably because the impact ejected super-heated debris at such high velocity that the debris almost went into orbit and could reach forests worldwide. The impact also generated 2-km-high tsunamis that inundated the shores of continents, and it generated a blast of superheated air."[44]

THE TRUTH ABOUT ENERGY

Robert DePalma, a professor in the Department of Geology at the University of Kansas, assembled an international team of geologists to study the most immediate effects of the Chicxulub impact. DePalma explored C-T boundary deposits at a place called Tanis, located in North Dakota. "Acipenseriform fish, densely packed in the deposit, contain ejecta spherules in their gills and were buried by an inland-directed surge that inundated a deeply incised river channel before accretion of the fine-grained impactite," DePalma and his scientific team reported in a 2019 article published by the *Proceedings of the National Academy of Sciences*.[45] An article published by the BBC in London explained in commonsense terms what DePalma and his team found. Fossil evidence revealed the tsunami from the Chicxulub reached North Dakota within the first few hours after the impact. DePalma's excavations in North Dakota revealed "fossils of fish and trees that were sprayed with rocky, glassy fragments that fell from the sky." The North Dakota site was "swamped with water—the consequence of the colossal sea surge that was generated by the impact." The scientists found fossil fish with impact-induced debris embedded in their gills. "There are also particles caught in amber, which is the preserved remnant of tree resin," the BBC article explained. "It is even possible to discern the wake left by these tiny, glassy tektites, to use the technical term, as they entered the resin."[46]

Steven Goderis, a geochemistry professor at Vrije Universiteit in Brussels, assembled an international team of geologists that reported their findings in 2021. Goderis and his team concluded that the iridium found in the Chicxulub crater corresponded with the iridium found in C-T sedimentary rock layers at fifty-two different sites worldwide. "This [iridium] layer is now formally the 'golden spike' that defines the end of the Cretaceous Period and the Mesozoic Era," Goderis and his colleagues declared.[47] The Goderis study analyzed the drill cores from the Yucatán site taken by the 2016 International Ocean Discovery Program (IODP)—International Continental Scientific Drilling Program (ICDP) expedition led by the University of Texas, Austin. That effort yielded a nearly 3,000-foot drill core taken from the rings of the Chicxulub impact structure. The continuous core samples taken from 505.7 to 1,334.7 meters below the seafloor allowed the scientists to date accurately the remaining iridium in the dust, which was all that remained of the Chicxulub meteorite. Isotope dilution mass spectrometry conducted at labs in Japan, Austria, Belgium, and the United States established that iridium-rich, airborne, microscopic dust and impact vapor were deposited in

C-T boundary sediments within a few minutes of the impact. The finer-grained material in the Chicxulub dust cloud was likely deposited over a more extended period, potentially circulating Earth in the atmosphere for as long as two decades after the impact.[48]

The Origin of the Chicxulub Asteroid

A group of scientists at the Department of Space Studies at the Southwest Research Institute in Boulder, Colorado, published an article in the November 2021 issue of the scientific journal *Icarus* in which they argued the meteorite that created the Chicxulub crater was a dark primitive asteroid. They claimed the asteroid came from the outer reaches of the solar system's main asteroid belt, situated between Mars and Jupiter. The Chicxulub asteroid impacted Earth with a force estimated at ten billion atomic bombs of the size used in World War II. The Chicxulub crater produced the third-largest impact structure on Earth, leaving a crater in Mexico's Yucatán Peninsula that spans ninety miles (145 kilometers) and goes twelve miles deep (19.3 kilometers). Geochemical analysis of the crater had led scientists to conclude that the Chicxulub asteroid was part of a class of carbonaceous chondrites, a primitive group of meteorites with a high ratio of carbon and an origin in the solar system's early history.[49]

The scientists created a computer model to track how often objects escape the main asteroid belt. They concluded that over eons, thermal forces allowed these carbonaceous chondrite asteroids, like the one that caused the Chicxulub impact, to drift into dynamical "escape hatches" where the gravitational pull from the planets in the solar system push them into orbits nearing Earth. Using NASA's Pleiades supercomputer, the scientists tracked 130,000 model asteroids that were evolving in this slow, steady manner for hundreds of millions of years. The scientists paid particular attention to asteroids located in the outer half of the asteroid belt, the furthest part from the sun. Before the publication of this study, scientists thought the escape of an asteroid from the upper half of the main asteroid belt between Jupiter and Mars was rare. To their surprise, the Southwest Research Institute found that six-mile-wide asteroids from this region strike Earth at least ten times more often than previously calculated. Yet, asteroids the size of the Chicxulub asteroid hit the earth only at long intervals. On average, the Southwest Research Institute scientists concluded that an asteroid more than six miles wide from the outer edge of the asteroid belt is flung into a collision course with Earth once every 250 million years.

"I had a suspicion that the outer half of the asteroid belt—that's where the dark primitive asteroids are—may be an important source of terrestrial impactors," said David Nesvorný, the lead scientist from the Southwest Research Institute team. "But I did not expect the results [would] be so definitive."[50] He concluded that some 60 percent of the large terrestrial impactors had come from the outer half of the asteroid belt, where the dark, primitive asteroids are found. He argued an escaping asteroid from this region of the asteroid belt is five times more common than previously thought, with a 60 percent probability the next asteroid to hit Earth will come from the same region. Simone Marchi, a member of the research team and a coauthor of the scientific paper published in *Icarus*, echoed these themes. "This result is intriguing not only because the outer half of the asteroid belt is home to large numbers of carbonaceous chondrite impactors, but also because the team's simulations can, for the first time, reproduce the orbits of large asteroids on the verge of approaching Earth," Marchi said. "Our explanation for the source of the Chicxulub impactor fits in beautifully with what we already know about how asteroids evolve."[51]

A few months earlier, in February 2021, Abraham (Avi) Loeb, the Frank B. Baird Jr. Professor of Science at Harvard, published a paper in *Scientific Reports* proposing an alternative theory for the origin of the Chicxulub asteroid. Professor Loeb argued that main-belt asteroids (known as "MBAs"), with diameters in the range of ten kilometers (approximately six miles) capable of producing Chicxulub impact events, strike Earth once every 350 million years or so—too rare for the C-T boundary extinction event.[52]

Loeb concluded that long-period comets (known as "LPGs" and typically called "sun grazers"), which can take as long as 200 years to orbit the sun, are also capable of producing Chicxulub-scale impacts on Earth. Loeb's theory was that an LPG originating from the Oort cloud—another ring of debris at the solar system's outer edge—can get bumped off-course by Jupiter's gravitational field. Loeb insisted that a fraction of these LPGs, disrupted by Jupiter after passing the sun, could get broken up into smaller fragments that cross Earth's orbit. Loeb's point was that his LPG comet theory predicted a larger proportion of Earth impactors with carbonaceous chondritic composition than would be expected from MPG meteorites impacting the earth. Loeb argued an LPG comet capable of causing a Chicxulub-magnitude event could occur every 250 to 750 million years instead of once every 250 billion years.

"When you have these sun grazers, it's not so much the melting that goes on, which is a pretty small fraction to the total mass, but the comet is so close to the sun that the part that's closer to the sun feels a stronger gravitational pull than the part that is farther from the sun, causing a tidal force," Loeb explained. "You get what's called a tidal disruption event and so these large comets that come really close to the sun break up into smaller comets. And basically, on their way out, there's a statistical chance that these smaller comets hit the earth." Loeb stressed that his calculations increase the chances of long-term comets impacting Earth by a factor of ten and show that about 20 percent of long-term comets become sun grazers. "Basically, Jupiter acts as a kind of pinball machine," said Loeb's coauthor, Amir Siraj, the copresident of Harvard Students for the Exploration and Development of Space. "Jupiter kicks these incoming long-term period comets into orbits that bring them very close to the sun."[53]

Soon after the Siraj-Loeb paper was published, a group of scientists at the School of Earth and Space Exploration and the School of Molecular Sciences at Arizona State University disputed their findings.[54] They began by noting that the debate over whether the Chicxulub impactor was an asteroid or a comet had raged for nearly four decades. The Arizona scientists argued that the distance pattern of the iridium layer that the Alvarez team found was associated with the Chicxulub impact crater. In other words, the iridium layer created by the Chicxulub impact was thicker closer to the impact, with the iridium layer thinner as the distance from the impact increased. The Arizona scientists concluded the iridium layer left by the Chicxulub impactor correlated better to the impact force of an asteroid. A comet, typically smaller than an asteroid, is about 50 percent ice. Therefore, the Arizona team argued, since a comet has a typically reduced impact force, then that meant a comet could not have caused the magnitude of the Chicxulub impact.

The Deccan Volcanism

Gerta Keller, a geologist on the faculty of the Geosciences Department at Princeton University, has been a leading advocate of the theory that massive volcanic activity in an area known as the Deccan Traps in west-central India at the end of the Cretaceous Period was what killed the dinosaurs. For scientists working to identify the causes of the mass extinction at the C-T boundary, the Deccan volcanism that occurred over some one million years is a candidate. The more uniformitarian theory that the Deccan volcanism led to the extinction of the dinosaurs over time is a direct challenge to the more cataclysmic theory that the Chicxulub asteroid was the impact event solely responsible.

In a 2014 paper published by the Geological Society of America, Keller explained that the Deccan volcanic eruptions once covered most of India, with an estimated 1.5 million km² (approximately 579,000 square miles in area) and 1.2 million km³ (approximately 290,000 cubic miles in depth) extruded lava, which even today, with about two-thirds eroded, covers an area the size of France or Texas. "Flow after flow of volcanic eruptions piled up horizontally layered sequences reaching several thousand meters, which today still form mountains up to 3,500 m [11,483 feet]," she wrote.[55] "Some massive eruptions reached over 1,500 km [932 miles] across India and out to the Bay of Bengal via intracanyon transport, forming the longest lava flows known on Earth."

Scientists have concluded there is no direct correlation between the volume of lava and the magnitude of a resulting extinction because there "is always sufficient recovery time between individual eruptions to negate any cumulative effect of successive flood basalt eruptions."[56] Instead, the trigger for a mass extinction event appears to be the rapid injection of vast quantities of volcanic gases (both CO_2, carbon dioxide, and SO_2, sulfur dioxide) into the atmosphere, causing a major biological catastrophe. The SO_2 can cause global cooling if it reaches the stratosphere and rapidly disperses around the hemisphere. In the lower atmosphere, SO_2 is converted to sulfuric acid by the sun's rays. The sulfuric acid reacts with water vapor to form sulfuric acid aerosol layers in the stratosphere, producing an atmospheric cooling effect.[57] A recent example of this happening involved the Mount Pinatubo eruption of 1991, which injected twenty megatons of SO_2 more than thirty km into the stratosphere. The result was a global temperature decrease approaching 0.5°C for approximately three years.[58] The other principal volcanic gas, CO_2, is a greenhouse gas that produces the opposite temperature effect. Scientists have speculated that the "cumulative effects of repeated, closely spaced flood basalt eruptions could potentially promote global warming."[59]

A recent study led by scientists from the Department of Earth and Environmental Sciences in the Graduate Center at the City University of New York (CUNY) concluded that volcanic eruptions were not the driver of dinosaur extinction. Analyzing the Deccan Traps' CO_2 budgets found that the amount of CO_2 outgassing from the Deccan volcanism lava volumes did not release enough CO_2 to cause the global warming event that occurred several thousand years before the C-T extinction.[60] "Our lack of insight into the carbon released by magmas during some of Earth's largest volcanic eruptions has been a critical gap for pinning down the role of volcanic activity in shaping Earth's past climate and extinction events," said Professor Benjamin Black, the study's principal investigator and a professor in the Earth and Environmental Science program at CUNY. "This work brings us closer to understanding the role of magmas in fundamentally shaping our planet's climate, and specifically helps us test the contributions of volcanism and the asteroid impact in the end-Cretaceous mass extinction."[61]

Researchers Courtney Sprain and Paul Renne at the Department of Earth and Planetary Science at the University of California, Berkeley, now believe the Deccan volcano eruptions began 400,000 years before the Chicxulub impact, releasing 75 percent of their total lava volume in the 600,000 years after the Chicxulub impact.[62] The Sprain-Renne research team has suggested the magnitude of the Chicxulub impact's seismic shock may have struck the planet so hard that it sent the Deccan Traps into "eruptive high gear."[63] The Berkeley researchers explained the possibly interrelated nature of the phenomena. "The close temporal coincidence of the impact and the accelerated volcanism makes it difficult to deconvolve the environmental perturbations attributable to each mechanism," they wrote. In other words, the Sprain-Renne research team has attempted to resolve the conflict between whether the Chicxulub impact or the Deccan volcanism killed the dinosaurs by saying the C-T boundary extinction event was an event that "probably resulted from the supposed effects of both phenomena."[64]

The Disappearance of the Dinosaurs

Another challenge to the theory that the Chicxulub impact caused a sudden and rapid extinction of the dinosaurs is the scientific evidence that the dinosaurs were in decline for tens of millions of years before their extinction. In 2016, three British scientists led by Manabu Sakamoto at the School of Biological Sciences at the University of Reading in the U.K. published an article applying statistical methodology to phylogenetic analysis to model the evolutionary dynamics of species extinction of Mesozoic dinosaurs through time.[65] "We find overwhelming support for a long-term decline across all dinosaurs and within all three dinosaurian subclades [major dinosaur groups]," the British scientists concluded. The study results showed that Mesozoic dinosaurs showed a marked reduction in their ability to replace extinct species with new ones, making them vulnerable to extinction and unable to respond quickly to and recover from the final catastrophic event of the Chicxulub impact. "Although Mesozoic dinosaurs undoubtedly dominated the terrestrial megafauna until the end of the Cretaceous, they did see a reduction in their capacity to replace extinct species with new ones making them more susceptible to sudden and catastrophic environmental changes, like those associated with the asteroid impact," the scientists concluded.[66] The British scientists stressed that the dinosaurs declined for a much more extended period than previously thought. The study results strongly suggested that the extinction rate of dinosaurs surpassed their ability to create new species over some forty million years before their final extinction.

The lack of a complete fossil record and the difficulties and continuing controversy over the dynamics of dinosaur species diversification have complicated the analysis. Various studies have disputed the evidence arguing for a global decline across dinosaur groups before extinction at the C-T boundary.[67] A decline of non-avian Mesozoic dinosaurs on the timescale of hundreds of thousands or millions of years would support the argument that the Deccan flood volcanism was a causative factor in the final extermination. Yet, an international study of six critical dinosaur families published in 2021 also found net dinosaur species diversification rates culminated in the middle Late Cretaceous Period, with the dinosaurs in decline for some ten million years before the final extinction at the C-T boundary.[68] An interesting finding of this study was that older dinosaur species were more vulnerable to extinction because they lacked "evolutionary novelty," or adaptation to changing environmental circumstances. Studies concluding the dinosaurs were in decline before their final demise supports the theory that a precipitating cause of the dinosaurs' death may well have been the sudden cataclysmic impact of the Chicxulub asteroid. But the extinction process included many factors, such as possibly the Deccan volcanism over millions of years, and most likely one additional cause that we will examine in chapter 8.

There is also a branch of earth sciences that insists the dinosaurs are not extinct. A modern pigeon or penguin does not appear to have much in common with a Tyrannosaurus rex. But during the Jurassic Period in the Mesozoic Era, about 150 million years ago, birds evolved from small, feathery, raptor-like dinosaurs, becoming another branch on the dinosaur family tree. "For more than 80 million years, birds of all sorts flourished, from loon-like swimmers with teeth to beaked birds that carried streamer-like feathers as they flew."[69] Birds can be characterized as "avian dinosaurs," while what the public imagines as dinosaurs, from the Brontosaurus to the T-Rex, are what paleontologists describe as "non-avian dinosaurs." A controversy continues seeking an explanation for why beaked birds survived the fifth extinction event at the C-T boundary, when winged avian dinosaurs like the Pterosaurs died along with some 75 percent of the species known to be alive during the Cretaceous Period. In the next chapter, when we discuss an additional possible cause for the dinosaurs' demise, we will return to this mystery.

The Permian-Triassic "Great Dying" Third Extinction

While the fifth extinction that killed the dinosaurs took place at the C-T boundary some sixty-six million years ago, the third extinction some 250 million years ago caused the most massive destruction of life on Earth that has ever occurred in geological time. The third extinction marks the boundary between the Permian Period (that ended the Paleozoic Era) and the Triassic Period (that began the Mesozoic Era). The event is known as the "Great Dying," occurring at the boundary in geological time known as the Permian-Triassic (P-T or P-Tr) boundary. Considered the most significant biological disaster in Earth's history, the third extinction killed some 95 percent of all species on the planet.

While the Chicxulub impact has achieved wide acceptance as the extinction event at the C-T boundary, earth scientists have not been able to identify any single cataclysmic event to mark the P-T boundary. Among the many causes speculated are the volcanic eruptions that lasted some one million years in the Siberian Traps. That explanation is like the Deccan volcanism in the Indian Traps, which we saw as a possible factor in producing the fifth extinction that killed the dinosaurs. A variation of the Siberian Traps' volcanism is that the CO_2 emitted by the volcanoes and the CO_2 emitted when the volcanoes burned forests four times the size of Korea caused global warming. As a result of that global warming, the theory asserts, frozen methane below the sea may have melted, producing a global warming effect twenty times more potent than the CO_2.[70]

Douglas H. Erwin, a senior scientist and curator in the Department of Paleobiology at the Smithsonian's National Museum of Natural History and chair of the faculty at the Santa Fe Institute, has spent much of his career since the 1980s searching for the cause of the P-T extinction. He has traveled extensively to China, South Africa, and Europe seeking geological evidence of an asteroid impact or other cataclysmic cause of the third extinction. In his 2006 book entitled *Extinction*, Erwin delivered his verdict as follows:

> So, what did cause the greatest mass extinction in the past 600 million years, and perhaps the greatest in the history of life? The short answer is that we do not know, or at least I do not know. Several of my less reticent

colleagues are sure they know but their answers are mutually contradictory and so cannot all be correct.[71]

As Erwin sorted through the possible causes, he was still left uncertain:

> We have growing evidence for an extraterrestrial impact, but evidence that is still, in my view, less than overwhelming; the evident coincidence of the Siberian flood basalts with the mass extinction, strong evidence for some degree of low oxygen and other changes in ocean chemistry, and a sudden spike in global temperatures. I have so often been wrong about the cause of the extinction that, in deference to my battered sense of scientific worth, I am tempted not to hazard an answer.[72]

When examining whether CO_2-caused global warming caused the third extinction, Erwin was again not convinced. "Does the change in the carbon cycle reflect the cause of the extinction, or was it the result of the extinction?" he asks.[73] His answer to that question is consistent with our earlier observation that changes in CO_2 lag global warming. Erwin considered:

> Photosynthesis partitions carbon on the Earth into two great reservoirs: the organic carbon reservoir of almost everything living as well as of coal, oil, and other organic remains, and a larger reservoir of inorganic carbon that has not passed through living things, including the limestone in a reef and carbon dioxide in the atmosphere. At the close of the Permian a massive amount of organic carbon was released into the atmosphere and oceans. Where it came from is the critical issue. Among the possibilities is that it reflects the carbon in all the plants and animals that died, or the burning of massive coal deposits in Siberia, or the release of methane gas trapped in sediments on the outer shelf of the continents.[74]

Erwin accepted the conventional explanation that hydrocarbon fuels are fossil fuels of organic origin. But the point here is that he concluded the CO_2 at the end of the Permian came from the plants and animals killed in the third extinction. That conclusion precludes reasoning that the CO_2 is the effect of the extinction event and not a cause of it happening.

An Ice Age Cause?

In 2017, Professor Urs Schaltegger from the Department of Earth Sciences at the University of Geneva in Switzerland and Hugo Bucher from the University of Zürich published a scientific paper reporting their research on sedimentary layers in the Nanpanjiang Basin in southern China. The research team led by Schaltegger and Bucher concluded that a short ice age preceded by global warming caused the P-T extinction.[75] A University of Geneva press release made clear the ice age conclusion the scientists discovered "completely calls into question the scientific theories regarding these phenomena, founded on the increase of CO_2 in the atmosphere, and paves the way for a new vision of the Earth's climate history."[76]

Through carbon-13 dating of the various sediment layers at the site in China, the researchers realized a gap in sedimentation represented the mass extinction of the Permian-Triassic boundary. This gap corresponded to a period when the seawater level decreased. The University of Geneva, when announcing the research, explained that the only explanation for this phenomenon is that there was an ice age that lasted 80,000 years. That was sufficient time to eliminate much of the existing marine life. Scientists attributed the volcanism in the Siberian Traps as being the cause of the global temperature. They argued the Siberian volcanos created a stratospheric injection of large amounts of SO_2 that reduced the intensity of solar radiation reaching Earth's surface. "We therefore, have proof that the species disappeared during an ice age caused by the activity of the first volcanism in the Siberian Traps," explained Urs Schaltegger. "This ice age was followed by the formation of limestone deposits through bacteria, marking the return of life on Earth at more moderate temperatures. The period of intense climate warming, related to the emplacement of large amounts of basalt of the Siberian Traps and which we previously thought was responsible for the extinction of marine species, happened 500,000 years after the Permian-Triassic boundary."

According to Schaltegger, one of the most important conclusions of the study is proof that climate warming is not the only explanation of global ecological disasters in the past on Earth.[77] The SO_2 emitted in the Siberian volcanism caused global cooling, while the CO_2 emitted in the same volcanoes failed to cause global warming. This research demonstrating the importance of volcanic SO_2 in driving an ice age extinction event at the Permian and Triassic boundary reinforces a point we made earlier in the chapter: water vapor is the most important greenhouse gas in the atmosphere. As volcano expert Richard Fisher pointed out, the "greatest volcanic impact upon the earth's short term weather patterns is caused by sulfur dioxide gas." Fisher continued to explain that in "the cold lower atmosphere it [the sulfur dioxide gas] is converted to sulfuric acid by the sun's rays reacting with stratospheric water vapor to form sulfuric acid aerosol layers." The aerosol remains in suspension long after solid ash particles from a volcano have fallen to Earth. A layer of sulfuric acid droplets forms between fifteen to twenty-five kilometers into the upper atmosphere, acting as a global cooling agent. Fisher added that "fine ash particles from an eruption column fall out too quickly to significantly cool the atmosphere over an extended period, no matter how large the eruption."[78] A 2008 study conducted by scientists at the University of Colorado at Boulder found that the lack of volcanic dust in the atmosphere in the past twelve years has reduced the amount of sunlight being refracted through Earth's atmosphere, allowing for brighter lunar eclipses. The study suggested that the relative lack of volcanic dust in the atmosphere could be responsible for as much as a .1–.2°C rise in the average Earth temperature since the 1960s. While this is a relatively small increase in Earth's temperature, the study points out that there are other climate drivers in the atmosphere than just anthropogenic CO_2 emissions.[79]

Ian Plimer, twice the winner of Australia's highest honor, the Eureka Prize, a professor of earth sciences at the University of Melbourne, and the author of more than 120 scientific papers, stressed that ice ages are proof CO_2 does not drive climate change. Consider this paragraph from his 2009 book entitled *Heaven and Earth: Global Warming, the Missing Science*:

> The proof that CO_2 does not drive climate is shown by previous glaciations. The Ordovician-Silurian (450-420 million years ago) and the Jurassic-Cretaceous (151-132 million years ago) glaciations occurred when the atmospheric CO_2 content was more than 4,000 parts per million by volume (ppmv) and 2,000 ppmv respectively. The Carboniferous-Permian glaciation (360-260 million years ago) had a CO_2 content of about 400 ppmv, at least 15 ppmv higher than the present figure. If the popular catastrophist view is accepted, then there should have been a runaway greenhouse when CO_2 was more than 4,000 ppmv. Instead, there was glaciation. Clearly a high atmospheric CO_2 does not drive global warming and there is no correlation between global temperature and atmospheric CO_2. This has never been explained by those who argue that human additions of CO_2 to the atmosphere will produce global warming.[80]

Plimer also dismissed the idea that global warming today will bring on the sixth extinction. He explained as follows:

> Some speculations suggest that a mere 0.8°C temperature rise over 50 years will result in extinction of 20% of the world's species. If this were the case,

we should have seen a mass extinction of life in the Minoan Warming, the Roman Warming, and the Medieval Warming. We did not. We may actually be living in a period of low extinctions, with relatively few species becoming extinct over the last 2.5 million years. Current projections of extinctions may be an overestimation as we focus on terrestrial vertebrates and not the spectrum of life on Earth.[81]

He argued that increased temperature would bring more species diversity "by extending the ranges of plants and animals." If a future warmer climate had a higher CO_2 content, "plant life would be far more vigorous because increased CO_2 enables plants to grow better in nearly all temperatures, especially at higher temperatures." He also dismissed the constant preoccupation of the IPCC that a doubling of CO_2 in the atmosphere caused by the burning of hydrocarbon fuels would cause catastrophic global warming hazardous to all life on Earth. Plimer pointed out that "the Cambrian explosion of life (542-520 million years ago) took place in the post-glacial warm times when atmospheric CO_2 was 25 times greater than today." He noted that satellite measurements of vegetation on a global scale between 1982 and 1999 showed "plant growth increased by 6% in response to slightly increased rainfall and slightly increased temperature but the major change was due to slightly increased CO_2." He projected that if the CO_2 atmospheric content doubled, "the net productivity rise of herbaceous plants is 30 to 50%, while of woody plants is 50 to 80%." He put the point succinctly: "To argue that increasing temperature and atmospheric CO_2 will result in extinction of plants is to argue that CO_2 is not plant food."[82]

Patrick Moore, Ph.D., the cofounder of Greenpeace in 1971, left the organization after some fifteen years. Moore, who received a Ph.D. from the University of British Columbia in 1974, wrote his dissertation on pollution control and the mining industry in Canada. In his testimony on May 22, 2019, before the Subcommittee on Water, Oceans, and Wildlife of the U.S. House Committee on Natural Resources, Moore also argued that we are not experiencing the much-touted sixth extinction. He explained he left Greenpeace when the organization began adopting policies that he did not believe were in accord with the sound scientific principles for the environmental issues we face today. In his congressional testimony, Moore argued the United Nations is using "extinction as a fear tactic to scare the public into compliance." He specifically took objection to how the United Nation's Intergovernmental Science-Policy Platform on

Biodiversity and Ecosystem Services (IPBES) accounts for the number of species on Earth. He pointed out the IPBES claims that there are eight million species, but only 1.8 million species are named and identified. "Thus, the IBPES [*sic*] believes there are 6.2 million unidentified and unnamed species," he testified. "Therefore, one million of the unknown species could go extinct overnight, and we would not notice it because we would not know they had existed." He claimed this was a highly unprofessional approach. "Scientists should not, in fact cannot, predict estimates of endangered species or species extinction based on millions of undocumented species."[83]

Conclusion

Throughout this chapter, we have seen climate scientists grapple with the sudden, dramatic changes that have occurred throughout history, as witnessed by the Chicxulub asteroid hitting the Yucatán. We have seen cataclysmic changes interacting to produce dramatic climate changes, as witnessed, for instance, by the possibility the Chicxulub impact triggered volcanic action that characterized the C-T boundary extinction event.

Psychologically, humans like to think of the earth as a stable environment that will be tomorrow as habitable for life as it is today. Yet, the geological record attests to Earth as a dynamic entity subject to sudden catastrophic events interspersed with relatively quiet and stable periods. Sedimentary rock outcrops show bands of different-colored earth accumulating through time. But the line demarcating one sedimentary rock level from another is generally sharp. The next level of sedimentary rock that accumulates is typically a different color, reflecting some dramatic change in that environment. Stephen Marshak's 2019 college geology textbook that we referred to earlier in this chapter, *The Essentials of Geology*, precisely makes this point: "Layers, or *beds*, of sedimentary rock are like the pages of a book, recording tales of ancient events and environments on the ever-changing face of the Earth."[84]

The U.S. Geological Survey (USGS) reports that 80 percent of Earth's surface, both above and below sea level, is of volcanic origin. "Over geologic eons, countless volcanic eruptions have produced mountains, plateaus, and plains, which subsequently eroded and weathered into majestic landscapes and formed fertile soils," the USGS proclaims on its website.[85] The USGS further estimates about 1,500 volcanoes are potentially active worldwide, aside from the continuous belts of volcanoes on the ocean floor at spreading centers like the Mid-Atlantic Ridge. Some 500 of the 1,500 volcanoes have erupted in historical time. The "Ring of Fire" around the rim of the Pacific Ocean consists of a 20,000-kilometer-long chain of volcanoes.[86] The Ring of Fire is the most seismically and volcanically active zone globally.[87]

While asteroids the size of the Chicxulub asteroid rarely hit Earth, smaller asteroids and comets hit Earth more frequently. In 1998, a paper published in *Astronomy and Astrophysics* solved the mystery that persisted over the previous ninety years: trying to find a scientific explanation for the explosion over Tunguska in central Siberia on June 30, 1908. Using a mathematical model of the hypersonic flow around a small asteroid in Earth's atmosphere, Luigi Foschini, a researcher at the National Institute of Astrophysics (INAF), Brera Astronomical Observatory, Rome, Italy, concluded that a stony asteroid with a diameter of about sixty meters (197 feet) caused the Tunguska event by reaching the lower atmosphere before exploding and fragmenting.[88] The explosion happened over sparsely populated forestland in present-day Krasnoyarsk Krai, Russia. The blast flattened an estimated eighty million trees over 2,150 square kilometers (830 square miles). Witnesses described seeing a fireball of bluish light in the sky, nearly as bright as the sun, that exploded in a flash with a sound like artillery cannon fire. The shockwave that followed the explosion broke windows hundreds of miles away.[89]

Much more recently, on February 15, 2013, an asteroid that observers caught on film broke up over Chelyabinsk, Russia. The blast, which witnesses said briefly outshone the sun and resembled a nuclear explosion, was detected by monitoring stations as far away as Antarctica. The shock wave shattered glass and injured some 1,200 people. Coincidentally, the same day, an asteroid designated 2012 DA14 passed within 27,000 kilometers (17,200 miles) of Earth. According to NASA, this second asteroid was traveling in a direction opposite to that of the meteorite that exploded over Chelyabinsk.[90] Bill Cooke of NASA's Meteoroid Environment Office at the Marshall Space Flight Center in Alabama identified the Chelyabinsk meteor as an asteroid coming from the asteroid belt, about 2.5 times farther from the sun than Earth. Peter Brown, a physics professor at the University of Western Ontario in Canada, estimated the asteroid was about seventeen meters in width (fifty-six feet), weighing 10,000 metric tons (11,023 tons). He estimated the Chelyabinsk asteroid struck Earth's atmosphere at 40,000 miles per hour and broke apart about twelve to fifteen miles above the planet. He further estimated the asteroid explosion exceeded 470 kilotons of TNT, thirty to forty times more powerful than the World War II atomic bombs the United States dropped on Hiroshima.[91] On September 16, 2021, another giant asteroid designated 2021 SG had a near-miss with Earth. The 2021 SG asteroid was four times larger than the asteroid that disintegrated over Chelyabinsk, Russia, in 2013. However, NASA failed to detect the asteroid until astronomers found it on September 17, 2021, the day after the asteroid's closest pass by Earth. NASA failed to see the asteroid because it came unexpectedly from the direction of the sun. They estimated asteroid 2021 SG had a diameter between 42 and 94 meters (138–308 feet). Its average diameter of 68 meters (223 feet) is four times the size of the Chelyabinsk asteroid. Astronomers using the large 1.2-meter (48-inch) telescope at Mount Palomar, California, were the first to detect the asteroid as it headed away from Earth. At its closest, 2021 SG passed Earth at about half the distance from it to the moon.[92]

In May 2021, journalists worldwide reported on a weeklong simulation exercise NASA conducted for a group of experts from the U.S. and European space agencies. The exercise involved a hypothetical scenario of an asteroid thirty-five million miles from Earth that could hit the planet within six months. NASA concluded existing technology was not sufficient to prevent the asteroid from impacting Earth.[93] They reported that at the start of 2019, the number of discovered Near-Earth Asteroids (NEAs) totaled more than 19,000, with thirty discoveries added each week.[94] The total of NEAs had grown to 26,442 according to a NASA report dated August 20, 2021.[95] Radio astronomer Gerrit L. Verschuur, in his 1996 book *Impact! The Threat of Comets and Asteroids*, noted that a total of some 10,000 tons of space debris hit Earth every year, mainly in the small meteoric form.[96] "But sometimes a larger object, a *meteorite*, survives heating in the atmosphere and lands intact," he wrote. "Meteorites weighing from ounces to tens of tons have been recovered."[97]

The Younger Dryas is one of the most well-documented examples of abrupt climate change. About 14,500 years ago, Earth's climate emerged from an ice age to a warmer interglacial state. Yet, the warming stopped suddenly, and conditions in the Northern Hemisphere returned to near-glacial conditions that took less than one hundred years and maybe only a decade, as suggested by an analysis of ice core data in Greenland.[98] The Younger Dryas is named after a flower (*Dryas octopetala*) known to grow in cold conditions that became common.[99] Some 11,500 years ago, the Younger Dryas ended equally abruptly. During the Younger Dryas, temperatures in Greenland rose 10°C (18°F) in a decade. A 2000 study published in *Quaternary Science Reviews* by Richard Alley, a professor at the Department of Geosciences and Environment Institute at Pennsylvania State University, found that the abrupt climate changes of the Younger Dryas occurred worldwide. "Near simultaneous changes in ice-core paleoclimatic indicators of local, regional, and more-widespread climate conditions demonstrate that much of Earth experienced abrupt climate changes synchronous with Greenland with thirty years or less," he wrote. Professor Alley also concluded that "post-Younger Dryas changes have not duplicated the size, extent, and rapidity of these paleoclimatic changes."[100]

Still, as we have noted, the IPCC issued a warning in August 2021 that anthropogenic CO_2 has already caused approximately 1.0°C of global warming above preindustrial levels. The IPCC warning cautioned that global warming is likely to reach 1.5°C between 2030 and 2052 if anthropogenic CO_2 emissions continue to increase at the current rate. This warning is equivalent to an alarm that human burning of hydrocarbon fuels will cause a sudden climate change with disastrous consequences for human life unless dramatically curtailed. David Sepkoski, who holds the Thomas M. Siebel Chair in History of Science at the University of Illinois at Urbana-Champaign, has published a 2020 book entitled *Catastrophic Thinking: Extinction and the Value of Diversity from Darwin to the Anthropocene*. In this book, Sepkoski interprets the causes and consequences of mass extinctions and their ensuing moral imperatives as deriving from and reflecting upon the cultural values of any given historical moment. Today, IPCC's thinking regards human beings as Earth's current climate catastrophe. The IPCC has judged we will bring about our own destruction in the sixth extinction, caused by our wanton burning of hydrocarbon fuels. Sepkoski asks whether "we have come to see humanity, not as some unpredictable external agent bring death from above—an asteroid—but rather as an implacable geological force capable of altering the basic conditions of life on Earth from within."[101]

In 2017, science writer Peter Brannen interviewed Doug Erwin, the Smithsonian paleontologist and expert on the Permian-Triassic mass extinction we mentioned earlier in this chapter.[102] Before interviewing Erwin at an annual meeting of the Geological Society of America, Brannen had corresponded with Erwin, asking him for his take on the contemporary idea that a sixth extinction is already underway on Earth today. Brannen described for Erwin his view that human beings have created a Frankenstein biosphere where we humans, together with our livestock and pets, take up 97 percent of Earth's biomass. Brannen detailed for Erwin a long list of ecological disasters that humans have caused, such as fishing trawlers obliterating some 90 percent of ocean predators since 1950, including "familiar staples of the dinner plate like cod, halibut, grouper, tuna, swordfish, marlin, and sharks." He lamented that "modern reefs are expected to collapse from warming and ocean acidification by the end of the century, and possibly much sooner." Brannen felt that Erwin was the person to ask since he "is one of the world's experts on the End-Permian mass extinction, an unthinkable volcanic nightmare that nearly ended life on earth 252 million years ago."

Brannen was surprised when Erwin explained his view that we are not in the sixth extinction. Erwin called that idea junk science. He explained that a mass extinction was much like the major power blackout the northeast United States suffered in 2003. From a mathematical point of view, Erwin reasoned, the failure of a regional power grid involves a complex in which a collapse in one part of the system can cascade into a complete system failure. "There's a very rapid collapse of the ecosystem during these mass extinctions," Erwin told Brannen. "If we're really in a mass extinction—if we're in the [End-Permian mass extinction 252 million years ago]—go get a case of scotch." Erwin's point was that a mass extinction involves a sequence of very complex events that include cataclysmic events and systematic climate disruptions that may occur over hundreds, thousands, or millions of years. Erwin conjured images of asteroids hitting the earth and massive volcanoes going into long-term, hyper-drive eruptions. Next, Earth's life-supporting ecosystem would collapse, causing an environmental shut-down like the collapse of the northeast U.S. power grid in 2003.

"People who claim we're in the sixth mass extinction don't understand enough about mass extinctions to understand the logical flaw in their argument," Erwin insisted. "To a certain extent, they're claiming it as a way of frightening people into action, when in fact, if it's actually true that we are in a sixth mass extinction, then there's no point in conservation biology," Erwin continued. Brannan explained that Erwin's point was that by the time a mass extinction starts, the world as we know it would already be over. "So, if we are really in the middle of a mass extinction," Erwin added, "it wouldn't be a matter of saving tigers and elephants." Erwin understood that Brannan was asking him whether groups like the IPCC were correct in asserting anthropogenic CO_2 would cause the sixth extinction. Yet, he dismissed the suggestion while being cautious to remain politically correct by agreeing that humans are capable of causing ecological damage to the environment.

Erwin returned to the analogy of a network power grid collapse. "Network dynamics research has been getting a ton of money from DARPA [Defense Advanced Research Projects Agency]," he said. "They're all physicists studying it, who don't care about power grids or ecosystems, they care about math. So, the secret about power grids is that nobody actually knows how they work. And it's exactly the same problem you have in ecosystems."

Radio astronomer Gerrit Verschuur made a similar point about comets and asteroids. He wrote the following in his 1996 book *Impact! The Threat of Comets and Asteroids*:

> The reality of the threat of comet impacts was brought home to everyone in July 1994 when fragments of a comet slammed into Jupiter's atmosphere to produce a stunning set of explosions that were seen from earth. Fortunately, we watched from a safe distance. If anything remotely similar had happened here, few human beings would have been left to think further about comets, asteroids, or anything else for that matter.[103]

The forces that caused the five mass extinctions we have experienced on Earth have been massive and complex. Understanding the dynamics of how the sudden impact of a gigantic asteroid could interact to intensify an ongoing period of volcanism requires understanding that the forces of nature are not necessarily within human control. Verschuur ended his book by contemplating NASA's plans to land an astronaut on an asteroid to conduct scientific exploration before returning to Earth safely. He commented that "this plan, if implemented would...represent the coming of age of the human species.... Walking on the moon was dramatic, but walking on a near-earth asteroid during its flight about the sun and past the earth would signify that our species had come to recognize that we do not live in splendid isolation from space," he wrote. "We need to understand near-earth objects, comets, and asteroids, if we are to live with them for a very long time into the future."[104]

CHAPTER 8

The Chaos Theory of Climate

The Butterfly Effect, Unpredictable Weather and Strange Attractors, the Unknown Precambrian Era, the Expanding Earth Theory, Another Explanation for the Extinction of the Dinosaurs, Milankovitch Cycles, Ice Ages, Shifting Magnetic Poles, and Catastrophe Theory

> *If a single flap of a butterfly's wing can be instrumental in generating a tornado, so also can all the previous and subsequent flaps of its wings, as can the flaps of the wings of millions of other butterflies, not to mention the activities of innumerable more powerful creatures, including our own species. If the flap of a butterfly's wings can be instrumental in generating a tornado, it can equally well be instrumental in preventing a tornado.*
>
> —Edward N. Lorenz, "Predictability: Does the Flap of a Butterfly's Wings in Brazil Set Off a Tornado in Texas," 1972[1]

> *But the scant two-week-long accuracy of weather forecasts reflects a fundamental problem described by Ed Lorenz at MIT in 1961. The weather is chaotic—small changes in how we start the model can lead to very different predictions after a few weeks. So no matter how precisely we might specify current conditions, the uncertainty in our predictions grows exponentially as they extend into the future. More computer power cannot overcome this basic uncertainty.*
>
> —Steven E. Koonin, former undersecretary for science, U.S. Department of Energy under the Obama administration, *Unsettled: What Climate Science Tells Us and What It Doesn't and Why It Matters*, 2021[2]

The Chaos Theory of Climate

The fact that the large computer models indicate such a temperature rise as a consequence of increased carbon dioxide cannot be taken as evidence of truth; for any such model is merely a formal statement of the modeller's opinion of how the atmospheric system works.
—Reid A. Bryson, "Simulating Past and Forecasting Future Climates," 1993[3]

Climate is related to Milankovitch Cycle wobbles— we just don't know how.
—Ian Plimer, *Heaven and Earth: Global Warming, the Missing Science*, 2009[4]

JOHN VON NEUMANN, BORN IN Hungary in 1903, was a brilliant mathematician who played an instrumental role in developing game theory and digital computing. During World War II, he worked on the Manhattan Project, developing the mathematical models needed for the explosive lenses in the implosion mechanism of Fat Man, the plutonium bomb dropped on Nagasaki. But von Neumann was not necessarily brilliant in everything. He was an early proponent of global warming who predicted (incorrectly) that the release of CO_2 from the burning of coal and oil would bring "a general warming of the world by about one degree Fahrenheit." He expected this to occur a little more than the end of his generation. He worried that the CO_2 content of the atmosphere would increase exponentially, and "another fifteen degrees of warming would probably melt the ice of Greenland and Antarctica."[5] John von Neuman died in 1957, and the ice remains in Greenland and Antarctica.

He also believed he could apply mathematics to predict the weather accurately. In a coauthored paper published in the geophysics journal *Tellus* in 1950, von Neuman explained using an early ENIAC computer how to find a numeric solution to weather prediction.[6] He believed that "our knowledge of dynamics of controlling processes in the atmosphere, together with the development of computing machines, was approaching a level that would permit weather prediction."[7] He also believed "one could understand, calculate, and perhaps put into effect processes ultimately permitting control and change of the climate." [8] In 1955, while serving as a member of the Atomic Energy Commission, von Neumann published an article in Fortune entitled "Can We Survive Technology?"[9] In this article, he discussed his concerns about anthropogenic CO_2 and global warming, but he also predicted that future mathematical analyses would allow us to control the global climate. In that article, he wrote the following fascinating paragraph, as follows:

> What could be done, of course, is no index to what should be done; to make a new ice age in order to annoy others, or a new tropical, "interglacial" age in order to please everybody, is not necessarily a rational program. In fact, to evaluate the ultimate consequences of either a general cooling or a general heating would be a complex matter. Changes would affect the level of the seas, and hence the habitability of the continental coastal shelves; the evaporation of the seas, and hence general precipitation and glaciation levels; and so on. What would be harmful and what beneficial—and to which regions of the earth—is not immediately obvious. But there is little doubt that one *could* carry out analyses needed to predict results, intervene on any desired scale, and ultimately achieve rather fantastic effects. The climate of specific regions and levels of precipitation might be altered. For example, temporary disturbances—including invasions of cold (polar) air that constitute the typical winter of the middle latitudes, and tropical storms (hurricanes)—might be corrected or at least depressed.[10]

In the same paragraph, von Neumann doubted we had a moral compass to make wise judgments about altering the climate. At the same time, he expressed complete confidence that our mathematical skills aided by advanced computers would be sufficient to make whatever changes in the environment we desired to make.

The Butterfly Effect

Edward Norton Lorenz was a mathematician and meteorologist who followed in von Neumann's footsteps. As a professor in the Department of Meteorology at MIT in the 1960s, Lorenz used an early digital computer, a Royal McBee LGP-30, to run a computer model he constructed to predict the weather. The computer model used twelve independent variables to measure various aspects of the weather, including temperature and wind speed. One day, in February 1961, Lorenz repeated a simulation he had run earlier. This time, he rounded off one variable from .056127 to .506, and then he went to get a cup of coffee. When he returned to his office, he was shocked to find "this tiny alteration drastically transformed the whole pattern his program produced, over two months of simulated weather."[11]

In 1963, Lorenz published a scientific paper in the *Journal of Atmospheric Sciences*. "Deterministic Nonperiodic Flow" was a highly technical study that triggered a scientific revolution.[12] The subject of the article was not specifically weather predicting. Instead, Lorenz focused on using finite systems of deterministic ordinary nonlinear differential equations to analyze hydrodynamical flow patterns. Cascading water has been problematic to describe in mathematical equations. Water cascades typically "vary in an irregular, seemingly haphazard manner, and, even when observed for long periods, do not appear to repeat their previous history."[13] Nonlinear differential equations are notoriously difficult to solve because a change in one variable does not produce the exact difference or reaction in related variables. But Lorenz felt utilizing nonlinear differential equations was appropriate for the dynamic system of water cascades he was trying to model mathematically. In a dynamic system, a change in one variable may not produce the exact change every time in other variables.[14] In his book *The Essence of Chaos*, Lorenz discussed why differential equations are the appropriate mathematical tool for handling flows, including water oscillations or the action of a pinball machine. "A system of differential equations amounts to a set of formulas that together express the rates at which all of the variables are currently changing, in terms of the current values of the variables," he explained.[15]

What he found out was that slight variations in the variables produced drastic changes in the results. He found that "slightly differing initial states can evolve into considerably different states."[16] He concluded the article by applying his findings to the atmosphere. "In view of the inevitable inaccuracy and incompleteness of weather observations, precise very long-range forecasting would seem to be non-existent," he noted.[17] Lorenz had concluded that the problem with long-range weather forecasting is that the variety of weather possibilities are so immense that even small changes can drastically affect weather outcomes. Lorenz's conclusion challenged the classical understanding of nature. Sir Isaac Newton published laws in 1687 that suggested a highly "predictable mechanical system—the 'clockwork universe.'"[18] Though Lorenz had just developed a mathematical proof that weather is unpredictable, his 1963 paper went largely unnoticed. Yet, the importance of Lorenz's work was enormous. "By showing that certain deterministic systems have formal predictability limits, Ed put the last nail in the coffin of the Cartesian universe and fomented what some have called the third scientific revolution of the 20th century, following on the heels of relativity and quantum physics," said Kerry Emanuel, a professor of atmospheric science at MIT.[19]

On December 29, 1972, Lorenz presented a paper at a meeting of the American Association for the Advancement of Science. The paper was entitled "Predictability: Does the Flap of a Butterfly's Wings in Brazil Set Off a Tornado in Texas?"[20] In the paper, Lorenz questioned whether a single flap of a butterfly's wings could generate a tornado somewhere else in the world. He was equivocal in that he also noted that a single flap of a butterfly's wing would have no more effect on the weather than any flap of any other butterfly's wings, not to mention the activities of other species, including our own.

The term "Butterfly Effect" gained notoriety in 1987 when science historian James Gleick published his book entitled *Chaos: Making a New Science*.[21] Gleick's first chapter, "The Butterfly Effect," was dedicated to giving Lorenz credit for discovering a new branch of mathematics now deemed "Chaos Theory." In his book *The Essence of Chaos*, Lorenz admitted he did not develop the Butterfly Effect name for chaos theory. He mused that the name might have come from "A Sound of Thunder," a short story written by science-fiction writer Ray Bradbury. Lorenz considered the reference appropriate. He noted that in the story the death of a prehistoric butterfly and its consequent failure to reproduce changed the outcome of a present-day election. "Perhaps the butterfly, with its seeming frailty and lack of power, is a natural choice for a symbol of the small that can produce the great," Lorenz wrote.[22]

But, as we will see in the next section, the strange attractor graphs Lorenz used in his 1963 paper to describe the action of a chaotic system resembled a butterfly. Lorenz resolved that perhaps that was the explanation after all.[23]

Unpredictable Weather and Strange Attractors

When applied to the atmosphere, the power of Lorenz's discovery is that any variations in measurement will produce enormous differences in outcomes. Mathematically this is impossible to avoid. To understand this, please consider that equally divisible numbers can be divided by other integers with no remainders. For instance, the number "4" is equally divisible by the number "2," while the number "7" cannot be equally divided by "2." An irrational number is a number that can be expressed as a decimal but not as a fraction. Irrational numbers are not the ratio of two integers.

For instance, π (Pi), the ratio of a circle's circumference to its diameter, is the irrational number 3.141592, etc. An irrational number has an infinite series of numbers after the decimal because an irrational number is never fully divisible. Temperature variables as simple as "degrees Celsius," to be precise, are of necessity measured in irrational numbers. Irrational numbers are characterized by infinite numbers lying to the decimal's right with increasingly more accurate measurements. Understanding the concept of infinity is mathematically tricky. No matter how many decimal places we calculate for an irrational number, there yet remain an infinitely greater number of additional decimal places that we have yet to figure. That irrational numbers are followed by an infinitely calculable number of decimal places means no human being will ever mathematically calculate the last remaining decimal place.

In contrast, nature does not need to calculate decimal places. Irrational numbers and their operation are intrinsic to the fabric of the mathematical logic by which nature operates just fine, without any assistance from human beings. The point is that the unavoidable use of irrational numbers in climate model measurements means Lorenz's problem with slight differences in initial calculations never goes away. This problem alone prevents any computer from accurately predicting weather or climate, no matter how powerful the computer might be.

In analyzing weather and climate, we must remember that nature does not worry about computing decimal points. But climate outcomes will be different because a computer model cannot compute infinite decimal places. Even computers set to calculate an enormous number of decimal places have limits. After calculating a vast number of decimal places, an endless number of decimal places still remains to be calculated. Human beings and computers cannot calculate infinite sets. But, again, nature operates without having to conduct human calculations and without needing computers. One of the leading reference books on chaos theory and fractal mathematics points out the problem as follows:

> In other words, even if the weather models in use were absolutely correct—that is, as models for the physical development of the weather—one cannot predict them for a long time period. The effect is nowadays called sensitive dependence on initial conditions. It is one of the central ingredients of what is called deterministic chaos.[24]

When we add positive and negative feedback effects into the climate model, the outcome is as follows:

Eventually the iterations of our feedback process become as trustworthy as if we had obtained them with a random number generator, or rolling dice, or flipping coins. In fact, the Polish mathematician Stan Ulam discovered that remarkable property when he constructed numerical random number generators for the first electronic computer ENIAC in the late forties in connection with large scale computations for the Manhattan Project.[25]

Yet another sort of measurement problem makes weather prediction impossible. Again, this problem derives from fractal mathematics. The problem is known as the "coastline paradox." As author James Gleick explained, "any coastline is—in a sense—infinitely long."[26] The length of the coastline depends on how the coastline is measured. Mathematician Benoit Mandelbrot elaborated the paradox this way in his 1973 book *The Fractal Geometry of Nature*:

> When a bay or peninsula noticed on a map scaled to 1/100,000 is reexamined on a map at 1/10,000, subbays and subpeninsulas become visible. On a 1/1,000 scale, sub-subbays and sub-subpenninsulas appear, and so forth. Each adds to the measured length.[27]

In a study of fractals applied to legal questions and lawsuits, David Post, a professor at the Beasley School of Law, Temple University, and his coauthor Michael Eisen at the Lawrence Berkeley National Laboratory, and the Department of Molecular and Cellular Biology, University of California, Berkeley, made an interesting comment about the fractal methodology for measuring coastlines. Post and Eisen noted that "true fractal objects" are "those for which (by definition) estimated length $L(\varepsilon)$ never converge." In Euclidian geometry, for instance, the length of the circumference of a circle is estimated by multiplying the diameter of the circle by π. As we have just seen, the answer will not be a perfectly precise number because π, 3.14, is an irrational number. But the point Post and Eisen wanted to make about fractal math is that in Euclidian geometry, the measurement of the circumference converges on the value of the circle's diameter multiplied by π. In Euclidian geometry, our measurement value will converge on the measurement estimated by the formula $C = \pi d$, where C is "circumference" and "d" is diameter, as the measuring rulers get smaller and smaller. Post and Eisen stated the point as follows: "Our estimating procedure will—must—converge on this value as our rulers get smaller and smaller; we may need an infinite number of infinitesimally small rulers to get it exactly correct, but Euclidian geometry is premised on the notion we can do just that." But true fractal objects, they point out, by contrast are those for which (by definition) estimated length L(ε) never converges. "Fractals appear to get longer and longer as the measuring stick gets smaller and smaller, and the estimated length of a true fractal diverges to infinity as ε approaches zero," they explained.[28]

Stated another way, a principle of fractal math is that measurement totals change as the dimension of the measurement is reduced. A map of the Dover coast on the English Channel gives one measure for the coastline length that differs from the measurements made walking along the Dover coast. Gleick explained the point as follows: "But Mandelbrot found that as the scale of measurement becomes smaller, the measured length of a coastline rises without limit, bays and peninsulas revealing ever-smaller subbays and subpeninsulas—at least down to atomic scales, where the process does finally come to an end. Perhaps."[29]

We are back to infinity again, and the same problem applies. We can never get a precise measurement because of the limitations in our measuring technology, whether the measuring technology is an ordinary ruler or an electron microscope. Thus, all our weather measurements are limited by the measurement instruments we utilize. Temperature measurements in Celsius or Fahrenheit are crude measurements compared to nature. Again, nature needs no ruler or any other measurement instrument to operate the weather. Our climate models are doomed to suffer from slight differences in initial states because the values we enter for critical variables like temperature, wind speed, etc. are inherently imprecise because of the coastline paradox in fractal mathematics.

So, as Lorenz demonstrated, weather prediction (and hence prediction of the climate) is impossible for several reasons. First, the environmental variables that need to be in a climate model approach being an infinite set. Second, the interaction between the independent and dependent variables in the computer model is inherently nonlinear. Third, the measurement of essential variables will involve irrational numbers, which inherently involve an infinite number of decimal place calculations. Fourth, the coastal paradox problem in fractal math means the measurement of weather variables generates results that approach infinity as our measuring instruments get increasingly fine, approaching zero, i.e., as we operate at increasingly more detailed levels of observation, down to the subatomic level. Nature suffers none of these problems. Weather and climate do not need human measurements or computer calculations to operate on Earth.

Lorenz understood from these calculations that the weather system lacks predictability that no computer can overcome, regardless of how powerful. What Lorenz demonstrated is that chaos is an inherent feature of the weather system. "Chaos wipes out every computer," the reference book on chaos and fractals insists. "The fact is that no matter how small a deviation in the starting values we choose, the errors will accumulate so rapidly that after relatively few steps the computer prediction is worthless."[30]

Yet, when Lorenz modeled the equations in his 1963 paper, he found the flow patterns of his twelve equations were not entirely random. Instead, seen in three-dimensional space, his observations graphed around two separate but related points that graphically look like a butterfly's wings. The two points around which the observations clustered are known as "Lorenz Strange Attractors." The point is that the calculations remained unpredictable, even though the calculations tended to cluster around two distinct points on the graph.

Some comments made by Steven Koonin, the former undersecretary for science in the Energy Department under the Obama administration, helps make the point. "Climate is not weather," Koonin stressed in his 2021 book *Unsettled*. "Rather, it's the average of weather over decades, and that's what climate models try to describe."[31] What Lorenz proved with the climate model he ran in his MIT office, on an early Royal McBee LGP-30 digital computer, was not that weather is random but that it is not predictable. The point is that even extreme weather in the form of severe floods, hurricanes, tornadoes, etc., is a natural feedback mechanism Earth uses to distribute heat around the planet. But usual weather features, including sunny days, rain, snow, etc., and extreme weather events like floods, hurricanes, and tornadoes are recurring weather patterns, even if not entirely predictable. In chaos theory, weather events are the strange attractors that constitute what we call climate.

A seven-day weather forecast can accurately predict the weather about 80 percent of the time, and a five-day forecast can accurately project the weather approximately 90 percent of the time.[32] But the farther out the prediction, the less reliable the weather forecast will be. A ten-day forecast is at the "practical predictability limit,"[33] right only about half the time. James Gleick understood this point when writing his 1987 book *Chaos*. Gleick imagined the earth covered with sensors spaced one foot apart, rising at one-foot intervals to the top of the atmosphere. He further assumed every sensor gave perfectly accurate estimates of every variable a meteorologist would want, including temperature, pressure, and humidity.

"Precisely at noon an infinitely powerful computer takes all the data and calculates what will happen at each point at 12:01, then 12:02, then 12:03....," Gleick imagined. "The computer will still be unable to predict whether Princeton, New Jersey, will have sun or rain on a day one month away," Gleick explained.[34]

Lorenz's scientific papers make clear that the inability to predict the weather accurately will not go away when our computers become quantum computers. The limits Lorenz identified do not derive from relativity theory. As weather observational technology advances, we will have an increased ability to predict the weather for the next few days. But even with highly advanced weather observational technology and quantum computers, the complexity of climate models, as demonstrated by Lorenz, means weather forecasting will never be accurate except for very near-future periods. As a rule of thumb, the best prediction for tomorrow's weather still is to know what the weather is doing today. But the fact that seasonal weather will change Earth's temperature necessitates the conclusion that the methodology of predicting tomorrow's weather from today's weather only works in the short term.

Those few IPCC climate experts who retain some shred of intellectual honesty and academic integrity have been willing to acknowledge the folly of IPCC climate models predicting CO_2 disasters from the burning of hydrocarbon fuels. In June 2007, James Renwick, a professor in the School of Geography, Environment, and Earth Sciences at the Victoria University of Wellington and a top IPCC scientist, stated on New Zealand Radio that "[t]he weather is not predictable beyond a week or two. He admitted the IPCC climate models were not reliable. "Climate prediction is hard, half of the variability in the climate system is not predictable, so we don't expect to do terrifically well."[35] Another high-profile IPCC lead author, Dr. Kevin Trenberth, a distinguished scholar at the National Center of Atmospheric Research (NCAR) in Boulder, Colorado, admitted IPCC climate models were merely "story lines." In a 2007 "Predictions of Climate" blog post appearing in the science journal *Nature*, Trenberth admitted the following: "None of the models used by IPCC are initialized to the observed state and none of the climate states in the models correspond even remotely to the current observed climate. In particular, the state of the oceans, sea ice, and soil moisture has no relationship to the observed state at any recent time in any of the IPCC models." In that blog Trenberth was particularly blunt. "In fact there are no predictions by IPCC at all. And there never have been," he wrote. "The IPCC instead proffers 'what if' projections of future climate that correspond to certain emissions scenarios."[36] IPCC reviewer and climate researcher, Vincent Gray, now deceased—the founder of the New Zealand Climate Science Coalition, a reviewer of every single draft of the IPCC reports going back to 1990 through 2014, author of more than one hundred scientific publications, and author of *The Greenhouse Delusion: A Critique of "Climate Change 2001"*[37]—declared in 2007 that "[t]he claims of the IPCC are dangerous unscientific nonsense."[38] Gray went on to claim that no IPCC climate model "has ever been properly tested, which is what 'validation' means, and their 'projections' are nothing more than the opinions of 'experts' with a conflict of interest, because they are paid to produce the models." Gray went on to say, "There is no actual scientific evidence for all these 'projections' and 'estimates.' It should be obvious that they are ridiculous."[39]

Yet, the IPCC insists on understanding Earth's weather and climate in linear terms, targeting CO_2 as the sole variable responsible for global warming. That the IPCC can confidently say that continued burning of hydrocarbon fuels will mean Earth's temperature will rise more than 1.5°C above preindustrial levels by 2030 or 2050 is nonsense. In a nonlinear climate system, any number of nonpredictable events may occur. The sun might flare or otherwise increase activity. Equally possible, the sun might already be entering a new minimum period. In time-series analysis, the variations of temperature recorded over a few years are too short a period to establish a new Earth temperature phenomenon statistically. In geological time, time-series climate analysis requires hundreds of thousands or even millions of years of accurate data to be meaningful.

The gases and dust a volcano throws into the atmosphere can have a global cooling effect regardless of how much CO_2 there is in the atmosphere. Thus, a cataclysmic event such as a massive volcanic eruption or a series of cataclysmic events such as a series of volcanoes erupting might dramatically cool Earth even if CO_2 concentrations were at prehistoric levels. Between June 12 and June 15, 1991, the sulfur dioxide emitted by the eruption of Mt. Pinatubo in the Philippines lofted an ash plume more than twenty kilometers into the atmosphere, throwing fifteen million tons of SO_2 into the stratosphere. The SO_2 reacted with water vapor to create upper atmosphere aerosols that reflected incoming sunlight. The Pinatubo eruption increased aerosol optical depth in the stratosphere by a factor of ten to one hundred times normal levels as measured before the eruption. Aerosol optical depth is a measure of how much light airborne particles prevent from passing into the atmosphere. As a result, the eruption of Mt.

Pinatubo caused Earth's mean surface temperature to decrease by about 0.6°C (1.0°F) over the next fifteen months.[40]

An extraterrestrial event may occur, such as another Chicxulub-size asteroid hitting Earth. Unusually high supernova activity could bombard Earth with cosmic rays. Recent scientific studies of supernovae and cosmic rays have stressed that historical and archeological evidence of global warming and cooling that occurred long before the Industrial Revolution "require natural explanations." Danish physicist Henrik Svensmark, whose work on supernovae we reviewed extensively in chapter 5, published a paper in 2015 in which he concluded that since "supernova variation reflects variation in star formation and the morphology of our Galaxy's spiral arms, one ends up with the surprising result that the conditions for life on Earth are a reflection of the shape of the Milky Way." Svensmark concluded that in geological time, supernova activity affected bio-productivity occurring in the cold intervals associated with the star-forming regions, regardless of how much CO_2 was in the atmosphere. "Biodiversity and bio productivity [as measured by studies of $\delta^{13}C$, i.e., the carbon-13 isotope] all appear so highly sensitive to supernova in our Galactic neighborhood that the biosphere seems to contain a reflection of the sky."[41] Or Earth's climate thermometer might simply continue to adjust the CO_2 level on Earth such that the interglacial period we are experiencing could extend another 10,000 or more years.

But Gleick understood the idea that weather and, consequently, the climate are inherently unpredictable would go against the grain of experienced meteorologists. One of Lorenz's oldest friends was Robert White, a fellow meteorologist at MIT who later became head of the National Oceanic and Atmosphere Administration. Lorenz told him about the Butterfly Effect. "White gave von Neumann's answer. 'Prediction, nothing,' he said. 'This is weather control.'" Gleick explained White's thought was that small modifications, well within human capability, could cause desired, large-scale changes. For White, small, unpredictable weather outcomes did not change the reality that weather follows a deterministic logic where climate processes can be modeled precisely such that human intervention could yet have outcomes that were predictable according to the climate laws that were fully understood. Lorenz not only disagreed, he felt White's conceptual weather paradigm was outmoded. Lorenz explained to White that we human beings can change the weather, but you would never be able to predict the outcome. "Yes, you could change the weather," Lorenz told White. "You could make it do something different from what it would otherwise have done. But if you did, then you would never know what it would otherwise have done," Lorenz explained. "It would be like giving an extra shuffle to an already well-shuffled pack of cards. You know it will change your luck, but you don't know whether for better or worse."[42] The Butterfly Effect, the logic of nonlinear equations, and the impact of fractal math were the reasons why mathematically based computer-driven climate models are doomed to failure, why long-term weather predictions are rarely correct, and why human climate interventions suffer from the laws of unintended consequences. "For small pieces of weather—and to a global forecaster, small can mean thunderstorms and blizzards—any prediction deteriorates rapidly," Gleick noted, explaining the Butterfly Effect. "Errors and uncertainties multiply, cascading upward through a chain of turbulent features, from dust devils and squalls up to continent-size eddies that only satellites can see."[43]

In his book *The Essence of Chaos*, Lorenz understood that our desire to control the climate is not founded on rational thinking. He wrote the following:

> We can readily disturb the existing weather, perhaps violently by setting off an explosion or starting a fire, or more gently by dropping crystals of dry ice into a cloud—or perhaps even by releasing a butterfly—and we can observe what will happen, but then we shall never know about what would have happened if we had left things alone.[44]

The unpredictability factor in Lorenz's chaos theory of weather was not limited to the Butterfly Effect. Yes, the mathematics of measurement made it impossible to avoid minor differences from nature in the starting values of the computer climate model. But there is also the Strange Attractor Effect that means no two weather events—no two rainstorms or snowstorms, no two hurricanes or tornadoes, no two volcanoes or asteroids hitting Earth—no two predictably operate the same. Yet, we can identify rainstorms, snowstorms, floods, hurricanes, tornadoes, and a host of other weather events as repeating patterns. That is how weather patterns form what we call "climate." But Lorenz was brilliant in using nonlinear differential equations in his original 1963 paper.

In a nonlinear equation, a change in one variable can produce various changes in other variables as the model is applied at different times or even as the model goes through multiple iterations in the analysis at hand. A nonlinear climate system is dynamic. To understand this point, we have to begin by explaining how linear equations work. That will set the stage for understanding how nonlinear equations are different.

In a linear mathematical model, we expect initial conditions, symbolized as I_1, to produce predictable initial results, represented as R_1. We go to the following period, such as $I_2 \rightarrow R_2$, with subsequent periods represented as $I_3 \rightarrow R_3$, and so forth through all model iterations. An example demonstrating why the IPCC logic regarding their case for global warming should make clear the IPCC predicates its case on linear equations. So, we assume that in I_1, we burn hydrocarbon fuels, with the result that in R_1, we get increased CO_2 emitted into the atmosphere. Then in I_2, the CO_2 in the atmosphere becomes a greenhouse gas, with the result that in R_2, we get an increase in global temperatures, i.e., global warming. The logic of global warming hysterics is that burning hydrocarbon fuels adds CO_2 to the atmosphere in at least an arithmetic manner, such that the CO_2 emitted in I_1 plus I_2 results in R_1 CO_2 plus R_2 CO_2. In other words, CO_2 builds as a percentage of the atmosphere arithmetically as we continue to burn hydrocarbon fuels. Similarly, global warming hysterics believe the global warming produced by the CO_2 is also an additive process, such that more CO_2 admitted increases global temperatures arithmetically.

Now let's apply the logic of nonlinear equations to demonstrate the logic flaw in the IPCC global warming model. Suppose Lorenz is correct and global climate, as is the case with global weather, is a nonlinear process. In a nonlinear mathematical model, we expect initial conditions, symbolized as I_1, to produce predictable initial results, represented as R_1. But when we go to the following period, we do not expect to get the same effect from the exact same cause. Thus, in a nonlinear model, instead of getting $I_2 \rightarrow R_2$, we get $I_2 \rightarrow R_x$, with "x" being an unpredictable outcome different from the expected R_2. Subsequent periods represented in the nonlinear model will produce $I_3 \rightarrow R_y$, and so forth through all model iterations. A nonlinear model is what screenplay author Winston Groom had in mind when Forrest Gump contemplated a box of chocolates. Forrest Gump explained, "My mama always said, 'Life was like a box of chocolates. You never know what you're gonna get.'"

Given the logic of nonlinear mathematical models, we can no longer assume that burning hydrocarbon fuels adds CO_2 to the atmosphere in an arithmetic manner. In a nonlinear model, I_1 may add R_1 CO_2 to the atmosphere. But in I_2, the amount of CO_2 emitted may be the same as in I_1, but the amount of CO_2 added to the atmosphere will be R_x, a different amount of total CO_2 in the atmosphere than what would be expected from an additive process. Why does this happen? The answer is that because the weather and the climate are dynamic systems. As I_1 transitioned to I_2, Earth may have begun absorbing more of the emitted CO_2 into the oceans or into Earth itself. We have noted earlier that weather events on Earth are part of a climate thermometer built into its weather/climate systems. The main purpose of weather events and climate patterns on Earth is to distribute heat as evenly as possible around the planet.

Please do not assume Earth is trying to maintain a temperature suitable or comfortable for human life. Earth, quite frankly, does not necessarily care about human beings. Remember, we have seen five near-total extinctions of life on Earth in geological time. To the planet, human beings are just another life form here for the time being. Earth's weather/climate system operates to distribute heat around the globe in an equal manner regardless what creatures happen to be living on the planet at any given time in geologic history. There are limits to how successful Earth is in distributing heat because of various events outside Earth's ability to control by the override and distribution functions built into Earth's naturally operating weather/climate thermostat. During both glacial and interglacial periods, for instance, Earth's weather/climate thermometer functions to distribute heat as evenly as possible around the globe. Even when Earth experiences an ice age, the planet's weather/climate system still functions to distribute heat around the planet. But when Earth is in the grip of an ice age, Earth temperatures are obviously at a lower average level than when Earth is in an interglacial warming period.

In a nonlinear model, global warming theorists are incorrect to assume CO_2 is also an additive process, such that more CO_2 admitted increases global temperatures arithmetically. By thinking that continued burning of hydrocarbon fuels will increase Earth's temperatures by more than 1.5°C above preindustrial levels by 2030 or 2050 at the latest, the IPCC makes the mistake of assuming Earth's climate is a linear system and that CO_2 emissions into the atmosphere are additive. The truth is that we do not know why CO_2 levels on Earth are where they are today, other than to conclude that CO_2 levels are where Earth's thermometer regulates CO_2 to be. As we have noted, the planet is currently in an interglacial period. Earth has no emotional reaction regarding whether it is good or bad to be in a glacial or an interglacial period. Similarly, Earth has no value position on whether having more or less CO_2 in the atmosphere is right or wrong morally. But the dynamic climate Earth has, the nonlinear mathematics of Earth's weather/climate system, can operate equally well with different results even when outside forces (like, for instance, variations in solar activity or intergalactic supernova activity) force dramatic warming or cooling changes in the overall average temperature of the planet. The bottom-line driving force result of higher CO_2 content in the atmosphere is today that plant growth is stimulated and Earth is greener. Why? Because we are in an interglacial warming period and there is more solar irradiance hitting Earth's surface to mix with more CO_2 in the atmosphere. These are the exact conditions needed to stimulate plant photosynthesis processes. If we were in a glacial period, higher atmospheric CO_2 concentrations would not equally stimulate plant growth because there is less TSI in an ice age.

Thus, the interaction of essential weather variables like temperature or CO_2 in the atmosphere can produce unpredictable reactions in other variables. Here is how Lorenz described the phenomenon:

> There remains the reasonably well established observation that weather variations are not periodic. Of course they have periodic components, the most obvious ones being the warming and cooling that occur with the passage of the seasons of the year or the hours of the day. Careful measurements have also detected weak signals with a lunar period, probably gravitational effects, and there is virtually no limit to the number of periods that investigators have *claimed* to have discovered. Some of these have been stated to several decimal places. Nevertheless, if we take an extended record of temperature or some other weather variable and subtract out all verified or suspected periodic components, we are left with a strong irregular signal. Migratory storms that cross the oceans and continents are still present in full force. These are presumably manifestations of chaos.[45]

Lorenz finally concluded the atmosphere itself is chaotic. He summed up by saying the following: "For one special complicated chaotic system—the global weather—the attractor is simply the climate, that is, the set of weather patterns that have at least some chances of occasionally occurring."[46] Earth's atmosphere permits it to have both weather and climate, with the climate being nothing more than weather patterns. First, the complexity of the planet's weather/climate systems is as such that there are so many different acting variables, including oceans, clouds, water vapor, etc. But also, the complexity of the systems are nonlinear, such that even the same weather/climate configurations of independent variables do not produce predictable outcomes of dependent variables. For example, the same factors that create a hurricane in the Atlantic Ocean at one time may not create a hurricane in the Atlantic in a different period. Why? The answer is because some other independent variable interacted. Perhaps the Atlantic currents were warmer or colder in one or other periods, and the ocean temperature affected the cloud and wind factors required to create a hurricane.

The consequences of Lorenz's work were to produce a paradigm shift in our understanding of weather and climate. The consequences of that paradigm shift are enormous. Yes, the governments of the world could force decarbonization according to the dictates of the IPCC. The governments of the world could even go so far as to criminalize the burning of hydrocarbon fuels. Such policy decisions would have an impact on Earth's weather and climate. But to assume we can accurately predict the outcome of decarbonization is ridiculous. IPCC adherents assume decarbonization will reduce CO_2 atmospheric concentrations and keep global temperatures from getting warmer. But it may not turn out that way, regardless of how many Hockey Stick graphs Michael Mann fabricates.

Before we move on, we need to stress one additional point. Lorenz's analysis also allows us to understand that extreme weather events like floods, hurricanes, and tornadoes are not only not entirely predictable but also not wholly avoidable. Extreme weather events are part of Earth's safety-valve mechanism for taking the drastic measures needed to regulate the planet's temperature, as dictated by the operation of its nonlinear climate system. Global warming climate models are typically doomed to produce predetermined results built into the model by the presumptions and prejudice of the climate scientists who created the model. Given the complexity of variables in Earth's climate system and the nonlinear manner in which the independent and dependent variables interact, it is hard to imagine that a computer model could ever predict Earth's weather or climate accurately. To most people, that result will appear counterintuitive. But when we realize that unpredictability is an inherent feature of nonlinear mathematical models, we can appreciate why the attempt to model the weather or the climate on a computer is futile, no matter how powerful the computer might be.

In August 2021, Ross McKitrick, the economics professor at the University of Guelph in Ontario whose criticism of Michael Mann's statistical methodology we reviewed extensively in chapter 6, published an important article in *Climate Dynamics*.[47] McKitrick, an economist specializing in environmental policy with an expertise in statistical analysis, demonstrated that studies conducted by Myles Allen, Simon Tett, and others following in their footsteps committed classic methodological errors that invalidated their conclusions.[48] The mistakes amounted to taking a "big smudge of data" and concluding "the fingerprints of greenhouse gas are on it," McKitrick explained. "When you do a statistical analysis, it's not enough just to crunch some numbers and publish the result and say, 'This is what the data tell us,'" he insisted. "You then have to apply some tests to your modeling technique to see if it is valid for the kind of data you are using."[49] The statistical technique in question, known as "optimal fingerprinting," is used by law enforcement officers to identify patterns of criminality. "Fingerprinting" is also used by social media Internet competitors who seek to profile people, for example, by tracking Internet search engine activity. Optimal fingerprinting statistical methodology is highly complicated. Optimal fingerprinting utilizes the statistical methods of Generalized Least Squares (GLS) regression analysis to discern patterns, applying a "Residual Consistency Test" (RCT) to check the GLS specification. McKitrick explained that global warming theorists have used optical fingerprinting techniques to link atmospheric CO_2 to everything from global temperature and forest fires to precipitation and snow cover.[50] In his 2021 scientific paper, McKitrick wrote the following:

> Their methodology [i.e., the methodology used by Allen, Tett, and others] has been widely used and highly influential ever since, in part because subsequent authors have relied upon their claim that their GLS model satisfies the conditions of the Gauss-Markov (GM) Theorem, thereby yielding unbiased and efficient estimators. But the AT99 [1999 paper published by Myles Allen and Simon Tett][51] stated the GM Theorem incorrectly, omitting a critical condition altogether, their GLS method cannot satisfy the GM conditions, and their variance estimator is inconsistent by construction. Additionally, they did not formally state the null hypothesis of the RCT ["Residual Consistency Test"—a measure used by Allen and Tett in AT99] nor identify which of the GM conditions it tests, nor did they prove its distribution and critical values, rendering it uninformative as a specification test.[52]

Again, in nontechnical terms, the Gauss-Markov (GM) theorem describes a statistical method for checking for bias in the data analysis.[53] McKitrick demonstrated that Allen and Tett, in their seminal 1999 paper, had failed to perform the required statistical tests. Thus, McKitrick argued that Allen and Tett developed a mathematical model that presumed as true what needed to be proven true. McKitrick argued they had created another tautology that was meaningless in the attempt to prove greenhouse gasses cause climate change.[54] Allen responded that a scientific paper he published in 2003 superseded the statistical methodology used in the 1999 paper.[55] McKitrick responded by pointing out that both Allen and Tett's 1999 paper and Allen's 2003 paper both traced back to a 1977 study where the methodological error regarding the Generalized Least Squares (GLS) regression analysis originated.[56]

To experts who understand the complicated statistical methodologies involved, McKitrick's analysis deconstructing optimal fingerprinting as applied to climate modeling was devastating. The various mathematical problems demonstrated in this chapter call into question the validity of any and all climate models to accurately predict the weather or the climate. In the earlier chapter 6 discussion of McKitrick's critique of Michael Mann, McKitrick explained why global warming theorists like Michael Mann distort legitimate historical climate data to argue their case. McKitrick's point here is subtly different. Another fatal error, as McKitrick demonstrated in this discussion, occurs when global warming theorists load their climate models with research and statistical methodologies inappropriately applied. Given the complexity of weather and climate on Earth, the most brilliant scientists, like Lorenz, abandon the linear thinking and single-factor explanations of those who bend science to demonize CO_2. What McKitrick so effectively demonstrates is that IPCC global warming scientists use the veneer of mathematical climate models to hide the reality that they presume the conclusion their climate models can never prove: namely, that CO_2 is the only global warming culprit about which we need to worry. As this chapter has demonstrated, Earth's weather and climate are complex, nonlinear, multivariate systems. Climate models constructed to demonize CO_2 as a single variable responsible for global warming are therefore inherently suspect, regardless how complicated their statistical methods may be.

The Unknown Precambrian Era

If we presume Earth is some 4.6 billion years old, over 80 percent of its history and possibly as much as 88 percent is mainly unknown. Geologists consider that early Earth experienced three eons: the Hadean, the Archean, and the Proterozoic. These three eons constitute the Precambrian Era, which lasted from 4.6 billion years ago until the Cambrian Period, some 540 million years ago. Stephen Marshak's 2019 college geology textbook that we referred to earlier in the previous chapter, *The Essentials of*

Geology, explains, as follows, the importance of the Cambrian Period that began the Paleozoic Era:

> The succession of fossils preserved in strata of the geologic column defines the course of life's evolution throughout Earth history. Simple bacteria and archaea appeared during the Archean Eon, but complex shell-less invertebrates did not evolve until the late Proterozoic. The appearance of invertebrates with shells defines the Precambrian-Cambrian boundary. At this time, a sudden diversification in life, with many new types of organisms appearing over a relatively short interval. Geologists refer to this event as the Cambrian explosion. Progressively more complex organisms populated the Earth during the Paleozoic.[57]

Geologists believe that during the Proterozoic Eon in the Precambrian Era, solid blocks of Earth's crust formed and sutured together to form continents and eventually supercontinents.[58] The subcontinents have been given different names. The Laurentia is a subcontinent believed to have consisted of North America and Greenland. Gondwana is another subcontinent that is considered to have included South America, Africa, Antarctica, India, and Australia.[59] About 250 million years ago, during the Late Paleozoic, geologists believe these subcontinents fused to form a giant subcontinent known as Pangaea. The subcontinent Pangaea was believed to have a northern part composed of Laurasia (North America, Europe, and Asia) and a southern part, Gondwana. At the end of the Paleozoic, geologists believe the "super ocean" Panthalassa surrounded the supercontinent Pangaea.

Alfred Wegener, a German who was born in Berlin in 1880, was a lecturer in meteorology and astronomy at the University of Marburg, Germany, when he published a paper in 1912 entitled "Die Entstehung der Kontinente" ("The Origin of the Continents.") In 1915, Wegener published a book-length treatment of the subject entitled *Die Entstehung der Kontinente und Ozeane* (*The Origin of the Continents and Oceans*). Wegener's theory of continental drift proposed that in the Triassic Period at the beginning of the Mesozoic Age, about 250 million years ago, Pangaea, a single landmass that included all the continents, began to break up. Wegener introduced the concept *Kontinentalverschiebung*, or of "continental drift," arguing that after the breakup of Pangaea, the continents moved to their current position, "closing oceans ahead of them, and opening oceans behind them."[60]

Wegener's continental drift concept led to a series of investigations in the twentieth century that developed a set of concepts known today as plate tectonics.[61] The idea is that Earth's outermost layer, the lithosphere, consisting of its crust and upper mantle, is broken into large plates. These plates lie on top of a partially molten layer of rock, the asthenosphere. Due to the convection of the asthenosphere and the lithosphere, the plates move relative to each other at different rates. The movement of the tectonic plates bumping into each other is considered responsible for orogens, the process of building mountains.[62] A convergent boundary forms when two plates, at least one of which is an oceanic plate, move toward each other. Rather than banging together, as is assumed in orogens, one oceanic plate is presupposed to bend, sinking down beneath the other plate, in a process geologists label "subduction." Convergent boundaries are known as subduction zones. Stephen Marshak's textbook explains this process as follows:

> Because subduction at a convergent boundary "consumes" old ocean lithosphere, geologists also refer to convergent boundaries as *consuming boundaries*, and because they are delineated by deep-ocean trenches, they are sometimes simply called *trenches*. The amount of oceanic plate consumption worldwide, averaged over time, equals the amount of sea-floor spreading worldwide, so the surface of the Earth remains constant over time.[63]

One of the fundamental concepts in continental drift and plate tectonics theories is that Earth has maintained the same size through geological history. Polish geologist Stefan Cwojdzinski noted "the idea that the planet was of a constant size was linked in a logical way to the nineteenth century concept of uniformitarianism."[64] Australian geologist James Maxlow explained that the process of subduction is an indispensable part of the continental drift and plate tectonic theories. Maxlow wrote:

> The tectonic plates are composed of two types of crust: thick continental crusts and thin oceanic crusts. One of the main points the theory proposes is that an equal amount of surface area of the plates must disappear into the mantle along the convergent boundaries by a process referred to as "subduction," more or less in equilibrium with the new oceanic crust that is formed along the divergent margins by seafloor spreading. This is also referred to as the "conveyor belt principle." In this way, it is assumed, but never acknowledged as a basic premise, that the total surface of the

Earth remains constant and hence the radius of the Earth also remains throughout time.[65]

Ironically, Wegener's theories began with cataclysmic events involving massive unexplained forces that caused Pangaea to break up and end with an idea at the heart of uniformitarianism. Despite the breakup movement of continents drifting to their positions today, adherents of continental drift theories insist Earth was the same size when Pangaea broke up at the end of the Paleozoic Age that it is today. Proponents of continental drift imagine the continents have just moved apart, shifting around oceans that were there in Precambrian time.

That Africa and South America appear to fit together is evident from even a cursory study of a global map. We can easily imagine all the other continents fitting together by examining the map with that observation as the starting point. When first introduced, the theory of continental drift drew ridicule in large part because Wegener could not explain how or why Pangaea split apart or how or why the continents began to drift apart. Although the forces required in breaking up Pangaea or drifting the continents had to be massive, Wegener just presumed these processes had to have happened because, on a map, the continents fit together.

Yet, there is at least one fundamental question in physics the continental drift theory leaves unanswered. Earth continues spinning because of inertia. In the vacuum of space, objects maintain their momentum and direction, including their spin, because no force intervened to stop them.[66] But would Earth, with all the land masses in one gigantic Pangaea mass and all the oceans separate in one big liquid Panthalassa, be sufficiently balanced to spin evenly? Oceans cover 70 percent of Earth's surface. If 30 percent of the planet consisted of one big landmass continent and 70 percent were water, would this constitute an uneven distribution of mass around the axis of rotation? Unbalance involves an unequal distribution of mass, causing the mass axis to differ from the bearing axis.[67] A rotor or a sphere that is unbalanced will cause vibration. Earth is not perfectly spherical, and, therefore, the sun exerts a torque on Earth's spin. The result is a slow shifting of Earth's axis, known as the precession of the equinoxes.[68] We will get to this later in the chapter when discussing Milankovitch cycles and the climate.

The point here is that while water covers 70 percent of Earth's surface, water represents only 0.05 percent of Earth's mass.[69] In chemical terms, the molar mass of water is 18.0153 g/mol, with a density of 997 kg/m^3.

We can assume a breakup of Pangaea into continents involved bedrock, not sedimentary rock. The molar mass of granite is 483.37 g/mol, with a density of 1.656 g/cm³. Thus, before Pangaea broke apart, Earth was 70 percent water with a significantly lower per unit mass and 30 percent land with a considerably higher per unit mass. Wegener's continental drift theory presumes that Earth, with an uneven distribution around the globe of landmass and oceans, would not begin vibrating so wildly that Earth would spin entirely out of orbit. Today, the seas divide the continents reasonably evenly. Did that just happen out of chance?

The Expanding Earth Theory

Samuel Warren Carey was an Australian geologist born in Campbelltown, New South Wales, in 1911. Early in his career, he was an advocate for Wegener's theory of continental drift. As he matured, he rejected Wegener's ideas in favor of what today is known as the "expanding Earth" theory. Here, in his 1976 book, *The Expanding Earth*, is how Carey described Wegener's ideas. Carey credited Wegener for breaking from the uniformitarianism that remains a geological presupposition even today:

> Wegener took the first big unorthodox step by concluding that at the time in question all the main land masses of the world had been a single supercontinent, Pangaea, which had disrupted into separate blocks that drifted apart like great tabular icebergs, opening the Arctic, Atlantic, and Indian Oceans and reducing the area of the Pacific by a like amount. All this in the latest twentieth of the earth's life![70]

In 1924, Carey got ahold of an English-language copy of Wegener's 1925 book. Reading that book initially convinced him that plate tectonics was a breakthrough concept. Yet, he did not as readily accept Wegener's ideas of Pangaea or continental drift. Carey correctly observed that from the 1930s until the early 1950s, Wegener's views "were generally reflected as a fantasy—fascinating but false." [71] He acknowledged Wegener's point that the contents could be "fitted together, jigsaw fashion, with remarkable success, into a supercontinent that he named Pangaea."[72] He also accepted that Wegener had proved that if "the continent's shapes had been cut from a newspaper, then reassembled, the print-lines should read across." [73] He granted that Wegener demonstrated paleontologically that vast oceans now separate closely related land plants and animals. Carey also acknowledged that Wegener had shown that similar strata sequences are separated, and progressive character trends continue across the join when reassembled.

The Chaos Theory of Climate

But by 1931, Carey had developed oblique stereographic projections of the global map that allowed him to display how the continents fit together on a round sphere considerably smaller than the current size of Earth. The bends, dislocations, and stretches he observed when viewing the continents together to form Pangaea went away once he mapped his stereographic projections of the continents onto his smaller Earth sphere. Carey concluded that the continents had once been joined, but not in a separate supercontinent; Pangaea surrounded by water. He had determined that the continents were the entire surface of a smaller planet Earth. In other words, Pangaea was Earth's surface before expansion.

"I knew by 1938 that if you straightened the obvious bends, restored the visible stretches, and reunited the dislocations, these processes alone reproduced a Pangaea essentially the same as Wegener had deduced from wholly different grounds," Carey explained. In his doctoral work, Carey studied oroclines, which are bends seen curved in a two-dimensional map-view showing a mountain bent or buckled around a vertical axis of rotation after the mountain was formed. His study of oroclines allowed Carey to appreciate how two-dimensional map-view distortions require shifting ground to see the phenomenon as a three-dimensional geological event occurring on a spherical Earth. "Most of what I published in the 1954 orocline paper was in the 1937 draft of my doctoral thesis, but had to be omitted at the eleventh hour because I realised that these concepts were too radical for acceptance then, and would have cost me my degree," he continued. "Even in 1954, publication was refused by referees of the Geological Society of Australia."[74]

Carey did additional research with the advent of paleomagnetism from the late 1940s to the early 1950s. He became convinced that paleomagnetic analysis had established "that the continents have separated and have relatively rotated to the extent claimed by Wegener."[75] But then, in the 1970s, as orthodox geologists began accepting plate tectonics, Carey made a final breakthrough. He developed an explanation for why the continents appeared to fit together that was compatible with plate tectonics but not with the version of continental drift Wegener had proposed. Carey explained that in the 80 percent of geological time that remains largely unknown, the continents split apart dramatically when Earth increased in size. Carey saw the earlier, smaller Earth as being without surface water.

He also rejected Wegener's idea that the continents splitting apart divided the seas. Carey accepted that tectonic plates formed when an expanding Earth broke apart the smaller crust to separate the continents.

But, as he rejected Wegener's ideas on the formation of the oceans, he also dismissed the plate tectonic idea that subduction was involved in ocean formation. "According to my views right from the thirties the ocean floors, except for part of the Pacific, had to be newly developed crust, risen from the mantle step by step as the continents moved apart," he wrote.[76]

Carey argued that the smaller Precambrian Earth had no oceans. Consider this paragraph from his book, *The Expanding Earth*:

> Most take it for granted that there have always been large oceans like those of today—at least since very early times. Such uniformitarianism seemed axiomatic. But is it? It is now known that *all* the floors of *all* the oceans, Pacific included, have been formed since the Paleozoic, so it is assumed that equivalent other ocean areas have been "consumed." Maxwell's demon is needed again at the helm to ensure so clean a sweep that no stable remnant of old ocean remained anywhere! Oceans like the Arctic, Atlantic and Indian, in any case, date only from the Mesozoic and have doubled their area since the Eocene. I suggest that the Pacific too was a fraction of its present size before the Mesozoic. Of course, there were extensive ancient seas, but oceans of the modern type are a new phenomenon with an antiquity no longer than the reptiles.[77]

Maxwell's demon is a reference to a thought experiment physicist James Clerk Maxwell performed in 1867. The demon opens and shuts a massless door between two gas chambers to test the second law of thermodynamics. Carey's point was that nobody opened a "door" that got rid of the ocean floors formed in Precambrian time. In subsequent writings, Carey insisted there were no great oceans on Earth's surface before the mid-Mesozoic Era.[78] The mid-Mesozoic was some 170 million years ago in the middle of the Jurassic Period. Carey placed the time of Earth's expansion, during which the oceans first formed, in the age of dinosaurs, a point we will consider more in-depth in the next section of this chapter. While Carey mentioned "extensive ancient seas," he failed to demonstrate geological evidence of their existence. He left open the question of when water first appeared on Earth. But he left no doubt that there was no water on the surface of a pre-expansion Earth.

Carey had concluded that plate tectonics became a factor only after the continents broke apart as Earth expanded. But he saw oceans forming and continents drifting not as the cause of Pangaea's breakup but as the effect of Pangea's breakup. He argued, for instance, that since the Cretaceous, North America has rotated a further 45° counterclockwise.[79] He also claimed the oceans have continued to expand, especially the Pacific, which he did not believe was initially nearly twice the size of the Atlantic Ocean. Many of Carey's writings are devoted to examining Earth's changes, including the ongoing movement of the continents and shifting of the poles.

Another Explanation for the Extinction of the Dinosaurs

Again, we turn to a mathematical concept to understand the argument of this subsection. The mathematical principle is that a difference in size means a difference in the phenomenon.[80]

We can understand the concept if we contemplate a bumblebee. A bumblebee can fly, given the size of the bumblebee and the dynamic construction of its wings. Yet, if a bumblebee is ten times or a hundred times its standard size, the bumblebee can no longer fly, even if we increase its wings proportionately such that the dynamic construction of its wings is still maintained. A giant bumblebee is not the same critter made larger. An enormous bumblebee is a different order of animal that requires different mathematical principles if the bumblebee is to fly in its larger form. In other words, being larger is not the same thing just made bigger. A difference in size typically involves a difference in the phenomenon. The famous Italian genius Galileo Galilei understood this mathematical principle as he made clear in his 1638 book *Discourses and Mathematical Demonstrations Relating to Two New Sciences*. Galileo realized that the skeletal structure of an animal was related to gravity. He argued the following:

> Nature cannot produce a horse as large as twenty ordinary horses or a giant ten times taller than an ordinary man unless by miracle or by greatly altering the proportions of his limbs and especially his bones, which would have to be considerably enlarged over the ordinary.[81]

In a modern-day example, Timothy Paul Smith, a research professor in physics and environmental studies at Dartmouth College, noted that in 1956, a hunter in Angola discovered the most giant savanna elephant ever measured. It was an elephant of the genus and species *Loxodonta africana* that weighed 12,000 kilograms. Smith reasoned that if an elephant were increased in size from three meters at its shoulders to thirty meters, it would weigh 6,000,000 kilograms and be big enough to fill the vault of the Notre Dame cathedral in Paris. An elephant that size would shatter its leg bone that would now be thirty centimeters in diameter.[82]

In his 1994 book *Dinosaurs and the Expanding Earth: Solving the Mystery of the Dinosaurs' Gigantic Size*,[83] British mechanical engineer

Stephen Hurrell proposed how the expanding Earth theory explained why the dinosaurs disappeared. Hurrell argued that Earth, before expansion, had less mass, hence less gravity. He discussed a Reduced Gravity Earth theory that argued animals as gigantic as dinosaurs would only be structurally viable in a reduced gravity environment. Near the end of his life, in the second edition of his book *Earth, Universe, Cosmos*, published in 2000, Carey agreed with Hurrell. In that book, Carey wrote the following:

> Reduced Earth radius with constant Earth mass implies higher surface gravity, but much reduced surface gravity is essential for dinosaurs to have existed. The mass of Earth must have been less. The size of dinosaurs peaked in the Jurassic with *Diplodocus, Brontosaurus*, and flying reptiles like *Quetzalcoatlus*. By the mid Cretaceous *Triceratops* and *Tyrannosaurus rex* were much smaller, although still huge. Oligocene animals were still smaller although very much larger than their modern relatives. Birds also became lighter from heavy-boned *Archeopterix* and the bird-like *Iguanadon* to much lighter modern birds.[84]

Hurrell made an important point when he challenged that the size of dinosaurs in the Jurassic Period required Earth's expansion, given the mathematical relationship between gravity and the structural scale of giant animals like dinosaurs. Hurrell explained the point as follows:

> The effect of gravity on life's scale is a distinct mathematical relationship that affects the basic building blocks of animals—bones, ligaments, muscles and blood pressure. A reduced gravity reduces the force on any animal's bones, ligaments and muscles so they can all be thinner and weaker for a particular scale of life. Blood pressure is also reduced in a weaker gravity since blood pressure is the hydrostatic weight of blood (mass x gravity).

Hurrell continued:

> This implies that the scale of ancient life was shifted toward a larger size in a reduced gravity. The most obvious result of this scale shift is gigantic dinosaurs with masses equal to several elephants but the effects are also plain on smaller animals as well. An elephant-sized dinosaur is noticeably more active and dynamic than any elephant because the dinosaur evolved to live in a reduced gravity.[85]

Hurrell estimated that Earth's gravity 300 million years ago was only 25 percent of today's gravity.[86] Valentin Sapunov, a professor at the Russian

State University of Natural Hazards in St. Petersburg, Russia, published a paper in 2015 entitled "On the nature of gravity and possible change of Earth mass in geological time." In this paper, Sapunov calculated that the size of Earth, during the last one hundred million years, had increased two times in linear scale and eight times in volume and mass.[87] Carey added the following:

> Dinosaur decline was spread over tens of millions of years, after they had dominated the lands for a hundred million years. But most dinosaur orders had reduced severely by Middle Cretaceous but all had vanished by the end of the Cretaceous. Dinosaurs were tropical animals and cooling global climates with narrowing of the humid tropics was a major factor in their extinction. Like every other group of organisms, they were replaced by animals more suitable to the environment, warm blooded mammals.[88]

Referring back to chapter 7, the death of the dinosaurs was a complex phenomenon that involved the cataclysmic impact of the Chicxulub asteroid and the Deccan volcanism, as well as most likely the reduced gravity of a smaller Earth as seen here. The extinction of the dinosaurs began tens of millions of years before the Chicxulub impact, and most likely, Earth expanded significantly as bigger, heavier dinosaurs gave way to still gigantic but smaller species of dinosaurs. Yet, global warming theorists want to argue that CO_2 is the only independent variable affecting climate throughout Earth's history. In the following subsection of this chapter, we will see global warming theorists struggle to explain the ice ages, given the significantly larger concentration of CO_2 that was in the atmosphere in prehistoric times.

But first, let's appreciate how threatening the expanding Earth theory is to orthodox science today. Paolo Sudiro, an Italian geologist who works in the oil industry as a reservoir navigation supervisor at Baker Hughes in Cepagatti, Italy, wrote a 2014 article entitled "The Earth expansion theory and its transition from scientific hypothesis to pseudoscientific belief." Sudiro objected to the expanding Earth theory because he viewed the amount of matter in the universe as a cosmological constant. "Expansionists have not been deterred by their inability to offer a plausible expanding mechanism; instead their failure in finding an explanation within the accepted physical laws had the effect of increasing their reliance on pseudoscientific solutions," Sudiro insisted.[89] In chapter 5, we saw that scientists today have abandoned the tenet of orthodox physics that maintained total solar irradiance (TSI) as a constant value. The idea that total mass in the universe is a constant is another legacy from uniformitarianism. Just as orthodox theory today embraces continental drift and plate tectonics but rejects Earth expansion, human beings are most comfortable believing Earth is a stable place. Carey commented on the creation of mass as follows:

> First, orthodoxy has always assumed that the universe was created with its complete inheritance of matter, which thereafter has remained constant. Likewise, that all the matter in the present solar system was present in the initial gaseous nebula that spawned the sun and its satellites. Similarly it is taken for granted that all the matter of the earth has been inherited from the time of its initial accretion. Each of these cognate assumptions is false: matter is created continuously and spontaneously at all levels.[90]

Reflect on all the dramatic Earth changes in geological time, including five significant extinctions of nearly all life on the planet that we have seen so far just in this book. Yet, global warming theorists insist our only concern about planetary survival should be that we will destroy the environment by burning hydrocarbon fuels.

Milankovitch Cycles

In 1904, Serbian-born Milutin Milankovitch received a Ph.D. at the University of Technology in Vienna, Austria. In 1920, Milankovitch published his first book calculating sun cycles and their impact on Earth's short-term weather and long-term climate. In what are known today as "Milankovitch Cycles," he identified how three variations in Earth's orbit around the sun affect weather and climate on Earth.[91] The astronomy and the mathematics of the Milankovitch Cycles are complex. However, the main point of this subsection should be clear. In understanding why glacial and interglacial periods have occurred throughout Earth's geologic history, we must acknowledge the central importance of the sun as a driving factor for Earth's climate. The following is a brief discussion of the three orbital variations that constitute the Milankovitch Cycles:[92]

Eccentricity: The Shape of Earth's Orbit

The gravitational pull of Jupiter and Saturn causes Earth's orbit around the sun to vary from nearly cyclical to slightly elliptical. Eccentricity measures how much the shape of Earth's orbit departs from a perfect circle. These orbital variations affect the distance between Earth and sun.

In what is known as perihelion, Earth's orbit is closest to the sun, and in the aphelion, Earth is farthest from the sun. The perihelion occurs each year on or about January 2, and the aphelion around July 4.

At present, Earth's Northern Hemisphere is at perihelion in the winter, 91,419,000 miles from the sun, and at aphelion in the summer, 94,581,000 miles from the sun. Currently, Earth's eccentricity is at its most circular, i.e., at its least elliptical orbit. Given Earth's average distance from the sun is 93,470,000 miles, Earth today is at approximately an eccentricity of 0.017. The eccentricity of an ellipse is measured by a number that varies between "0" and "1.0." The value "0" represents a perfect circle, and "1" represents a flattened ellipse. With a near-circular orbit, Earth at perihelion is about 2,051,000 miles closer to the sun than the average distance, and at aphelion, approximately 1,111,000 miles farther away.

NASA explains Earth's current orbit as follows:

> The difference in the distance between Earth's closest approach to the Sun (known as perihelion), which occurs on or about January 3 each year, and its farthest departure from the Sun (known as aphelion) on or about July 4, is currently about 5.1 million kilometers (about 3.2 million miles), a variation of 3.4 percent. That means each January, about 6.8 percent more incoming solar radiation reaches Earth than it does each July.[93]

Thus, at eccentricity 0.017, the Northern Hemisphere receives approximately 6.8 percent more solar energy in winter and about 6.8 percent less solar energy in summer. But at this eccentricity, the Northern Hemisphere is still 7.5°C cooler in winter, despite receiving more solar energy. Remember, when the Northern Hemisphere is experiencing winter, the Southern Hemisphere is experiencing summer. The Northern Hemisphere contains more continents, and the Southern Hemisphere has more oceans. Water absorbs more sunlight than land. Otherwise, the Northern Hemisphere would be even cooler in winter if it were not for the seas in the Southern Hemisphere experiencing summer.

At 0.017 eccentricity, with the Northern Hemisphere at aphelion in summer, Earth gets approximately 6.8 percent less energy from the sun. Although Earth receives less solar energy in the current orbit in summer, at 0.017 eccentricity, the Northern Hemisphere is still 7.5°C warmer in summer at aphelion. Again, remember that summer in the Northern Hemisphere is winter in the Southern Hemisphere, and the southern oceans get less sunlight to absorb. So, at 0.017 eccentricity, at aphelion, the Southern Hemisphere in winter cools the Northern Hemisphere in summer, making the summer in the Northern Hemisphere cooler than it would otherwise be.

Another consequence of a nearly circular orbital eccentricity is Kepler's second law of planetary motion that states a planet moves fastest when nearer to the sun. Thus, at 0.017 eccentricity, with the Northern Hemisphere closest to the sun in winter, the winter there is six days shorter. In turn, the summer in the Northern Hemisphere is six days longer.

In the next future cycle, with Earth's orbit at its most elliptical, the Northern Hemisphere is at perihelion in the summer, 86,685,000 miles from the sun, and the Northern Hemisphere is at aphelion in winter, 99,315,000 miles from the sun. In this configuration, Earth's eccentricity is at 0.0679. So, at perihelion in the near-circular orbit, Earth is roughly 6,785,000 miles closer than the average distance to the sun, and at aphelion, some 6,315,000 farther away. Note, in the most elliptical orbit, Earth is closer to the sun at perihelion than in the nearly circular orbit and farther from the sun at aphelion.

At eccentricity 0.0679, the Northern Hemisphere will experience winter in December at aphelion and it will experience summer in June at perihelion. In this configuration, the Northern Hemisphere will receive 31 percent more solar energy in summer and 31 percent less solar energy in winter. The result is that Earth's temperature will be approximately 20°C warmer in summer and 20°C colder in winter. Another consequence given Kepler's law is that the summer in the Northern Hemisphere will be fifteen days shorter and the winter in the Northern Hemisphere will be fifteen days longer.

In the elliptical orbit at eccentricity 0.0679, winters in the Northern Hemisphere will probably start in September or October and last until April or May, while summers may be short but hot, probably starting in June and ending in August. The key is that with greater distance from the sun during a longer-lasting Northern Hemisphere winter in the elliptical orbit, the larger ice caps in the Arctic than in the Antarctic will melt less in the summer. Again, because winter starts fifteen days earlier and ends fifteen days later in the most elliptical orbit, the Arctic icecap is more likely to expand into Canada, Northern Europe, and Siberia, possibly pushing Earth into the next ice age.

Earth's eccentricity cycle spans about 100,000 years, the time it takes the planet to transition from its most circular orbit to its most elliptical. In the next few years, Earth's orbit will begin transitioning into a more elliptical one. When Earth reaches maximum eccentricity, the Milankovitch

Cycle predicts that Earth will be deep into the next ice age. Yet, we should not relax, confident that a new ice age is tens of thousands of years in the future. Only 12,000 years from now, Earth's increasingly elliptical orbit will cause temperature swings between winter and summer, which could be twice what we experience today.[94] Shorter summers will melt less of the Northern Hemisphere ice. Longer winters will cause more of the Northern Hemisphere ice to expand.

Scientists at the Woods Hole Oceanographic Institution have also warned that a sudden global cooling period like the Younger Dryas that we discussed in chapter 7 could occur again. These scientists reference climate chaos to note a Younger Dryas–like quasi period of around 1,500 years that could cause abrupt global cooling to happen today. The Woods Hole analysis stresses that our climate promotes cold, deep-water formation around Antarctica and the northern North Atlantic Ocean. The cold, deep-water formations in the high latitudes of the Northern and Southern Hemispheres are a consequence of Earth's complex hydrologic cycle that acts as a moderating mechanism for climate on the planet. The oceans balance excess heating at the equator and cooling at the poles by using the atmosphere and the oceans to reduce equator-to-pole temperature differences.

The concern is that during millennial periods of cold climate, North Atlantic Deep Water (NADW) formation either stops or is seriously reduced. The Woods Hole scientists warn that Earth could abruptly enter another cycle with the potential to cool the oceans and the atmosphere some 3–5°C. A cooling of that magnitude is a third to a half the temperature change experienced during significant ice ages. The scientists warned that Earth could experience global cooling of the magnitude of the Little Ice Age, with Northern Hemisphere glaciation persisting possibly for centuries. The scientists also specifically commented that these ocean effects on climate involve dynamics that "lie beyond the capability of many of the models used in IPCC reports." The scientists stressed that the consequences of a new Younger Dryas would be particularly severe for those living around the edge of the northern Atlantic Ocean. They concluded by suggesting that "we may be planning for climate scenarios of global warming that are opposite to what might actually occur."[95]

Obliquity: The Angle of Earth's Axis Is Tilted with Respect to Earth's Orbital Plane

Earth's obliquity is the tilt of the planet measured from its axis. The axis is currently tilted at 23.4°. Over the last million years, Earth's obliquity has ranged between 22.1° and 24.5° perpendicular to the planet's orbital plane. The tilt accounts for the seasons. During winter at eccentricity 0.017, the Northern Hemisphere in perihelion tilts from the sun, causing the North Pole to be dark. Other places in the Northern Hemisphere experience shorter days and colder winter temperatures. When it is winter in the Northern Hemisphere, the Southern Hemisphere tilts toward the sun, experiencing summer.[96]

Solstices occur when the axis of Earth points directly at the sun, which happens twice each year. The axis is currently tilted at 23.5°. On or near June 21, the North Pole tilts 23.5° toward the sun and the Northern Hemisphere experiences the longest day of the year in the Northern Hemisphere, the summer solstice. On that same day, the Southern Hemisphere tilts 23.5° away from the sun and the Southern Hemisphere experiences the shortest day of the year in the Southern Hemisphere, the winter solstice. The second solstice date is December 21 or 22, when the process reverses and the Northern Hemisphere tilts 23.5° away from the sun, while the Southern Hemisphere tilts 23.5° toward the sun. December 21 or 22 is the shortest day in the Northern Hemisphere, the winter solstice, and the longest day in the Southern Hemisphere, the summer solstice.[97]

Twice each year, Earth experiences the equinoxes, with days and nights of equal length. During the equinoxes, Earth's axis tilts perpendicular to the sun, and every location on Earth (except the extreme poles) experiences twelve hours of sunlight and twelve hours of darkness. On or about March 21 or 22, the Northern Hemisphere experiences the vernal or spring equinox, and the Southern Hemisphere experiences the autumnal or fall equinox. On September 22 or 23, the Northern Hemisphere experiences the fall equinox, and the Southern Hemisphere experiences the spring equinox. In contrast, Earth tilts toward the sun during the summer solstice in the Northern Hemisphere. On the summer solstice in the Northern Hemisphere, the sun hits perpendicular to the surface of the Tropic of Cancer that lies at 23.5° north latitude, corresponding to the tilt of Earth's axis. During the winter solstice in the Northern Hemisphere, the sun's rays hit Earth perpendicularly at the Tropic of Cancer located at 23.5° south latitude.[98]

Earth's axis, currently tilted at 23.5°, is about halfway between the extremes. But the planet is presently heading toward decreased obliquity. As a result, the seasons moderate into cooler summers and warmer winters. Cooler summers allow snow and ice to build into large ice sheets. As Earth's obliquity increases, the seasons become more extreme, with hotter summers and colder winters. With increased obliquity, the ice sheets in the Northern Hemisphere grow and Earth's global cooling intensifies. The ice sheets have a positive forcing effect by reflecting more of the sun's radiant energy into space, causing additional cooling.[99]

According to NASA, Earth reached maximum tilt about 10,700 years ago and will achieve minimum tilt some 9,800 years in the future. Earth's obliquity cycles between maximum and minimum tilt approximately every 41,000 years. The shorter obliquity cycle and the longer eccentricity cycle affect Earth's temperature. Why? Earth's obliquity cycles every 41,000 years and its eccentricity cycles every 100,000 years. Thus, Earth's tilt toward the sun will vary at perihelion and aphelion. The variance depends on the obliquity of Earth in each orbit. Consequently, the solar radiation hitting Earth varies in each orbit, depending on the angle of obliquity.

Precession: The Direction Earth's Axis of Rotation Is Pointed

As Earth rotates, it wobbles, much like a toy top spinning off-center. NASA points out that the wobble is due to tidal forces caused by the gravitational influences of the sun and moon that cause the planet to bulge at the equator, affecting its rotation. The trend in the direction of the wobble relative to the fixed seasons of stars is known as "axial precession." Precession in astronomy refers to changes in the rotating body's axial orientation.

In astronomy, the ecliptic is the apparent great-circle path of the sun in the celestial sphere. The celestial sphere is like a great dome sphere in the sky that shows the apparent position of the stars in the sky as seen from Earth. The celestial equator is the projection of Earth's equator on the celestial sphere. The celestial equator is currently at 23.5°, Earth's current angle of obliquity. The two points at which the ecliptic (the apparent path of the sun in the sky) intersect the celestial equator are the vernal equinox on or about March 21 and the autumnal equinox on or about September 23. As we noted above, the equinoxes do not happen at the same ecliptic points every year because the ecliptic plane and the equator plane revolve in opposite directions. The two planes make a complete revolution concerning each other every 25,878 years. As we also noted above, precession in astronomy refers to changes in a rotating body's axial orientation. The movement of the equinoxes along the ecliptic has been known for ages as the "precession of the equinoxes."[100]

The name "precession of the equinoxes" comes from an ancient perception of the sky. Ancient observers of the stars saw that the equinoxes were moving westward along the ecliptic plane relative to their position. These sky watchers had trouble explaining this since they believed the stars were in fixed positions in the celestial sky. In 129 BCE, the Greek astronomer Hipparchus noticed that the positions of the stars did not match up with Babylonian measurements. Hipparchus concluded that what was moving was Earth as a frame of reference, not the stars. Early astronomers believed the stars were motionless, opposite to the motion of the sun along the ecliptic.[101]

So, after a complete cycle of 25,878 years, the celestial North Pole and the equinoxes have made one complete revolution, and the celestial North Pole again points to Polaris.[102] Presently Polaris, the brightest star in Ursa Minor, appears close to the North Celestial Pole (NCP), but as Earth goes through the 25,878-year cycle, the NCP displacement will point away from Polaris. Scientists explain Earth's precessional wobbling by the action of a spinning gyroscope. As the gyroscope rotates, the top repeatedly describes a circle moving in an arc due to the wobbling. Precessional wobbling causes the North Pole to trace a 47° arc through the sky during the 25,878-year cycle. Because of precession, in about 13,000 years, the NCP will be closest to Vega, the brightest star in Lyra, the harp constellation. As Earth completes the 25,878-year cycle, the NCP will again return to Polaris.[103]

NASA also points out that axial procession makes seasonal contrasts more extreme in one hemisphere and less extreme in the other. As we noted above, currently, in the Northern Hemisphere, Earth is tilted toward the sun when it is farthest from the sun, at aphelion, in June, and we experience summer. Today, in the winter, Earth tilts away from the sun at perihelion, when the sun is closest to Earth, and we experience winter. In an estimated 13,000 years, axial precession will cause this to flip.[104] In 13,000 years, the Northern Hemisphere will tilt toward the sun by 23.5° in June when Earth is closest to the sun at perihelion. At that time, the Northern Hemisphere will be tilted away from the sun in December, at aphelion, when Earth is farthest from the sun, and we will experience winter. What will change will be the constellation that the summer solstice occurs in. In 13,000 years, the Northern Hemisphere will still experience winter in December and summer in June. Still, Earth will have traveled one-half of a complete cycle around the zodiac, and we will see Orion as the summer solstice constellation. Precession affects the background constellations around which the Earth-sun dynamics of motion occur.[105]

A related phenomenon is known as apsidal precession. In celestial mechanics, the term "apsides" refers to the nearest and farthest points reached by an orbiting planetary body, i.e., the perihelion and aphelion of Earth's orbit. Apsidal precession involves a complete rotation of the imaginary line connecting the perihelion and the aphelion rotating around the sun in a complete 360° cycle. Thus, apsidal precession involves the tendency of Earth's elliptical orbit to rotate around the sun through an entire period of 360°. In technical terms, the apsidal precession is also known as the perihelion precession.[106] NASA points out that not only does Earth's axis wobble, its entire ellipse also wobbles irregularly due primarily to the gravitational interactions with Jupiter and Saturn. The imaginary line connecting the perihelion and aphelion slowly rotates around the sun. A complete 360° cycle of apsidal precession spans about 112,000 years.[107] All these changes in the complex movement of Earth through space impact climate, with some Earth movements as defined by Milankovitch processes taking thousands of years to pass through cycles.

Ice Ages, Shifting Magnetic Poles, and Catastrophe Theory

We begin this subsection by explaining in more detail that in order to understand how complex processes like the Milankovitch Cycles "cause" ice ages, we require, once again, to deepen and reinforce the discussion of dependent versus independent variables and how each operates in scientific analysis. As noted earlier in the chapter, the equations we need to understand the complexities of weather and climate are nonlinear. Also, given the feedback mechanisms of the weather and climate systems, variables like CO_2 can be independent variables (i.e., causes) or dependent variables (i.e., effects) in the nonlinear equations. This is another mistake IPCC adherents make in assuming increasing atmospheric concentrations of CO_2 cause increases in the earth's average temperature. The IPCC and their global warming adherents tend to see CO_2 as the only cause of climate change throughout history. But CO_2 is also a dependent variable (i.e., an effect) when we consider that increases in CO_2 in the atmosphere lag global warming, as discussed in the previous chapter. When Earth warms, plant and animal life flourish, and the resulting impetus, given the carbon cycle, emits more CO_2 into the atmosphere. In this instance, increased CO_2 emissions are an effect (i.e., a dependent variable), not a driver (i.e., a cause) of the climate change phenomenon.

In the discussion of ice ages that follows, please keep in mind that while the warming currently experienced with Earth in interglacial cycle has caused CO_2 levels to increase, life on the planet has blossomed as a result of increased TSI. In the distant billions of years of earth history, atmospheric concentrations of CO_2 were dramatically higher during both glacial and interglacial periods. With the abundance of life on the planet today, CO_2 levels are lower than Earth experienced in Precambrian time, when Earth was largely devoid of life—a time period that comprised some 80 percent of Earth's history as we previously noted. In other words, the study of Milankovitch cycles and the impact of those cycles on ice ages through geologic time demonstrates that CO_2 cannot be understood as the only independent variable that drives global warming. Moreover, not only is CO_2 one of many independent variables affecting Earth's climate, CO_2 also operates as a dependent variable in the climate equation, as for instance now, when the interglacial global warming and the blossoming of life on the planet causes more CO_2 to be emitted from natural causes into the environment. The refusal of CO_2 to behave only as an independent variable in geologic time is why Michael Mann and his desperate cohorts went to such great lengths to falsify data producing their fraudulent Hockey Stick graphs.

Again, there are credible climate scientists who agree. In 2018, Philip Stott, a professor emeritus at the University of London and editor of the *Journal of Biogeography*, refuted the claim that CO_2 is the main climate change driver. "As I have said, over and over again, the fundamental point has always been this: Climate change is governed by hundreds of factors, or variables, and the very idea that we can manage climate change predictably by understanding and manipulating at the margins one politically selected factor (CO_2), is as misguided as it gets," he said. "Climate is the most complex coupled nonlinear chaotic system known to man. Of course, there are human influences in it, nobody denies that. But what outcome will they get by fiddling with one variable (CO_2) at the margins? I'm sorry but it's scientific nonsense."[108] Atmospheric scientist Hendrik Tennekes, a pioneer in the development of numerical weather prediction and former director of research at the Royal Netherlands Meteorological Institute, declared, "I protest vigorously the idea that the climate reacts like a home heating system to a change setting of the thermostat: just turn the dial, and the desired temperature will soon be reached."[109]

After we discuss the impact of Milankovitch Cycles on the ice ages and before we examine the mathematics of catastrophe theory, we will briefly comment on the shifting of Earth's magnetic poles currently happening. The discussion of Earth's shifting magnetic poles once more illustrates the prejudice that IPCC adherents have by insisting CO_2 is the only forcing agent causing climate change. The discussion of shifting of the magnetic poles will also demonstrate that changes that rarely occur in Earth's geological history can happen quickly, as well as over thousands, millions, and even billions of years in geologic time. Even more, the discussion of the shifting magnetic poles will illustrate how discontinuous changes from one state to another without a smooth or gradual transition also can and do happen in Earth's environment.

Ice Ages

Milankovitch had begun working on the Earth-sun cycles because he wanted to develop a methodology for mathematically calculating Earth's past climate in geological time. With the assistance of German climatologist Wladimir Köppen, Milankovitch created a graph mapping solar radiation at latitudes 55°, 60°, and 65° North over the past 650,000 years. Köppen realized Milankovitch's graph matched reasonably well with the history of the Alpine glaciers that German geographers Albrecht Penck and Eduard Brückner had constructed fifteen years earlier.[110] Milankovitch saw that the periodicity of the ice ages in Earth's history matched the periodicity of Earth reaching maximum eccentricity at aphelion. As we have already discussed, when Earth is in its most circular orbit, the Northern Hemisphere cools at perihelion because the Southern Hemisphere is in winter. When Earth is in its most elliptical orbit, the Northern Hemisphere reaches its maximum distance from the sun at aphelion. Those studying glaciers since the nineteenth century realized the ice ages involved the spread of the Arctic ice cap over North America, Europe, and Asia, but not the spreading of the Antarctic ice cap.

The problem is that the history of Earth's ice ages does not match up precisely with the Milankovitch Cycles. Scientists have identified five significant ice ages in Earth's history. Approximately a dozen significant glaciations have occurred over the past one million years. The most extensive glaciation occurred 650,000 years ago and lasted 50,000 years. The most recent glaciation peaked some 18,000 years ago before giving way to the current interglacial Holocene Epoch approximately 11,700 years ago.[111] Earth has alternated between long ice ages and shorter interglacial periods for around 2.6 million years. For the last million years, approximately, ice ages have occurred roughly every 100,000 years, with 90,000 long years of glaciation regularly followed by approximately 10,000 years of relatively interglacial global warming.[112] Given this cycle, Earth is due for another ice age.

Australian geoscientist Ian Plimer, whom we quoted in the previous chapter, acknowledged that the Milankovitch Cycles are related to the cause of Earth's ice ages. Still, he struggled with the match not being perfect. In his 2009 book, *Heaven and Earth*, Plimer wrote the following:

> The problem with the Milankovitch Cycle theory may be that orbital wobbles are not the main driver of climate. The annual amount of solar energy between the hemispheres is on a 21,000-year cycle. Yet, this is not the major glacial cycle. The 41,000-year cycles affect the amount of solar radiation entering the tropics and the poles. This is a glacial cycle and the glacial cycles switched from 41,000 to 100,000-year cycles about 1 Ma [1 million years ago]. The 41,000-year cycle is no longer a glacial cycle. Why? Despite the most uniform amount of solar energy striking the Earth in the 100,000-year cycle, it is the 100,000-year cycle that is the major glacial cycle. Why? The 100,000-year cycle is the weakest of the three Milankovitch Cycles and hence may be hardly enough to drive climate change. The Earth's orbital eccentricity also shows a 400,000-year as well as a 100,000-year cycle. The two cycles are of comparable strength yet the 400,000-year cycle is not recorded in climate records. A warming climate predates by about 10,000 years the change in incoming solar radiation that supposedly had been its cause.[113]

Plimer refers to a 400,000-year eccentricity cycle apparent only in climate records older than the last million years. Plimer's point is that the scientific investigation of Milankovitch Cycles has many gaps, like the 400,000-year cycle that ceased to be detected a million years ago.[114] Plimer continued this thought, stressing the following:

> What is remarkable and cannot be explained well by Milankovitch Cycle theory is that the transition from an interglacial to a glaciation occurs at peak temperature when the melting of ice is at a maximum. There are great variations in temperature during the transition. At present, the Earth is close to the level at which past transitions occurred.[115]

Plimer conceded that the Milankovitch Cycle theory is correct in that incoming solar radiation drives Earth's ice ages. However, he commented that "the mechanism whereby very slight changes in incoming radiation drive major climate change is unclear."[116] Later he repeats this thought. He added the following: "It is widely accepted that Earth's orbit affects glaciation but a better and more detailed understanding of this process is needed."[117]

The U.S. government's National Centers for Environmental Information of the National Oceanic and Atmospheric Administration (NOAA) explain what causes glacial-interglacial cycles in terms friendly to Milankovitch Cycles. Consider NOAA's following explanation for what causes those cycles:

> Variations in Earth's orbit through time have changed the amount of solar radiation Earth receives in each season. Interglacial periods tend to happen during times of more intense summer solar radiation in the Northern Hemisphere. These glacial-interglacial cycles have waxed and waned throughout the Quaternary Period (the past 2.6 million years). Since the middle Quaternary, glacial-interglacial cycles have had a frequency of about 100,000 years. In the solar radiation time series, cycles of this length (known as "eccentricity") are present but are weaker than cycles lasting about 23,000 years (which are called "precession of the equinoxes").[118]

But at least NOAA acknowledges that understanding nonlinear equations is a vital prerequisite for determining what causes glacial-interglacial cycles. Consider NOAA's next paragraph:

> Interglacial periods tend to occur during periods of peak solar radiation in the Northern Hemisphere summer. However, full interglacials occur only about every fifth peak in the precession cycle. The full explanation for this observation is still an active area of research. Nonlinear processes such as positive feedbacks within the climate system may also be very important in determining when glacial and interglacial periods occur.[119]

NOAA also introduced an essential concept to catastrophe theory mathematics when commenting on how abruptly glacial periods end. "Warming at the end of glacial periods tends to happen more abruptly than the increase in solar insolation." Next, the organization wrote a few sentences that properly position CO_2 feedback mechanisms as both a cause and effect of the climate dynamics that cause glaciers to come and go. They continued with the following:

> Direct measurement of past CO_2 trapped in ice core bubbles shows that the amount of atmospheric CO_2 dropped during glacial periods, in part because the deep ocean stored more CO_2 due to changes in either ocean mixing or biological activity. Lower CO_2 levels weakened the atmosphere's greenhouse effect and helped to maintain lower temperatures. Warming at the end of the glacial periods liberated CO_2 from the ocean, which

strengthened the atmosphere's greenhouse effect and contributed to further warming.[120]

As noted earlier, prehistoric ice ages are a problem for the IPCC and global warming theorists because the high concentration of CO_2 in the prehistoric atmosphere did not stop repeated glacial periods from happening. We find NASA arguing the IPCC's current argument that the Milankovitch Cycle solar radiation factors are not sufficiently powerful to produce the ice ages. But, at the same time, NASA explains CO_2 is a sufficiently robust climate forcing factor to prevent the predicted next ice age from occurring on Earth. To find IPCC minimizing the Milankovitch Cycle's reliance on solar radiation as a cause of glacial-interglacial periods, we must return to NASA's website.

On their website, NASA issued IPCC-like warnings that postindustrial anthropogenic CO_2 will raise Earth's temperature by another 0.5°C by 2030 or 2040 at the latest. Next, they explained how anthropogenic CO_2 outpowers the predicted coming of a Milankovitch Cycle global cooling and near-term ice age:

> This relatively rapid warming of our climate due to human activities is happening in addition to the very slow changes to climate caused by Milankovitch cycles. Climate models indicate any forcing of Earth's climate due to Milankovitch cycles is overwhelmed when human activities cause the concentration of carbon dioxide in Earth's atmosphere to exceed about 350 ppm [parts per million].[121]

Other studies have suggested that current and anticipated near-future levels of anthropogenic CO_2 will postpone the next ice age for at least 100,000 years.[122] A separate, published scientific paper estimated that anthropogenic CO_2 might delay the next ice age by half a million years.[123] Another published scientific article termed CO_2 as "the principal control knob governing Earth's temperature."[124] The IPCC and their global warming adherents refuse to abandon the argument that CO_2 is the sole important causative forcing mechanism to the climate that deserves any attention at all. While the view puts CO_2 back in the driver's seat as a causal temperature forcing element, we must ask if postponing the next ice age might be among the first, if not only, positive benefit global warming theorists have posited.

On May 8, 2013, Harrison Schmidt, a geologist and a retired NASA astronaut who was a former U.S. senator from New Mexico, and Dr. William Happer, then a professor of physics at Princeton University who was a former director of the office of energy research at the U.S. Department of Energy, coauthored an editorial in the *Wall Street Journal* entitled "In Defense of Carbon Dioxide." In the editorial, they wrote:

> Of all of the world's chemical compounds, none has a worse reputation than carbon dioxide. Thanks to the single-minded demonization of this natural and essential atmospheric gas by advocates of government control of energy production, the conventional wisdom about carbon dioxide is that it is a dangerous pollutant. That's simply not the case. Contrary to what some would have us believe, increased carbon dioxide in the atmosphere will benefit the increasing population on the planet by increasing agricultural productivity.
>
> The cessation of observed global warming for the past decade or so has shown how exaggerated NASA's and most other computer predictions of human-caused warming have been—and how little correlation warming has with concentrations of atmospheric carbon dioxide. As many scientists have pointed out, variations in global temperature correlate much better with solar activity and with complicated cycles of the oceans and atmosphere. There isn't the slightest evidence that more carbon dioxide has caused more extreme weather.[125]

On December 8, 2015, Princeton's Professor Happer, a distinguished research physicist and author of two hundred peer-reviewed papers, testified to the U.S. Senate Subcommittee on Space, Science, and Competitiveness. "Few realize that the world has been in a CO_2 famine for millions of years, a long time for us, but a passing moment in geologic history," Happer explained to the subcommittee. "Over the past 550 million years since the Cambrian, when abundant fossils first appeared in the sedimentary record, CO_2 levels have averaged many parts per million (ppm) not today's few hundred ppm." Happer noted that preindustrial levels of 280 ppm were not far above the minimum level, around 150 ppm, when many plants die from CO_2 starvation. "Thousands of peer review studies show that almost all plants grow better (and land plants are more drought resistant) at atmospheric CO_2 that are two or three times larger than those today."[126]

Magnetic Pole Shifts

Starting in 1990, Earth's magnetic North Pole began shifting, accelerating northwards. In 2017, the magnetic North Pole passed the geographic North Pole and began heading toward Siberia. In 2019, the movement of the magnetic North Pole was moving so unexpectedly fast that scientists issued an irregular update in advance of the scheduled five-year update of the magnetic North Pole's position.[127] The geographic North Pole is a fixed point in the Arctic. The magnetic North Pole is where a compass needle points to align with Earth's magnetic field. The current shifting of the magnetic North Pole moving in a sprint at about thirty to nearly forty miles a year is mainly a problem for navigation systems requiring a recalibration of the World Magnetic Model.[128] The difference between the true geographic north and the true magnetic north is magnetic declination. Philip Livermore at the University of Leeds in the U.K. believes the shifting is related to two patches of relatively strong magnetic fields below Canada and Siberia. Livermore has explained that these two patches are involved in a tug-of-war that pulls the magnetic North Pole back and forth.[129]

Earth's magnetic field varies over time, such that the positions of the North and South magnetic poles have changed continuously over time. Scientists have constructed a reliable history of geomagnetic patterns with paleomagnetic techniques for reading magnetic patterns in rock layers.[130] NOAA maintains an interactive website that shows the locations of the magnetic poles and historical declination lines calculated for the years 1590-2020.[131] Geomagnetic pole reversals, in which the North and South Poles flip, has happened occasionally throughout geological history. Paleomagnetic records reveal Earth's magnetic poles have reversed 183 times in the last eighty-three million years and at least several hundred million times in the past 160 million years. The time intervals vary widely but average about 300,000 years. The last magnetic pole reversal occurred 780,000 years ago.[132]

NASA reports that geophysicists are pretty sure that the reason Earth has a magnetic field is that the planet's center consists of an iron core surrounded by a fluid ocean of hot liquid metal.[133] But while NASA scientists hesitate to give a firm explanation for why Earth has magnetic fields, NASA is entirely confident the current shift of the magnetic North Pole has no impact whatsoever on climate change. They insist "changes and shifts in Earth's magnetic field polarity don't impact weather and climate for a fundamental reason: air isn't *ferrous*." In other words, NASA maintains there is not sufficient iron in the atmosphere for Earth's magnetic field to impact climate. Here is their explanation:

> While iron in volcanic ash is transported in the atmosphere, and small quantities of iron and iron compounds generated by human activities are a source of air pollution in some urban areas, iron isn't a significant

component of Earth's atmosphere. There's no known physical mechanism capable of connecting weather conditions at Earth's surface with electromagnetic currents in space.[134]

NASA's comment about electromagnetic currents in space refers to what is known as the magnetosphere that NASA describes as "a system of magnetic fields" that shield Earth "from harmful solar and cosmic particle radiation." They also note that Earth's magnetosphere can change shape in response to sun activity.[135]

Not all geoscientists are equally sure Earth's magnetic field has no impact on the climate. The discussion in chapter 5 regarding a correlation between cosmic rays hitting Earth and the effects on cloud formation suggests Earth's magnetic field may play a role in climate. Several scientific studies have noted NASA's point that the variations in the magnetic field on Earth and the magnetosphere surrounding Earth can affect the number of cosmic rays that reach the planet.[136] A scientific study published in *Atmospheric Chemistry and Physics* in 2014 stressed that "atmospheric deposition of iron (Fe) plays an important role in controlling oceanic primary productivity." The study also pointed out that "the sources of Fe in the atmosphere are not well understood." It noted that the higher than previously estimated Fe from coal combustion places a more significant atmospheric anthropogenic input of Fe into the northern Atlantic and northern Pacific Oceans, "which is expected to enhance the biological carbon pump in those regions."[137] The IPCC should find this study interesting. However, the IPCC might have concerns that acknowledging the importance of atmospheric anthropogenic Fe could elevate the question that Earth's magnetic field may play a role in climate change.

Researchers from the University of Liverpool published a study in the August 2021 issue of the *Proceedings of the National Academy of Sciences*, providing evidence of an approximately 200-million-year cycle linked to the strength of Earth's magnetism.[138] The Liverpool researchers performed thermal and microwave paleomagnetic analysis on rock samples from ancient lava flows in Eastern Scotland to measure the strength of the magnetic field during periods for which there is no reliable preexisting evidence. They found that between 332 and 416 million years ago, the strength of the magnetic field in these ancient rocks was only one-quarter of what it is today. But the low strength of Earth's magnetic field some 332–416 million years ago was comparable to another period of Earth's low magnetic field 120 million years ago.

A separate study published in *Science Advances* in 2020 suggested the terrestrial mass extinction at the Devonian-Carboniferous boundary (some 359 million years ago) was linked to elevated UV-B radiation [type B ultraviolet]. The study found that the second mass terrestrial extinction, also known as the Late Devonian extinction (typically dated at 365 million years ago), occurred around the same time the Liverpool scientists measured the weakest magnetic field measurements in their study published in 2021. The second extinction killed approximately 75 percent of all life on Earth. The researchers in the 2020 study believed a reduction in the ozone layer caused elevated UV-B radiation. These two studies support the theory that Earth's magnetism affects climate. The 2020 study noted that the Devonian-Carboniferous boundary's mass extinction coincided with significant climatic warming that ended the intense final glacial cycle of the latest Devonian Ice Age.[139]

Catastrophe Theory

French mathematician René Thom (1923–2002), a specialist in topology, is the founder of catastrophe theory. Topology involves the mathematical study of non-Euclidian shapes created by bending, stretching, twisting, and otherwise distorting objects in three-dimensional space. With his 1975 book *Structural Stability and Morphogenesis*,[140] Thom applied topology mathematics to understand a host of real-world events that demonstrated discontinuities. He defined a catastrophe as "any discontinuous transition that occurs when a system can have more than one stable state, or can follow more than one stable pathway of change."[141] The catastrophe involves the "jump" from one state or pathway to another. In mathematical terms, the equations of catastrophe theory are nonlinear. In non-mathematical terms, the idea is that a situation or event may proceed in a relatively stable and somewhat predictable way until a tipping point is reached. At the tipping point, the system flips into a dramatically different mode of operation.

Like topology, the equations of catastrophe theory define the discontinuities intrinsic to complex physical phenomena like the weather and climate. Catastrophe theory graphs fold, bend, drop off, change shape, and generate dramatically new configurations, much like the non-Euclidian surfaces in topology bend, twist, curve, and in general change shape. A few examples help to make the idea of catastrophe theory mathematics easier to understand. The water heats up, then suddenly its energy reaches a tipping point, and the water begins to bubble up and boil. Volcanoes develop magma underground for considerable periods until suddenly the volcano experiences a discontinuity, and the phenomenon turns into an eruption. A counterintuitive principle of catastrophe theory is that constants defining stable states are suspect. Thus, a solar constant gives way when we find out the total solar irradiance (TSI) varies. A few paragraphs earlier, we saw the assumption that all the matter that ever will be in the universe already exists.

Conclusion

Analyzing concepts like global warming and climate change without appreciating the mathematics of catastrophe theory is like trying to fly to the moon with Euclidian geometry as your understanding of spacetime. Earth's atmosphere, its weather, and climate are not only chaotic and unpredictable, but they are also subject to quick and sudden changes, reversals, and different states of operation. Catastrophe theory in geology explains the limitations of uniformitarianism mathematically. For millions of years, Earth may be in a glacial period, then suddenly, the glacial period ends, and an interglacial period begins. With catastrophe theory, we understand Earth's geoscience rules and laws are subject to change.

Milankovitch, for instance, assumed Earth's orbit, while subject to the dynamic changes of its orbit cycles, operated the same today as 4.6 billion years ago. But, if Earth had changed its mass or size, we should understand it would also have changed its orbit. Gravity is a function of mass, as reflected by the formula $g = GM/r^2$, where "g" is gravity, "G" is the gravitational constant, "M" is the mass of Earth, and "r" is the radius of Earth. Again, assuming gravity on Earth is a constant value is a mistake. Suppose the mass or the size of Earth changes, then Earth's gravity changes. Again, Earth's climate is a dynamic system. If Earth's orbit changes, all the values in the Milankovitch Cycles change over time. Suppose the sun's irradiance that hits Earth changes, then Earth's climate changes, either because the its orbit has changed as a function of changing mass or because the sun is switching between active and passive cycles. The Milankovitch Cycle calculations may not match the ice age cycles in geologic time in part because Earth's orbit has changed in geologic time, the sun's irradiance has changed, or both. That is a catastrophe theory idea. We have constantly criticized the IPCC's global warming theories, not because CO_2 is not a greenhouse gas, but because CO_2 is not a significant enough independent variable to be the single driving cause in as complex and dynamic a system as Earth's climate.

Professor Jan C. Schmidt, from the Department of Social Sciences at the Darmstadt University of Applied Sciences in Darmstadt, Germany, explained how challenging catastrophe theory is to stable-state thinkers. In an article entitled "Challenged by Instability and Complexity," Schmidt observed the following:

> Throughout history, stability metaphysics has always played a major role in science, beginning in ancient times with Plato's stability concept of the cosmos. In modern times, stability metaphysics can be found in the works of outstanding physicists such as Newton and Einstein. For instance, in his *Opticks* Newton did not trust his own nonlinear equations for three- and n-body systems which can potentially exhibit unstable solutions. He realized God's frequent supernatural intervention in order to stabilize the solar system. In the same vein, Einstein introduced ad hoc—without any empirical evidence or physical justification—the cosmological constant in the framework of General Relativity in order to guarantee a static and stable cosmos, "Einstein's cosmos." Both examples, from Newton and Einstein, illustrate that metaphysical convictions—what nature is!—can be incredibly strong, even if they are in conflict with what is known about nature at that time.[142]

Schmidt correctly insisted that instabilities are not exceptions in a stable world. Instead, instability is inherent to complex systems like Earth's weather and climate. The complexity inherent to catastrophe theory demands we realize no climate factor, including CO_2, can be expected to produce the same result every time. In an intrinsically nonlinear environment, we should not be surprised to find CO_2 at extremely high levels during the prehistoric ice ages. Milankovitch failed to add into his calculation of cycles that sun activity is also an independent variable. In the 100,000-year cycle, when Earth reaches aphelion of the highest eccentricity orbit, the sun might be in a highly active phase or a minimum. The variance in sun activity clarifies that orbital differences of the Milankovitch Cycles are not singular driving causes of Earth's temperature. Nor does a particular sun activity cycle occurring in the same orbit configuration of a Milankovitch Cycle produce the same temperature effect on Earth each time it happens.

In the multivariant, nonlinear system of equations needed to understand Earth's climate, the same independent variables do not necessarily produce the same outcome in the dependent variables. Consider a particular sun activity cycle occurring at a specific configuration of Earth's orbit

The Chaos Theory of Climate

in a Milankovitch Cycle. In $Time_1$, these conditions may arise when volcanoes are superactive on Earth. In $Time_2$, the same sun activity and orbit configuration may occur when a Chicxulub-size asteroid has just impacted the planet. In $Time_3$, possibly the Chicxulub-size asteroid that hit Earth has caused a significant increase in Earth's volcanism. There is no single cause that explains a phenomenon as complex as why ice ages have occurred, why the continents separated, or why the dinosaurs died.

Catastrophe theory may also help explain how the expanding Earth theory remains compatible with plate tectonics. Suppose Earth in Precambrian time expanded according to catastrophe theory mathematics. Conditions within early Earth may have grown to a discontinuity. Suppose Precambrian Earth was the size of a tennis ball that suddenly popped open to the size of a softball. There is no reason to assume the relatively slow rate Earth is expanding today is the same rate at which Earth expanded in the Precambrian Era. If that happened, we could imagine the fuzz on the tennis ball popping apart to be the continents on an expanded planet. This paradigm would view plate tectonics as the effect of an expanded Earth rather than the cause of a supercontinent Pangaea breaking apart. Plate tectonics may explain continental drift experienced today quite adequately. But even if today's plate tectonics are the cause of continued continental drift, plate tectonics do not have to be the cause of a Precambrian Earth expanding according to the laws of catastrophe theory.

The theory of continental drift assumes rather than proves that Pangaea simply broke up. Catastrophe theory is more compatible with the expanding Earth theory. Forces within Earth could have reached a tipping point where conditions changed, and Earth grew suddenly. The sudden expansion of the planet would have separated the continents while opening up on its new surface the new crust for the oceans to form.

Plate tectonics could operate much as modern structural geology contemplates once we account for how and why the continents separated. After the continents separated, plate tectonic processes like orogeny help explain how mountains formed. After the continents separated, plate tectonic processes like subduction explain the continuing submersion of the oceanic lithosphere into Earth's mantle at boundaries where ocean and continental plates converge. Processes of plate tectonics thus are compatible with an expanding Earth theory.

Comprehending the mathematics of chaos theory and catastrophe theory makes the IPCC's determination to explain all the various phases of Earth's temperature and climate by reference to CO_2 alone a fool's errand at best. For IPCC adherents, rigging the data to fit a predetermined, politically motivated conclusion that anthropogenic CO_2 causes Earth-threatening global warming is necessary and justified. Similarly, for this group, building computer-run climate models that present CO_2 as predetermined independent variables that drive global warming is also necessary and justified. If a linear cause-and-effect relationship existed throughout geologic time and established beyond doubt that CO_2 causes global warming, then an event like Climategate would never have happened.

In the attempt to make human beings the catastrophe of the Holocene, IPCC adherents have fallen into the trap of conceptualizing Earth's climate in an anthropocentric interpretation. Earlier in this chapter, we stressed that geologists know very little of the 80 percent of Earth's history that constituted Precambrian time. Geologists John J.W. Rogers and M. Santosh, in their 2004 book *Continents and Supercontinents*, stressed how much remains unknown about Wegener's Pangaea theory even today. After Wegener's various publications around World War I, Rogers and Santosh commented that geologists first recognized Gondwana as a supercontinent. However, Rogers and Santosh stressed the following:

> Three or more supercontinents may have existed before the assembly of Gondwana at ~500 Ma [about 500 million years ago]. A supercontinent named Rodina or Palaeopangaea almost certainly accreted at ~1 Ga [about 1 billion years ago] and broke up a few hundred million years later. An earlier supercontinent, named Columbia, may have been accreted by ~1.6 Ga [1.6 billion years ago) and broken up 100 million years later. One or more supercontinents may also have existed in the latest Archean, although they have not yet been named.[143]

Yet, the IPCC never dares to suggest that CO_2 may have contributed to global warming. At the same time, the IPCC's Sixth Climate Assessment issued in 2021 has not a single reference to the Milankovitch Cycles.[144] All human existence on the planet has occurred during the present Holocene Epoch that began some 11,600 years ago.

If we reduce Earth's geologic 4.6-billion-year history to a twenty-four-hour clock, the Holocene has occurred in less than the last minute. Ancient Greek philosopher Protagoras of Abdera (485–415 BCE) is famous for arguing "πάντων χρημάτων μέτρον ανθρωπος" (Man is the measure of all things). Diogenes Laertius recorded the complete version of what Protagoras wrote as follows: "Man is the measure of all things; things which are, that they are, and things which are not, that they are not."[145]

As we noted earlier, Earth does not operate by calculating mathematics or by using a computer. To Earth, we humans are just the most recent inhabitants of the tiny planet—a planet that has existed for 80 percent of its eons without life beyond the microbial state. If we human beings extinguish ourselves in a thermonuclear war, Earth will likely survive, recover, possibly flourish, but most certainly go on to the next era of its existence.

For the IPCC to presume that we are now in the Anthropocene Epoch awaiting the destruction of Earth because we put a slightly more significant quantity of a trace chemical compound into the atmosphere is the height of human hubris. As we have made evident in this second section of the book, the science of Earth's climate is complex beyond our ability to calculate the causes or effects completely or to predict the future. We also clarified that Earth's environment had endured a chaotic climate and gone through cataclysmic change for billions of years before we ever got here.

Yet clearly, serious scientists are still left who understand all too well that the IPCC's driving purpose is not geoscience. As we have also pointed out, that purpose is neo-Marxist, and its raison d'être is eliminating capitalism. Plato, in writing *Laws*, book 4, section 716c, gave his interpretation of Protagoras. Plato correctly restructured the comment to note that God truly is the measure of all things in the highest degree, a degree much higher than any human being Protagoras may have had in mind. A central theme of this book is that the global warming movement grew from the 1950s' concerns that we were overpopulating Earth.

In part III of this book, we will review the economic consequences of abandoning hydrocarbon fuels. We will conclude part II by joining Julian Simon to celebrate that we human beings have a mission on this planet, as specified in Genesis 9:7, to be fruitful, bring forth abundantly on Earth, and multiply.

PART III

The Economics of Energy, Global Warming, and Climate Change

CHAPTER 9

Abiotic Oil

Biogenic versus Abiogenic Oil, The Fischer-Tropsch Process and Synthetic Oil, J.F. Kenney and the Russian-Ukrainian School, Thomas Gold and the Deep Hot Biosphere, Deep-Sea Hydrothermal Vents, Resolving the "Petroleum Paradox," Fossil Fuel, Kerogen, and Biogenic Oil Chemistry, Fractures in Basement Rock, Offshore Drilling and Deep-Earth Hydrocarbons, Extraterrestrial Hydrocarbons, and CO_2 Emissions from Deadwood Decomposition

It is generally accepted that petroleum is derived from the remains of organic life, but many uncertainties exist concerning the processes involved.
—**Hollis D. Hedberg,** Petroleum Geologist, 1964[1]

The capital fact to note is that petroleum was born in the depths of the earth, and it is only there that we must seek its origin.
—**Dmitri Mendeleev,** *L'Origine du pétrole,* 1877[2]

The theory of the abyssal abiogenic origin of petroleum confirms the presence of enormous, inexhaustible resources of hydrocarbons in our planet and allows us to develop a new approach to methods for petroleum exploration and to reexamine the structure, size, and location of the world's hydrocarbon reserves.
—**Vladimir Kutcherov** and **Vladilen Krayushkin,** 2010[3]

THE TRUTH ABOUT ENERGY

THERE ARE TWO PARADIGMS CENTRAL to understanding the IPCC's argument demonizing anthropogenic CO_2. The first is that CO_2 is a greenhouse gas humans must bring under control, despite being a trace chemical compound in the atmosphere. The second is the idea that hydrocarbon fuels are fossil fuels that we must stop burning, even though the IPCC views hydrocarbon fuels as depleting resources that will soon be exhausted by increased human consumption. The two concepts are linked. We must stop burning hydrocarbon fuels because oil, coal, and natural gas all emit CO_2, and CO_2 is a significant cause in producing catastrophic climate change. IPCC adherents insist a consensus of scientists holds both ideas as accurate. IPCC adherents, therefore, ridicule and scorn those of us who argue that we can safely and productively burn hydrocarbon fuels, as well as those who say CO_2 is not a significant driver of Earth's temperature.

The purpose of this chapter is to argue that natural hydrocarbon fuels found fully formed in the earth are abiotic, i.e., not made up of organic, biological material.[4] This chapter will also clarify that hydrocarbon fuels degassing through deep-sea hydrothermal vents and deep-earth fissures are an essential feature of Earth's carbon cycle long before the first signs of life appeared on the planet in the form of deep-earth microorganisms. The carbon cycle starts in Earth's core, not with biogenic decay. Conventionally trained petroleum geologists will find the idea of abiotic oil hard to accept, much as petroleum geologists demonized for decades anyone who dared suggest that peak oil was a tautology, not proven science. So far in this book, we have demonstrated the complexity of Earth's weather and climate. Because we understand that Earth's weather and climate are both chaotic and subject to catastrophic change, then we must also recognize that we live on a planet whose environment is beyond our control. We will conclude this chapter by arguing that decarbonizing by restricting or significantly reducing the use of hydrocarbon fuels will have detrimental effects on the life-supporting nature of the carbon cycle itself.

This chapter is mainly about the science of abiotic oil. But the chapter is placed within part III on economics because the realization that the mantle of Earth abiogenically creates the hydrocarbon fuels that we burn as oil, coal, and natural gas has significant economic consequences. First, this chapter will show abiotic hydrocarbons are plentiful on Earth. Next, abiotic hydrocarbons are being created by Earth continuously. Finally, burning hydrocarbon fuels is one of the many natural and nonharmful processes in Earth's natural carbon cycle. Decarbonizing may have unanticipated severe and harmful effects on the planet's carbon, climate, and food cycles.

As famously pointed out by Thomas Kuhn, paradigms in science shift when incontrovertible scientific evidence begins to challenge orthodox beliefs.[5] In previous chapters, we have argued the IPCC's fears, which are the burning of hydrocarbon fuels since the dawn of the industrial age will cause catastrophic global warming and climate change, are unfounded, given the history of Earth's changes in geologic time. A paradigm shift away from the concept that hydrocarbon fuels are fossil fuels will open the mind to the possibility that hydrocarbon fuels are a natural product of Earth, not inherently harmful. If we assume what Earth produces naturally may be used safely, we can also consider continuing the advances made so far in burning hydrocarbons more cleanly.

Biogenic versus Abiogenic Oil

Dmitri Mendeleev, the Russian chemist, who in 1896 first arranged the sixty-three known elements into a periodic table based on atomic mass, also studied the origin of petroleum. Mendeleev was one of the first to suggest that oil is primordial material that arises from great depths from within Earth.[6] He reasoned that oil moves upward toward Earth's surface along structures he theorized were "deep faults" within Earth's crust. Philosophy of science professor Clifford Walters published a historical survey of views on the origin of petroleum. In his survey, Walters observed that Mendeleev's abiotic theory "was viewed initially as particularly attractive as it offered an explanation for the growing awareness of the widespread occurrence of petroleum deposits that suggested some sort of deep, global process."[7]

Michael D. Gordin, a professor of history at Princeton University, was less kind in discussing Mendeleev's theory of abiotic oil. In his 2004 biography of Mendeleev entitled *A Well-Ordered Thing*, Gordin commented as follows:

> There has been understandable emphasis on Mendeleev's consulting for the growing Baku oil industry and its relation to the Witte system. Mendeleev was affiliated with the oil industry (one of the most important sectors of the economy) from the start of his career, beginning with private work for V.A. Kokorev in Baku in 1863, and he published on oil throughout his life—including a monograph comparing the Pennsylvania and Baku oil fields as well as newspaper pieces advocating the elimination of the

government lease system and the excise tax. Oil became a focus of his scientific work as well: Mendeleev argued (incorrectly) that oil was not derived from microorganisms, but rather from metallorganic reactions with water.[8]

Gordin's politically correct assertion that Mendeleev's abiotic oil theory was wrong is typical of the intolerance orthodox scientists and historians of science have when alternative explanations challenge their beliefs. Note also Gordin's subtle suggestion that Mendeleev's oil industry financial ties influenced his incorrect thinking.

Geoscientists Mark A. Sephton and Robert M. Hazen, in their 2013 article "On the Origin of Deep Hydrocarbons," expressed accurately that geology textbooks continue to view the organic theory of the origin of oil as the scientific consensus. They summed up as follows, making it clear they understood the economic importance of the abiotic versus organic debate:

> Deep deposits of hydrocarbons, including varied reservoirs of petroleum and natural gas, represent the most economically important component of the deep carbon cycle. Yet despite their intensive study and exploitation for more than a century, details of the origins of some deep hydrocarbons remain a matter of vocal debate in some scientific circles. This long and continuing history of controversy may surprise some readers, for the biogenic origins of "fossil fuels"—a principle buttressed by a vast primary scientific literature and established as textbook orthodoxy in North America and many other parts of the world—might appear to be settled fact. Nevertheless, conventional wisdom continues to be challenged by some scientists.[9]

A brief survey of college geology textbooks revealed that even in the 1940s, geology professors writing textbooks expressed discomfort with the organic theory of the origin of oil. In 1949, Cecil G. Lalicker, professor of geology at the University of Kansas, published his textbook entitled *Principles of Petroleum Geology*. While not abandoning the organic theory, Lalicker clarified that the exact organic processes by which decaying organic material turned into oil remained somewhat mysterious. He wrote the following:

> The organic theory of the origin of petroleum is now generally accepted by most scientists, but there remain many problems which are yet unsolved. It is generally believed that petroleum originated by a series of complex

processes, from plant and animal substances. The exact nature of the original organic material is not yet known, although many valuable data have been assembled on this problem. The complex biological, chemical, and geological processes necessary in converting the organic material of plants and animals into hydrocarbons are not completely known.[10]

Lalicker noted that French chemist Pierre Eugène Marcellin Berthelot in the 1800s believed Earth contained free alkaline metals that could react with CO_2 at deep levels to produce hydrocarbons. In 1886, Berthelot demonstrated that acetylene, when heated to approximately 900°C, polymerizes into benzene, C_6H_6. Scientists then considered benzene to be only of organic origin.[11] Lalicker also discussed Mendeleev's proposition in 1877 that iron carbides within the earth could generate hydrocarbons by reacting with percolating waters. Lalicker was intrigued that someone of Mendeleev's evident genius with chemistry considered hydrocarbons to be of abiotic origin. Yet, he had his doubts. "The existence of iron carbides within the earth has not been proven," Lalicker concluded.[12]

Eminent geologist A.I. Levorsen, in his 1954 textbook *Geology of Petroleum*, also expressed hesitation about the organic theory. Levorsen noted that "only organic matter that can be used as a source material for petroleum is that which has become buried and preserved in the sediments before it has been destroyed by oxidation." Yet, Levorsen openly discussed his concern that organic material decomposes too quickly to turn into oil. He wrote the following:

> A key question, then, in the origin of petroleum, is how to preserve the decomposition products within the sediments short of complete decay. The presence of a reducing environment seems the most likely method of arresting decomposition; in fact, it is probably essential, as indicated by the presence in petroleum of the readily oxidized porphyrins. At some stage in a cycle from organic matter to petroleum, the material must become buried and preserved within the sediments. Whether the burial is closer to the primary organic end of the cycle or closer to the ultimate petroleum end is still unknown. Theories have been advanced to explain the burial at all stages in the cycle.[13]

Kenneth K. Landes, a professor of geology at the University of Michigan, published in 1951 another very prominent textbook, *Petroleum Geology*. Again, Landes proclaims his certainty that petroleum is of organic origin. However, he raised an interesting concern about biomarkers. These are the remnants of organic life commonly found in oil deposits that petroleum geologists interpret as proving organic material formed the oil. Landes appears to have had his doubts in the following passage:

> Sanders [J. McConnell Sanders] and Waldschmidt [W.A. Waldschmidt] have described such microscopic objects as Foraminifera, diatoms, plant remains, insect scales, and spines, as well as fragments of other materials, present in crude oils. These discoveries may or may not be significant in terms of the origin of oil. It may be that the petroleum in the course of its travels through sedimentary rocks rich in organic materials picked up the microscopic objects.[14]

Yet Landes embraced orthodoxy with this passage:

> The geological associations of oil and gas have led practically all geologists to reject the inorganic theories as entirely inadequate. Over 99 percent of the world's oil and gas so far produced has come from sedimentary rocks. Furthermore, in every oil-producing region the sedimentary rock section includes beds which either contain or have contained considerable organic material.[15]

In recent years, petroleum geology textbooks have acknowledged the possibility of abiotic oil but have dismissed its commercial importance. The 2015 third edition of Richard C. Selley and Stephen A. Sonnenberg's textbook *Elements of Petroleum Geology* weighs the debate as follows:

> There is now clear evidence for the origin of abiogenic hydrocarbons in the deep crust or mantle, and for its emergence along faults and fractures, notably in midoceanic ridges and intracontinental rifts. Geologists will not fail to note, however, that commercial accumulations of oil are restricted to sedimentary basins.[16]

Selley, a professor at the Royal School of Mines, Imperial College London, and Sonnenberg, a professor at the Colorado School of Mines in Golden, Colorado, stressed that "the apparently unquestionable instances of indigenous oil in basement are rare and not commercially important."[17] Later in the textbook, Shelby and Sonnenberg commented that approximately "99 percent of the world's oil and gas occur in sandstone or carbonite reserves."[18] Although a few sentences later, they acknowledge that several fields are producing oil from basement reserves, including the Hugoton field of Texas and Oklahoma, the Augila field of the Sirte Basin in Libya, and various fields in Long Beach, California.[19] The persistence of the organic theory among petroleum geologists today strongly suggests that the peak oil adherents of M. King Hubbert's "peak oil" theory will resurface if the Biden administration succeeds in curtailing the production of hydrocarbon fuels in the United States.

Yet *Hydrocarbon Chemistry*, a definitive treatment of hydrocarbon chemistry now in a two-volume 2018 third edition, leaves no wiggle room insisting hydrocarbon fuels are all organic in origin. The three distinguished, internationally recognized chemists coauthoring *Hydrocarbon Chemistry* wrote the following in volume I:

> All fossil fuels (coal, oil, gas) are basically hydrocarbons, varying, however, significantly in their H/C ratio [Hydrogen/Carbon ratio]. These are formed over eons by the anaerobic decay of living organisms that is, they are fossilized solar energy. Consequently, all hydrocarbons available for mankind are of biologic origin.[20]

In the following sentence, the coauthors fully endorse the IPCC's theories that anthropogenic CO_2 causes catastrophic global warming.

> When burned they [i.e., hydrocarbon fuels] undergo oxidation to form carbon dioxide and water and, consequently, they are not renewable on a human timescale. Furthermore, their burning (oxidation) results in a large anthropogenic CO_2 emission causing harmful effect in the environment (global warming, rising sea levels, acidification of the oceans, etc.).[21]

That such a definitive scientific treatment of hydrocarbon fuels has become this politically correct is a strong indication the fossil fuel paradigm has transitioned from orthodoxy to the status of a secular, religious-like doctrinal belief. Ironically, this is the exact moment when Thomas Kuhn would predict that the fossil fuel belief system has stretched to the point of breaking.

As we pointed out in chapter 1, the peak oil theory reduces to a tautology if we assume hydrocarbon fuels are organic fossil fuels. Tautological thinking runs deep into the core of chemical science worldwide. The distinction between organic and inorganic chemistry did not last long. In 1828, the German physician Friedrich Wöhler synthesized urea from inorganic starting materials. Urine is produced by many animals, including us, and urea is a significant component of urine. Wöhler conducted his experiment by evaporating a water solution of ammonium cyanate that he had prepared by adding silver cyanate to ammonium chloride. Instead, he got a crystalline material identical to urea, a chemical easily isolated from urine.

This result blew apart any theory that only living organisms can produce organic chemicals. Wöhler clearly showed that no mysterious vital force was needed to synthesize organic chemicals.

Wöhler's experiment should have destroyed the traditional basis for distinguishing between organic and inorganic chemicals. Marc A. Shampo, Ph.D., and Robert A. Kyle, M.D., noted the importance of Wöhler's discovery as follows:

> Urea is important in physiologic chemistry because it is the principal end product of the metabolism of nitrogenous foods in the body and is found in the urine. Before this experiment, most investigators believed that a "life force" motivated or influenced all substances found in plants and animals. They therefore contended that any substance produced by a life process could not be made from inorganic chemicals. Wöhler's experiment ushered in the era of synthetic organic compounds, which led to the production of dyes, dynamite, plastics, sulfa drugs, and synthetic fibers and to the process of petroleum cracking.[22]

Still, even today, the fields of organic and inorganic chemistry are studied separately. The American Chemical Society acknowledges that organic chemistry was originally "limited to the study of compounds produced by living organisms." Yet, the American Chemical Society recognizes organic chemistry as "the study of the structure, properties, reactions, and preparation of carbon-containing compounds. Most organic compounds contain carbon and hydrogen, but they may also include any number of other elements (e.g., nitrogen, oxygen, halogens, phosphorus, silicon, sulfur)."[23] The tautology remains in place in the academic study of chemistry. Organic chemistry is where hydrocarbons are studied. Why? Because dating back to before Wöhler, geologists assume hydrocarbon fuels had an organic origin. Subconsciously the classification of oil as an organic chemical continues to reinforce the fossil fuel theory. If hydrocarbons are organic chemicals, it implies an origin from chemicals that somehow once lived.

Mendeleev's genius was that he did not categorize the elements by distinguishing them by "organic elements" and "inorganic elements." Mendeleev organized his periodic table by ranking the elements according to their atomic weights.[24] He understood that the elements are just substances in the final analysis, and chemical compounds are formulations in which elements combine. Hydrocarbon fuels are not organic because they are compounds of hydrogen and carbon. Hydrocarbon fuels are, by definition, organic only if decaying organic materials are required to produce them. If Earth creates hydrocarbon fuels without decaying organic materials, then fossil fuels as a synonym for hydrocarbon fuels is inappropriate.

The Fischer-Tropsch Process and Synthetic Oil

"Germany has virtually no petroleum deposits," observed Anthony N. Stranges of the Department of History at Texas A&M University, noting a resource reality even today. "Prior to the twentieth century, this was not a serious problem because Germany possessed abundant coal resources. Coal provided for commercial and home heating; it also fulfilled the needs of industry and the military, particularly the navy."[25]

In the opening decade of the twentieth century, however, Germany's fuel requirements began to change. Germany became increasingly dependent upon gasoline and diesel oil engines to fuel automobiles and trucks. Then, the development of commercial airlines made producing aviation fuel another requirement. Germany's ocean-going ships, including their navy, converted from coal to diesel oil as their energy source. "Petroleum was clearly the fuel of the future," Stranges noted, and Germany had a problem. How would twentieth-century Germany develop the abundant gasoline and diesel fuel supplies needed to propel a competitive national industrial economy and mount a world-class military operation second to none in Europe without ample petroleum resources?

Then, in the 1920s, German scientists needed to solve this energy problem to fuel Germany into economic recovery following the devastating loss in World War I and the equally disastrous aftermath of the Treaty of Versailles in 1919.

The solution came from two German chemists, Franz Fischer (1877–1947) and Hans Tropsch (1889–1935), working at the Kaiser Wilhelm Institut für Chemie (Kaiser Wilhelm Institute for Chemistry) in Berlin. In the 1920s, Fisher and Tropsch developed a series of equations that became known as the "Fischer-Tropsch process." Defining a methodology for producing synthetic petroleum from coal, Fisher and Tropsch were aware of the chemistry of carbon monoxide reactions developed since 1900. The hydrogenation of carbon dioxide, producing "water gas" from hydrogen and carbon monoxide, was key to the synthetic production of methane in the early 1900s. In 1923, Fisher and Tropsch realized that alkalized iron turnings (i.e., iron filings resulting as debris from manufacturing processes) at 100–150 atm of hydrogen (i.e., standard atmosphere, with 1 atm being Earth's atmospheric pressure at sea level), plus carbon monoxide, and 400°–500°C produced synthetic hydrocarbons in a catalyzed reaction.[26]

In 1925, Fisher and Tropsch began using an iron-zinc oxide preparation as their first catalyst. They went into commercial operation with a cobalt catalyst. In 1937, the Kaiser Wilhelm Institut für Kohlenforschung (Kaiser Wilhelm Institute for Coal Research) developed, on a laboratory scale, alkalized precipitated iron catalysts that ultimately became the standard for commercial Fischer-Tropsch operations. From 1935–1940, Ruhrchemie (Ruhr Chemical) A.G. in Germany developed the Fischer-Tropsch process on a large commercial scale using synthetic gas containing two volumes of hydrogen per volume of carbon monoxide, compressed to about 7 atm through a granular bed of cobalt catalyst at 185°–205°C. The major products of the synthesis were wax, oil, water, gaseous hydrocarbons, and a small amount of carbon dioxide. The hydrocarbons were largely straight-chained alkanes, i.e., saturated hydrocarbons, with the chemical formula C_nH_{2n+2}. As German scientists refined the Fischer-Tropsch process through World War II, the process produced twenty different hydrocarbon chemical compounds, ranging from propane (C_3H_8) to n-butane (C_4H_{10}) to benzene (C_6H_6) to n-octane (C_8H_{18}) to n-eicosane ($C_{20}H_{42}$). With the Fischer-Tropsch process, German scientists could produce synthetic gasoline, diesel fuel, and aviation fuel from coal.[27]

So, by the 1920s, Fisher and Tropsch had developed a process to produce synthetic hydrocarbons. Their process involved passing hydrogenated carbon gas (H + CO) through an iron (Fe) catalyst at high pressure and intense heat. The result produced methane (CH_4) synthetically. For our discussion here, the importance of the Fischer-Tropsch process was the demonstration that hydrocarbons could be produced synthetically on a commercial basis without the involvement of any organic materials— no microbes, no dead plants, no decaying animals. The Fischer-Tropsch process alone proves hydrocarbon fuels are not necessarily organic in origin, a point Wöhler's synthesis of urea had prefigured a century earlier, in 1828. Some may object that the coal used in the Fischer-Tropsch process is a fossil fuel. We would add that biomass can also be used in the Fischer-Tropsch process to produce synthetic hydrocarbons. But the critical chemical reaction, as German scientists began proving in the early 1900s, was the catalytic hydrogenation of carbon monoxide (CO and H_2) to form C_1 hydrocarbons like methane (CH_4) and methanol (CH_3OH).[28]

During the early 1930s, the Luftwaffe, Germany's military air force, contracted with German industrial giant IG Farben to produce a synthetic high-quality aviation fuel. Germany's military arm, the Wehrmacht, followed suit by hiring IG Farben to produce synthetic diesel fuel. By 1936, IG Farben was no longer an independent company, but a government-private enterprise partnership run by the Nazi government. When Hitler attacked Poland on September 1, 1939, Nazi Germany had fourteen synthetic fuel plants in operation and six more under construction, producing approximately 95 percent of the aviation fuel used by the Luftwaffe. By 1943, using synthetic oil production defined by the Fischer-Tropsch process, Germany had almost three million metric tons of gasoline by hydrogenation of coal. Adding to this diesel fuel, aviation fuel, and various lubricants produced synthetically from coal, Nazi Germany was able to satisfy up to 75 percent of its fuel demand through coal conversion processes made possible by the equations developed in the Fischer-Tropsch process.[29]

Also constrained by lacking extensive national petroleum reserves, imperial Japan followed Nazi Germany into synthetic fuel production. In 1936, Japan calculated that the nation would have had a 400-to-500-year fuel reserve by converting coal to liquid fuel. Japan's seven-year plan of 1937 called for the construction of eighty-seven synthetic fuel plants by 1944, all of them using the Fischer-Tropsch process. The imperial Japanese government set a goal of producing 6.3 million barrels annually of synthetic gasoline and the same quantity of synthetic diesel fuel. While the economic demands of waging war in China and across the Pacific ultimately thwarted Japan's ambitions to produce synthetic oil, Japan still managed to construct fifteen synthetic fuel plants that reached peak production of 717,000 barrels of synthetic fuel in 1944.[30]

Another measure of oil's economic value involves the United States and the Allies' bombing over Germany during World War II. On November 3, 1944, well before the end of the war, President Roosevelt issued a directive calling for a government study to determine whether or not all the bombings served any purpose.[31] What precisely did the dropping over 2.7 million tons of bombs on Europe accomplish?

The resulting *United States Strategic Bombing Study* produced some surprising results. The bombing attack on the German airplane industry culminated in the last week of February 1944, when the U.S. dropped 3,636 tons of bombs on German airframe plants. In that week and the days following, the U.S. and the Allies bombed every known aircraft factory in Germany. But, surprisingly, in 1944, the Nazis manufactured a total of 39,807 aircraft of all kinds. The number in 1942 before the bombing attacks began had only been 15,596. The German aircraft production had increased despite the massive bombing of Nazi aircraft plants.

As the U.S. and the Allies destroyed Germany's aircraft manufacturing plants, the Germans adapted to recover the machinery and disperse the manufacturing. Why? The bombing devastated the buildings, but the machines "showed remarkable durability." The Germans reorganized the management of the aircraft plants and subdivided production into many small units that were immune to massive bombing raids. The result was clear—bombing the plants had not slowed down the Nazis' ability to make new airplanes. The Allied bombing of German oil and chemical production plants told a similar story. By the end of the war, the Germans could produce Messerschmitt fighter planes, but they had no airplane fuel with which to fly them. The output of aviation gasoline from synthetic plants fell from 316,000 tons per month, when the air attacks began in 1943, to 5,000 tons in September 1944, when the U.S. and the Allies had bombed every primary airplane manufacturing plant. Without fuel, the Nazi war machine came to a grinding halt.

In his 2021 book *Stalin's War*, Bard College history professor Sean McMeekin noted that in the early stages of World War II, the British and French developed a plan for waging war on the Soviet Union.[32] On January 4, 1940, the British war cabinet discussed bombing the Baku oil fields. The British knew that three-quarters of Russia's petroleum production came from the Baku oil fields in Azerbaijan, then a part of the Soviet Caucasus. On March 28, 1940, at a Supreme War Council in Paris, the British and French formalized their plans for bombing Soviet oil installations in Baku, later code-named Operation Pike. On April 1, 1940, the British Air Ministry ordered four squadrons of Bristol Blenheim Mk IV bombers, a total of forty-eight bombers, to redeploy and reinforce Britain's Middle East command in Iraq. The British Air Ministry acted after a military reconnaissance flight over the Baku oil fields reported the wooden oil derricks along the Caspian Sea were only seventy yards apart. The ministry realized "incen diary bombs could easily ignite a conflagration of the entire petroleum-saturated area."[33] The bombing raid never happened. Still, McMeekin allowed himself an exercise in hypothetical history. He mused that a British air attack on the Baku oil fields could have created "an alternative world in which the war machines of Stalin and Hitler might have slowly ground to a halt for the lack of oil in the weeks after May 15, 1940."[34]

The Fischer-Tropsch Equations and Inner Earth Processes

The Fischer-Tropsch equations to produce synthetic hydrocarbons describe how Earth could manufacture hydrocarbons synthetically. Current technical descriptions of the Fischer-Tropsch process acknowledge that coal is not necessary to make the synthetic creation of hydrocarbon fuels. The following, for instance, is a contemporary Stanford University 2015 chemistry course description of the Fischer-Tropsch process:

> The Fischer-Tropsch process is a gas-to-liquid (GTL) polymerization technique that turns a carbon source into hydrocarbon chains through the hydrogenation of carbon monoxide by means of a metal catalyst.

The Stanford chemistry course summary continues as follows:

> The carbon source is converted to *syngas*, a combination of carbon monoxide (CO) and hydrogen (H_2) gas, through a process of gasification ($C + H_2O \rightarrow CO + H_2$) [carbon plus water produces carbon monoxide plus hydrogen] where a controlled flow of steam and oxygen is maintained through the source at high temperature and pressure (1200–1400°C and 3MPa [Megapascal] ~ 30 atm) without enough oxygen for complete combustion.[35]

Note: the definition specifies that the process begins with a "carbon source," specified only as "C" in the chemical equation above. The chemistry coursework does not identify that coal or any organic material, such as biomass, is the source of carbon that the Fischer-Tropsch process requires. The Stanford course material suggests that using CO_2 removed from the ocean or the atmosphere as the carbon source would allow the Fischer-Tropsch process to be carbon-neutral.

This subsection will examine the chemicals in Earth's core and mantle and the chemical reactions in Earth's mantle. We will argue that all the chemical reactions that the Fischer-Tropsch equations specify for the synthetic creation of hydrocarbon fuels have been present in inner Earth in geologic time and are still operating today. The main point of this subsection is that the inner Earth is fully capable of manufacturing abiotic hydrocarbon fuels. In other words, Earth's lower mantle acts to produce abiotic oil as if it were a Fischer-Tropsch plant producing synthetic hydrocarbon fuels.

A secondary theme is that the IPCC's concentration on anthropogenic CO_2 emissions fails to appreciate the complexity of Earth's carbon and hydrogen cycles. We will demonstrate that the largest reservoirs of both carbon and hydrogen on Earth reside in its core. Recent scientific studies have established that oxygen is also present in the core, challenging generations of geoscience that have considered it to be oxygen depleted. The carbon and hydrogen cycles in inner Earth have mechanisms whereby carbon and hydrogen emissions reach the surface from the core and mantle. These findings challenge the IPCC calculations that the only significant additions to CO_2 concentrations in the atmosphere are due to humans burning hydrocarbon fuels.

Concentrations of Hydrogen, Carbon, and Oxygen in Earth's Core

Earth's core is composed of two parts: an inner and an outer core. The outer core is about 2,900 km (1,802 miles) below the surface, and the inner core is approximately 5,150 km (3,200 miles) below the surface. The distance from Earth's surface to its center is 6,730 km (4,182 miles). The inner and outer cores combined constitute approximately 15 percent of Earth's volume and 32.5 percent of Earth's mass.[36]

Geoscientists have traditionally believed Earth's inner core consists primarily of solid iron alloyed with a small amount of nickel and lighter elements. A scientific study published in 2014 found a surprisingly large amount of carbon in Earth's inner core, demanding a revision of the traditional view. Seismic studies showed that shear wave (S-wave) travels through the inner core at an anomalously low speed, challenging the notion of the inner core's solidity. The study proposed iron carbide (Fe_7C_3) was the leading candidate component of the inner core. Adding carbon to the inner core provided an excellent match to account for the shear wave anomaly. The 2014 study produced the following surprising conclusion:

> Current estimates of carbon in the mantle ranges between 0.8×10^{20} kg and 12.5×10^{20} kg. If the inner core is made up of Fe_7C_3, its carbon inventory amounts to 60×10^{20} kg, which considerably exceeds the average mantle budget. In this case, the inner core would be the largest carbon reservoir in Earth, accounting for two thirds of its total carbon inventory estimated on the basis of volatility systematics. This model challenges the conventional view that the Earth is highly depleted in carbon, and therefore bears on our understanding of Earth's accretion and early differentiation. Carbon in the core may exchange with shallower reservoirs through mantle convection in combination with grain boundary diffusion through the D" zone. Through Earth's history, this process may have

Abiotic Oil

played a significant role in the outgassing of CO_2 from the interior and the carbon cycle involving the surface and internal reservoirs.[37]

The D" zone is an area at the bottom of Earth's mantle. The D" zone thus lies at the boundary between the mantle and the core.

A scientific study published in *Nature Reviews Earth and Environment* in August 2021 found evidence for a surprising amount of hydrogen to be present as one of the low atomic number "light" elements to be in the core in addition to iron. In Japan, a group of scientists led by Kei Hirose, from the Earth-Life Science Institute at the Tokyo Institute of Technology, and Department of Earth and Planetary Science at the University of Tokyo, began by observing seismic data suggesting a range of light elements were in Earth's core. Traditionally earth scientists have assumed the core was predominately solid iron. But the Japanese scientists concluded that the seismic readings suggested otherwise. Hirose suspected that a substantial amount of hydrogen was in the core, in addition to sulfur, silicon, oxygen, and carbon. Using a diamond anvil apparatus to simulate the temperature and pressure of the young Earth's core, Hirose and his team demonstrated for the first time that hydrogen can bond strongly with iron in these extreme conditions.[38]

Using high-resolution imaging in a secondary ion mass spectroscopy technique, Hirose confirmed that hydrogen under the conditions of the core of the early Earth is iron-loving, or siderophile. "This finding allows us to explore something that affects us in quite a profound way," Hirose said in an interview written by the University of Tokyo. "That hydrogen is siderophile under high pressure tells us that much of the water that came to Earth in mass bombardments during its formation might be in the core as hydrogen today. We estimate there might be as much as 70 oceans' worth of hydrogen locked away down there. Had this remained on the surface as water, Earth may never have known land, and life as we know it would never have evolved." The finding that there could be seventy times more hydrogen in the core than in the oceans should confound IPCC adherents who focus on elements in the atmosphere but not within Earth.[39]

The amount of carbon in the core is also surprising. A team of geoscientists from Florida State University and Rice University published an article in August 2021 that estimated that Earth's outer core may be the largest terrestrial carbon reservoir on the planet.[40] Using seismic data readings from the core, the scientists estimated between 93 percent and 95 percent of all the carbon on Earth resides in the inner and outer cores at the planet's center. "Understanding the composition of the Earth's core is one of the key problems in the solid-earth sciences," said research scientist Mainak Mookherjee, an associate professor of geology in the Department of Earth, Ocean, and Atmospheric Sciences at Florida State University. "We know the planet's core is largely iron, but the density of iron is greater than that of the core. There must be lighter elements in the core that reduce its density. Carbon is one consideration, and we are providing better constraints as to how much might be there."[41]

As recently as 2011, ScienceDaily reported that geoscientists had determined Earth's core was deprived of oxygen.[42] Ten years later, in 2021, a team of scientists from the School of Earth and Environment at the University of Leeds and their colleagues at the Department of Earth Sciences and Thomas Young Centre at the University College of London reported oxygen was present in Earth's core after all.[43] "Chemical interactions between metal and silicates at the core-mantle boundary (CMB) are now thought to lead to transfer of oxygen into Earth's liquid core," the geoscientists wrote. "Previous models of FeO [iron oxide, also known as ferrous oxide] transfer have considered a solid mantle; however, several lines of evidence suggest that the lowermost mantle could have remained above its solidus long after core formation was complete, which would allow much faster mass transfer." In chemistry, the solidus defines a temperature range above which solids melt and below which substances are solid (i.e., crystallized). The authors identified that the main power for Earth's magnetic field involved the release of light elements, including oxygen, to the liquid outer core "due to the ongoing growth of the solid inner core." They also noted that mass exchange between Earth's core and mantle "depends on the nature of the light elements and the physical conditions at the CMB." The focus of their research established oxygen transferred into the core as ferrous oxide (FeO). Their bottom-line conclusion stated more simply was that "chemical interactions at the core-mantle boundary lead to oxygen transfer to the core."[44]

A 2020 study led by geoscientist Jung-Fu Lin at the University of Texas, Austin, used a diamond anvil apparatus that allows the application of high temperature and pressure suggested that Earth's core did not form early in Earth's history.[45] The study placed the date of creation for Earth's solid core at between one billion to 1.3 billion years ago, while the planet itself is estimated to be some 4.6 billion years old. Lin's scientific team noted that the 1–1.3 billion-years-ago estimate for forming the core coincided with paleomagnetic studies of ancient rock formations that revealed

Earth's magnetic field strengthened suddenly between one billion and 1.5 billion years ago. The late-date formation of the core adds a dimension to the expanding Earth theory. It adds the building of a series of conditions within Earth that could produce rapid and dramatic changes consistent with predictions deriving from catastrophe theory mathematics.

Carbon, Hydrogen, and Oxygen Reactions in Earth's Mantle

The mantle lies between the superheated core and the thin outer layer, the crust. It is about 2,900 km (1,802 miles) thick and makes up approximately 84 percent of Earth's volume and 67 percent of its mass.[46]

A leading college textbook on structural geography describes this earth structure as containing two outer shells distinguished by their chemical composition and mineralogy. The text notes the following:

> The upper shell, referred to as the crust, is made up of various sedimentary, metamorphic, and igneous rocks that are rich in silica, and are composed primarily of the minerals feldspar and quartz. The lower shell, called the mantle, is primarily peridotite, an igneous rock that is relatively poor in silica and is composed mostly of the minerals olivine and pyroxene. Due in large measure to the different densities of these common minerals, crustal rocks typically are less dense than mantle rocks. The oceanic crust is up to about 10 km thick and the continental crust is about 50 km thick, whereas the mantle is about 1,800 km thick.[47]

As we saw in the previous chapter, subduction zones are areas where a plate with oceanic crusts descends beneath a plate with continental crust. In the process of subduction, organic and inorganic carbon are both drawn into the mantle. A 2019 study published in *Nature* demonstrated that subducting carbon resides in the ocean in the form of carbonate shells and remains of marine organisms, as well as carbonite in the oceanic lithosphere. The lithosphere is one of the top layers of the earth, composed of the crust and the upper part of the mantle. Not all the subducted carbon in the mantle is recycled to Earth's surface. Deeply subducted carbon potentially forms diamonds.[48] A 2016 study published in *Nature Geoscience* studied boron isotopes in subducted carbonatites from forty million to 2.6 billion years ago. The study provided evidence for carbon of primordial origin to be in Earth's mantle. But the study also found that the subduction rate has varied over geologic time. During the first two billion years or so, the mantle was much hotter than today, a phenomenon that prevented subduction plates from penetrating the mantle as deeply as today. During the last two billion years, a cooler mantle has allowed subduction plates to move at greater depths, possibly to Earth's core-mantle border.[49]

Earth's mantle also contains abundant hydrogen. One of the chemical processes that release hydrogen into the mantle is serpentinization. We also know that seawater percolates through tectonic fractures in subduction zones. Serpentinization occurs when this percolation water transforms ultramafic rocks into the crystal structure of the minerals found in the rock. Ultramafic rocks are high in magnesium and silica, such as igneous olivine and peridotite. Ultramafic rocks are igneous rocks abundant in the mantle composed of magnesium and silicon. An example of the process is the serpentinization of peridotite into the mineral serpentine. Peridotite is an ultramafic igneous rock. Peridotite consists primarily of olivine and other iron- and magnesia-rich minerals (generally pyroxene).[50] In serpentinization, water acts as a catalyst. The chemical reaction causes the iron and magnesium minerals in the ultramafic rocks to transform into serpentine-group minerals (e.g., antigorite, chrysotile, and lizardite). The chemical reaction involved in serpentinization releases hydrogen through a dehydroxylation process.[51] Dehydroxylation consists of a heating process in which the hydroxyl group (OH) is released from the ocean water (H_2O) involved in the chemical reaction.[52]

Another complexity IPCC adherents need to consider in comprehending the intricacy of Earth's hydrogen cycle is discussed in a 2017 scientific study published in the *Proceedings of the National Academy of Sciences*. Using advanced diffraction tools, the scientific team found that hydrogen freed from goethite ($FeO2H$) can rise through Earth's mantle to the surface.[53] Goethite is a hydrous compound found in subduction slabs deep in the lower mantle. Subducted ocean water also acts as a catalyst in freeing hydrogen from goethite in the mantle.[54]

Oxygen is the most abundant element in Earth's mantle. The top three elements in the mantle are oxygen (45 percent), magnesium (23 percent), and silicon (22 percent). Scientists from the University of Bonn in Germany demonstrated in laboratory studies that majorite, at a depth of several hundred kilometers underground, stores oxygen under high pressures and temperatures. Majorite is a granite-like mineral found in the mantle that acts as an oxygen reserve. The Bonn scientists also stressed that the majorite acts as an oxygen elevator. Professor Christian Ballhaus from the Mineralogical Institute at the University of Bonn emphasized the importance of majorite's oxygen mechanism in the mantle to life on Earth. "According to our findings, planets below a certain size hardly have

Abiotic Oil

any chance of forming a stable atmosphere with a high water content," explained Arno Rohrbach, a doctoral student at the Mineralogical Institute. "The pressure in their mantle is just not high enough to store sufficient oxygen in the rock and release it again to the surface." Nearing Earth's surface, the pressure in the mantle becomes too weak to maintain the majorite. As the majorite decomposes, the oxygen is released. The released oxygen bonds with hydrogen to form water. Ballhaus stressed that without this mechanism, Earth would not be known as the Blue Planet.[55]

Earlier in the chapter, we identified that gasification produced hydrogenated carbon monoxide in the Fischer-Tropsch process. The gasification chemical equation is $C + H_2O \rightarrow CO + H_2$. As just shown, hydrogen, carbon, and oxygen are all present in Earth's mantle. So too, carbon monoxide is present.[56] Geoscientists have established that there is possibly as much water in the mantle as in the oceans.[57] Moreover, research published in 2013 found that deep in the mantle, at the high temperatures and pressures, the hydrogenation of carbon takes place relatively easily. Thus, all the chemical reactions, temperature, and pressure conditions necessary to make hydrocarbon fuels are present within the mantle.[58]

Earth's mantle forces are also sufficient to cause hydrocarbon fuels to rise to the surface. Convection in the mantle is a process that occurs when materials near the core heat up and rise to the surface. Suppose a Fischer-Tropsch–like process occurs within the mantle of Earth. As hydrocarbons form and heat up, the process of convection could carry the deep-Earth abiotic hydrocarbons so formed to Earth's surface. We now have the required conditions for abiotic oil created in the mantle to pass through tectonic fractures in the bedrock crust and pool in sedimentary rock at Earth's surface. In the process of migrating into sedimentary rock reservoirs, the abiotic hydrocarbons created in the mantle pick up organic biomarkers. The organic biomarkers in Earth's abiotic oil have tricked Western petroleum geologists into thinking hydrocarbon fuels are organic in origin.[59]

As we continue to learn more about the deep-Earth cycles of hydrogen, carbon, and oxygen, the IPCC's assumption becomes increasingly suspicious. While conventionally trained petroleum geologists refuse to accept that Earth produces abiotic hydrocarbon fuels continuously, Russian and Ukrainian petroleum geologists have understood that the creation of hydrocarbon fuels are a natural part of the planet's hydrogen, carbon, and oxygen cycles, not a process depending upon abundant organic material already being here to decay. The IPCC assumes humans bear the primary responsibility for adding extremely high CO_2 into the atmosphere since the Industrial Revolution. But seeing humans as the cause of an imminent Anthropocene climate crisis and possible sixth extinction requires the IPCC to assume that Earth's hydrogen, carbon, and oxygen cycles cannot absorb the CO_2 produced by our use of hydrocarbon fuels as our primary energy source. More CO_2 in the atmosphere causes Earth to be greener. Plant photosynthesis absorbs solar radiation, acting as a negative feedback mechanism for temperature by keeping solar radiation from directly hitting Earth's surface. This plant temperature feedback mechanism is similar to clouds acting as shields to bounce solar radiation back into space.

Plants also cool the temperature through a process of transpiration. When the surrounding atmosphere heats up, plants release excess water into the air from their leaves. As NASA explains, a forest canopy or a vast expanse of grassland releases large amounts of transpiration as temperatures heat up.[60] The increased water vapor in the atmosphere from transpiration causes more precipitation and cloud cover. Transpiration then is another way plants act as a negative feedback mechanism cooling Earth. While NASA has left this discussion on the NASA website, the organization has added the following comment: "This page contains archived content and is no longer being updated. At the time of publication, it represented the best available science."[61] Since NASA has taken the politically correct path of supporting the IPCC on global warming, the inconvenient scientific principles discussed in the article appear headed for the government's Orwellian "memory hole."[62]

More abundant plant life also removes a greater quantity of CO_2 from the atmosphere. NASA has retained on its website a 2016 article reporting on a study published that year in *Nature Climate Change*. It was run by a group of thirty-two climate scientists from twenty-four institutions in eight countries. The study used satellite data from NASA's Moderate Resolution Imaging Spectrometer and the National Oceanic and Atmospheric Administration's Advanced Very High Resolution Radiometer to help determine the leaf area index, i.e., the amount of leaf cover, over Earth's vegetated regions. The scientists found that from a quarter to a half of Earth's vegetated lands has shown significant greening over the previous thirty-five years due to rising levels of atmospheric CO_2. The study concluded that the greening of Earth represented an increase in leaves on plants and trees equivalent to an area two times the continental United States. NASA had the following quotation from one of the researchers:

Results showed that carbon dioxide fertilization explains 70 percent of the greening effect, said coauthor Ragna Myneni, a professor in the Department of Earth and Environment at Boston University. "The second most important driver is nitrogen, at 9 percent. So we see what an outsized role CO_2 plays in this process."[63]

NASA appears to have liked the professor making CO_2 the culprit. But once again, why is CO_2 causing the greening of Earth such an environmental negative? Neither the professor nor NASA appears to have appreciated that plants act as a negative feedback mechanism cooling the planet. Nor did the professor or NASA bother to consider the obvious benefits a warmer, greener Earth has for life, including human life, on the planet.

Ironically, by forcing decarbonization, the IPCC could cause global warming by reducing plant life on Earth. Now that we see the extent to which the hydrogen, carbon, and oxygen cycles operate within the inner Earth, IPCC-caused decarbonization could adversely affect hydrogen, carbon, and oxygen cycle processes running in the mantle as well. Are we certain a human goal to decarbonize on a global scale will not adversely affect these complex temperature management systems Earth utilizes naturally? Patrick Moore, Ph.D., the cofounder of Greenpeace, emphasized the importance of CO_2 in an essay he published in 2018. "During the past 150 million years CO_2 had steadily declined to such a low level that plants were seriously threatened with starvation during the peak of the last few glacial cycles," he wrote. "Thankfully, our CO_2 emissions have inadvertently reversed that trend, bringing some balance back to the global carbon cycle. All of this can be verified yet the narrative of 'climate catastrophe,' which has no basis in science, is hollered from rooftops around the world."[64]

The IPCC appears indifferent to whether hydrocarbon fuels are abiotic or organic in origin. Either way, the IPCC seems to have no problem with Earth creating hydrocarbon fuels, provided we humans stop using them. The central point here is that a natural outcome of the hydrogen, carbon, and oxygen cycles on Earth is oil being produced deep within the planet without the assistance of human beings or any other living organism. Once hydrocarbon chemists concede that the earth makes hydrocarbon fuels abiotically, we intend to stress that Earth's natural processes only create substances it needs.

J.F. Kenney and the Russian-Ukrainian School

In the West, we owe our appreciation of the Russian-Ukrainian theory of abiotic petroleum to petroleum geologist J.F. "Jack" Kenney, who owned and operated the Gas Resources Corporation in Houston, Texas. Kenney was a hands-on international oil entrepreneur affiliated with the Institute of Physics of the Earth (IPE), a part of the Russian Academy of Sciences. The IPE, founded in 1928, one of the oldest scientific institutes in the Russian Academy of Sciences, is a prominent research center of global and national geophysics. The academic side of the IPE has built university-level schools in geophysics, seismology, experimental geophysics, and geo-electromagnetic research.[65] In the 1960s, Kenney began coauthoring scientific publications with geologists at the IPE who were doing leading research into the abiotic theory of the origin of oil. Kenney coauthored these articles with credentials both from the IPE and from his Houston corporation. Though he passed away in 2004, Kenney's website, https://www.gasresources.net/, contains an excellent description of his work and an archive of the essential papers he coauthored with Russian IPE geoscience colleagues.[66]

After World War II, Stalin determined that the Soviet Union would never be vulnerable again because of a dependence on foreign oil. In 1947, the U.S.S.R. had limited oil resources, the largest of which were still the Baku oil fields in present-day Azerbaijan. Russian petroleum geologists were convinced that the Baku oil fields were depleting and near exhaustion after World War II. During the war, Russia had occupied Iran, but U.S. President Harry Truman was determined to force Russia out of Iran, believing the Soviets were bent on expansion. Truman took the case to the United Nations and accused the Soviets of interfering with a foreign nation. On March 25, 1946, the Soviets announced they would begin withdrawing their military forces from Iran within six weeks.[67] As Kenney described, Stalin responded by initiating a "Manhattan Project"–type effort after the pullout from Iran to study every aspect of petroleum to ascertain if Russia had any commercially exploitable petroleum reserves within the country. By 1951, Russian petroleum geologist Nikolai Kudryavtsev articulated what today has become known as the Russian-Ukrainian Theory of Deep, Abiotic Petroleum Origins. Between 1940 and 1995, Russian scientists published more than 300 scientific publications on the Fischer-Tropsch process while obtaining some 170 Fischer-Tropsch patents.[68] Since 1951, Russian and Ukrainian geoscientists have published hundreds of scientific papers rigorously exploring the abiotic theory. However, except for Kenney's efforts, this body of geoscience has remained largely unknown in the West, primarily due to language.

In recent years, Swedish geophysicists have advanced much of the early work Russian and Ukrainian geoscientists did to promote the abiotic oil theory. In 2010, Vladimir Kutcherov, from the Division of Heat and Power Technology at the Royal Institute of Technology, in Stockholm, Sweden, and Vladilen Krayushkin from the Laboratory of Inorganic Petroleum Origin at the Institute of Geological Sciences in the National Academy of Sciences, in Kyiv, Ukraine, coauthored an essential article in *Reviews of Geophysics*. The paper bore the title "Deep-Seated Abiogenic Origin of Petroleum: From Geological Assessment to Physical Theory." Kutcherov and Krayushkin presented their reasons for concluding the abiogenic theory had reached a new level of scientific proof. They proclaimed the following:

> Experimental results and geological investigations presented in this article convincingly confirm the main postulates of the theory and allow us to reexamine the structure, size, and locality distributions of the world's hydrocarbon reserves.[69]

Kutcherov and Krayushkin detailed experiments conducted in Russia that resulted in 1990 with the patenting of a high-pressure chamber that, when fully sealed, reached pressures of 50 kbar and temperatures of 1,200°C for several hours. Using 99 percent pure solid iron oxide, FeO, calcium carbonate, $CaCO_3$, and double-distilled water, H_2O, the repeated experiments produced alkanes, i.e., saturated hydrocarbons, with the chemical formula C_nH_{2n+2}. The chemical synthesis closely followed the Fischer-Tropsch formula, using calcium carbonate for the source of carbon, water for hydrogen, and zinc for the catalyst in a reaction that called for the hydrogenation of carbon monoxide. The remainder of the published thirty-page scientific paper analyzed oil fields worldwide, arguing the oil fields' characteristics are not consistent with organic oil expectations. For instance, Kutcherov and Krayushkin document natural gas and petroleum resources in Precambrian crystalline shields where there was no involvement of the sedimentary source rock.

Scientists in the United States have replicated the laboratory experiments that Kutcherov and Krayushkin discussed. In 2004, Henry Scott of Indiana University in South Bend organized a research team to see if they could produce methane in a laboratory without using organic materials of any kind. Simply put, the scientists were trying to see if iron oxide, calcium carbonate, and water would produce methane under pressures and temperatures comparable to those experienced in Earth's upper mantle. The research team included Dudley Herschbach, a Harvard University research professor of science and recipient of the 1986 Nobel Prize in Chemistry, and other scientific colleagues from Harvard University, the Carnegie Institution of Washington, and the Lawrence Livermore National Lab.[70]

The research protocol called for generating hydrocarbons in a chemical reaction involving iron oxide, calcium carbonate, and water at temperatures as hot as 500°C and under pressures as high as 11 GPa (gigapascals). One gigapascal is equivalent to 10,000 atmospheres. To experiment, the scientists designed a "diamond anvil cell" mechanism consisting of two diamonds, each about three millimeters high (about one-eighth inch). The tips of the diamonds pointed together to compress a small metal plate. The plate held the iron oxide, the calcium carbonate, and the water that the scientists wanted to force together. The scientists then conducted a variety of highly accurate spectroscopic analyses on the sample material that resulted. Herschbach explained the diamonds were ideal material for the experiment because, as one of the "hardest substances on earth, they can withstand the tremendous force, and because they're transparent, scientists can use beams of light and X-rays to identify what's inside the cell without pulling the diamonds apart."[71]

Again, the experiment worked. The scientists found they could synthetically produce methane, CH_4, the principal component of natural gas. Laurence Fried of Livermore Laboratory's Chemistry and Minerals Science Directorate summed up the importance of these findings as follows:

> The results demonstrate that methane readily forms by the reaction of marble with iron-rich minerals and water under conditions typical in Earth's upper mantle. This suggests there may be untapped methane reserves well below Earth's surface. Our calculations show that methane is thermodynamically stable under conditions typical of Earth's mantle, indicating that such reserves could potentially exist for millions of years.

Fried continued:

> At temperatures above 2,200 degrees Fahrenheit, we found that the carbon in calcite formed carbon dioxide rather than methane. This implies that methane in the interior of Earth might exist at depths between 100 and 200 kilometers. This has broad implications for the hydrocarbon reserves

of our planet and could indicate that methane is more prevalent in the mantle than previously thought. Due to the vast size of Earth's mantle, hydrocarbon reserves in the mantle could be much larger than reserves currently found in Earth's crust.[72]

In a separate scientific paper published in *Nature Geoscience*, Kutcherov and his colleagues at the Geophysical Laboratory, Carnegie Institution of Washington, reported on additional diamond-anvil experiments. Here Kutcherov exposed methane to pressures higher than 2 GPa and to temperatures in the range of 1,000°K (726.85°C) to 1,500°K (1,226.85°C). It formed saturated hydrocarbons containing two to four carbons (ethane, C_2H_6; propane, C_3H_8; and butane, C_4H_{10}), molecular hydrogen, and graphite. Conversely, the research scientists found the exposure of ethane C_2H_6, to similar conditions resulted in methane CH_4. Kutcherov and his colleagues concluded that their experiments with ethane suggested the synthesis of saturated hydrocarbons is reversible. "Our results support the suggestion that hydrocarbons heavier than methane can be produced by abiogenic processes in the upper mantle."[73] In an interview with the Swedish Research Council, Kutcherov spoke clearly about the results of this experimental laboratory research.[74] "There is no doubt that our research proves that crude oil and natural gas are generated without the involvement of fossils," Kucherov said. "All types of bedrock can serve as reservoirs of oil."[75] He added there is no way that fossil oil, with the help of gravity or other forces, could have seeped down to a depth of 10.5 km (6.5 miles) in the state of Texas, for example, that he noted is rich in oil deposits.

Thomas Gold and the Deep Hot Biosphere

Thomas Gold, a brilliant scientist, was born in Vienna on May 22, 1920. His father, Max Gold, was a doctor of law who served as CEO of ÖMAG, a large industrial mining and metal corporation. In the 1930s, Max Gold moved the family to Berlin to become CEO of another large company in the same industry. In 1933, when Hitler assumed power as Germany's chancellor, Max Gold, a Jew, decided to flee Germany to return to Austria. Thomas Gold spent his early years in Switzerland, attending the Lyceum Alpinum Zuoz boarding school near St. Moritz. When he graduated at age seventeen, he joined his family, who had fled back to Austria by then.

On March 12, 1938, at the Anschluss, Germany invaded Austria and annexed it into Nazi Germany. Gold and his family went to the U.K. with stateless papers. In September 1939, when war was declared, the

British government, fearing invasion by Hitler's Nazi Germany, rounded up hundreds of other Germans and Austrians who were technically enemy aliens. Gold had entered Trinity College in 1939, where he began studying mechanical sciences. But when the internment began, the British government picked up Gold and shipped him to Canada with some 800 others.

Fifteen months later, when the U.K. released most of these detainees, Gold returned to the U.K., and in 1942 he resumed his studies at Cambridge University, switching from mechanical sciences to physics. Gold left Cambridge in 1952 to serve as the chief assistant to Astronomer Royal at the Royal Greenwich Observatory in Sussex, England. In 1956, he moved to Cambridge, Massachusetts, to serve as professor of astronomy at Harvard University. In early 1959, Cornell hired Gold away from Harvard, offering him a professorship in the Department of Astronomy, with an assignment to set up an interdisciplinary unit for radiophysics and space research. Gold chaired the astronomy department and directed the Center for Radiophysics and Space Research, where he hired many prominent astronomers, including Carl Sagan. But he had to wait until 1969 to get his doctorate when Cambridge University finally decided to bestow upon him an honorary degree. Gold was a Fellow of the Royal Society in London and a member of the U.S. National Academy of Sciences.[76]

Gold has a long list of scientific achievements to his credit. In April 1942, Gold joined the British naval research establishment to become a member of the radar establishment at the Admiralty. He worked with the theory group headed by two scientists, Fred Hoyle and Hermann Bondi, also famous. There, Gold directed the development of new radar devices, which played a significant role in defending the U.K. against Nazi air attacks. In 1946, when he was a graduate student in astrophysics at Cambridge University, he was intrigued with a problem that had stumped auditory physiologists for years. How did the human ear manage to distinguish so finely the subtleties of musical notes? The conventional scientific wisdom then was a nineteenth-century idea first proposed by the German physicist Hermann von Helmholtz. The accepted theory was that the inner ear functioned as a series of "strings" vibrating at different frequencies. The conventional wisdom held that the ear was the rough instrument that took in the noise. The brain was the vital organ of hearing, distinguishing between the tones to identify individual notes and combinations. Gold disagreed. His work on radar convinced him that the ear as a detecting instrument had to function more finely, truly adding something to distinguish the sounds. He believed the ear amplified incoming noise by adding energy to the detected frequencies before the ear transmitted the signals to the brain. Not until the 1970s did physiologists finally conclude that the inner ear contained a series of fine hairs that did act as amplifiers, exactly as Gold had suggested years before.

Then, in 1955, Gold suggested, contrary to conventional wisdom, that a fine rock powder covered the moon's surface. Again, he was vindicated, but not until Apollo 11 touched down on the moon in 1969 and the world watched as Neil Armstrong hopped around, kicking up a fine-grain powder of rock as he moved along. Gold was one of the 110 scientists worldwide who received moon soil samples from the Apollo 11 mission to test. In 1967 Gold advanced the theory of pulsars by suggesting that pulsars are neutron stars that spin out radio waves as they rotate. Traditional astronomers ridiculed the idea until astronomers discovered a pulsar in the Crab Nebula, a neutron star that spun out radio waves as it revolved.

Gold's thinking about oil began with his primary academic discipline, astronomy. As an astronomer versed in spectroscopy to determine the chemical composition of stars and planets, Gold understood that hydrocarbons are abundant in the universe. Since the early part of the twentieth century, spectrographs that analyzed wavelengths have permitted astronomers to determine with certainty that carbon is the fourth most abundant element in the universe, right after hydrogen, helium, and oxygen. Furthermore, among planetary bodies, "carbon is found mostly in compounds with hydrogen—hydrocarbons—which, at different temperatures and pressures, may be gaseous, liquid, or solid. Astronomical techniques have thus produced clear and indisputable evidence that hydrocarbons are major constituents of bodies great and small within our solar system (and beyond)."[77] In other words, Gold understood that hydrocarbons are not organic chemicals resulting from life processes on Earth, as proponents of the fossil fuel theory commonly assume. Instead, hydrogen and carbon are elements readily available in the universe, elements that combine with carbon to form hydrocarbons, whether life is present or not.

"Power from Earth:" Deep-Earth Gas

Thomas Gold published his first of two books on abiotic hydrocarbons in 1987 entitled *Power from Earth: Deep Earth Gas—Energy for the Future*.[78] In chapter 7, "Where Are Oil and Gas Found?" Gold puzzled that the giant oil fields in the Middle East did not have the geological characteristics that Western petroleum geologists presumed were necessary to produce oil biologically. Gold wrote:

In detail, the oilfields of the area have little in common. Some are in the folded mountains of Iran, some in the flat deposits of the Arabian desert. The oil and the underlying gas fields span over quite different geological epochs, have different reservoir rocks and quite different caprocks. The search for organic source-rocks responsible for the world's largest oilfields has not led to any clear consensus. Sediments of quite different type and age have been suggested here and there, and evidently quite different materials serve as caprocks. The quantities of organic sediments have been regarded as inadequate for the production of all the oil and gas, and would probably be seen to be much more inadequate still if one allowed for natural seepage in the area.[79]

He speculated that the reason the Middle East has such giant oil fields is presumably "the mantle of the earth in that area happens to be particularly hydrocarbon rich."[80] He also observed that oil and gas producing wells in Indonesia and Burma were the same regions in those countries that coincided with volcanic and seismic belts, i.e., volcanoes and earthquakes.[81] He concluded that the mantle produces hydrocarbons in an abiogenic process and that hydrocarbons escape the mantle through fractures in the crystalline basement bedrock of Earth's crust.[82] He was particularly fascinated with the amount of methane formed at a depth of 150–300 km (62–186 miles). He again stressed deep-Earth methane gas made its "way to the surface through the crust."[83] He concluded that if deep-Earth methane is plentiful, then "the economic outlook would be greatly improved and many countries would breathe more freely when they are relieved from the obligation to build or expand a nuclear power industry in a hurry, or from the need to depend on imported fuels."[84]

The Deep, Hot Biosphere

In his second book published twelve years later, in 1999, entitled *The Deep Hot Biosphere: The Myth of Fossil Fuels*,[85] Gold advanced his thoughts on abiogenic deep-Earth methane vents. What fascinated Gold was the abundant life scientists had begun discovering at great ocean depths with no sunlight. He puzzled at how this life at great ocean depths could survive without sunlight. He became fascinated that life at these ocean depths were huddled around deep-sea vents that were exuding menthane, and he came up with the concept of a "deep, hot biosphere." He explained the following:

> In these ocean vents, a borderland between the surface and the deep biospheres, there may be some atmospheric oxygen available that was

carried down in solution in the cold ocean water. If this were sufficient for converting all the methane supplied from the vents into carbon dioxide and water, then this borderland province would be dependent on surface biological processes, and it would not be an outpost of what I suggest is an independent realm of life stretching down into the rocks below. It seems doubtful that the prolific life at these concentrated locations on the ocean floor could receive enough waterborne atmospheric oxygen, but a firm answer is not yet known. However, this issue is not of central importance. We now know of many cases where we can probe so far down into the deep biosphere that atmospheric oxygen has absolutely no access, and we observe generally similar metabolic processes taking place there. Where does the necessary oxygen come from?[86]

Around these ocean vents, marine scientists discovered many living creatures, from simple organisms (such as bacteria) to more advanced microorganisms (such as tubeworms). How could these macrofauna live in seawater so deep no light could penetrate that far? He realized that the microbes were living off the food base of the deep-sea methane vents, and more advanced macrofauna survived by feeding on the microbes. Gold explained this phenomenon as follows:

> Two decades of studies have revealed that these microbes feed on molecules gushing from the vents: hydrogen (H_2), hydrogen sulfide (H_2S), and methane (CH_4), each of which can supply energy only if oxygen is available. No known animal can feed on any of these chemicals directly, but animals can feed on microbes that do. What is particularly remarkable about the deep-ocean vent communities is that many of the macrofauna seem to be dependent on symbiotic partnerships with the microbes.[87]

From there, Gold came to another startling conclusion. The presence of methane in the output of deep-ocean vents assumed primary importance because the methane could be the source of the carbon required for the microbes to live and the source of chemical energy. He explained his realization as follows:

> Hydrocarbons bear a structural resemblance to foods we eat that are derived from photosynthesizers. For example, the only material difference between a molecule of hexane (a six-carbon form of petroleum) and a molecule of glucose (a six-carbon sugar, common in foods at the surface) is that hydrogen atoms surround the chain of carbon in hexane, whereas water molecules surround the chain of carbon in sugar. The hexane

C_6H_{14} is a *hydrocarbon*, whereas the sugar $C_6H_{12}O_6$ is a *carbohydrate*. The terminological difference is subtle but important. For us animals, the carbohydrate is a food, the hydrocarbon poison. Nevertheless, the biological idiosyncrasies of our own tribe of complex life should not be allowed to constrain our judgment as to the possibilities—indeed preferences—among the multi-talented microbes. They might well have a metabolism that requires an input of petroleum.[88]

Gold contemplated that the deep-ocean microbes feeding on deep-Earth methane were the beginning of life on Earth. This perception helps explain why 80 percent of Earth's history, the entirety of Precambrian time, was needed to develop life on Earth. Complex structures like trilobites, a now-extinct form of marine arthropods, only emerged in the Early Cambrian period some 521 million years ago. Gold proposed that the source of energy for life on Earth was not photosynthesis. Instead, he suggested that the degassing of hydrocarbons in deep-sea hydrothermal vents was crucial to life-forming Earth.[89] Gold realized the carbon cycle on Earth began with deep-Earth hydrocarbons. He rejected the conventional notion "that hydrocarbons present within the earth's upper crust are derived strictly from plant and animal debris transformed by geological processes," insisting instead that hydrocarbons exuding from these deep-sea hydrothermal vents played a critical role in the *origin* of life on Earth by feeding the microorganisms that live there.[90]

Equally startling, Gold's contemplation of the deep, hot biosphere led him to an essential perception that atmospheric CO_2 was a by-product of Earth's carbon cycle, not a central feature. Gold was fascinated by ocean methane hydrate structures. Ocean methane hydrates are white, ice-like solids in which microscopic cages composed of water molecules trap methane molecules.[91] Methane hydrate structures also exist under the Arctic permafrost. Gold puzzled why more ocean hydrates were not made up of CO_2 if CO_2 was such a central part of Earth's carbon cycle. He analyzed the following:

> Hydrates made up of CO_2 rather than methane can exist also, though over a smaller range of temperature and pressure than methane hydrates. Nevertheless, there are substantial areas of ocean floor that could support CO_2 hydrates, but few—if any—such samples have been found.[92]

He noted that often "there is more carbon in the methane atoms trapped in a deposit of hydrate than in all the sediments associated with that deposit." He explained his reasoning:

In such instances the conventional explanation of its [carbon's] source (biological materials buried within the sediments) cannot account for the production of so much methane. The methane embedded in the ice lattices must have risen from below, through innumerable cracks in the bedrock. Once a thin, capping layer of the solid forms, the genesis of more such hydrate underneath becomes an inevitability, provided methane continues to upwell.[93]

He observed the conclusion that the source of methane in ocean hydrate lies beneath, not within, the sedimentary layers of Earth, was strengthened by evidence of pockets of free methane gas beneath some regions of hydrate ice and also beneath permafrost layers of Arctic tundra. He noted that in ocean methane hydrate, downward migration of methane gas from overlying sediments did not seem conceivable. "Gases, after all, do not migrate downward in a liquid of greater density," he wrote. "If there is any flow, it is in the reverse direction."[94] His study of ocean methane hydrates reinforced his conclusion that deep-sea methane vents were fundamental to producing life on Earth and how the carbon cycle on Earth operates. He further concluded "that 'gentle' but widespread addition of carbon to the atmosphere is a global phenomenon of diffusion from the ground of methane and other hydrocarbons, no doubt at different rates at different locations and at different times."[95] He combined this with his understanding of the chemical composition of meteorites to conclude that "hydrocarbons and not CO_2-producing compounds will have been the principal input of carbon in the forming earth."[96]

Deep-Sea Hydrothermal Vents

In chapter 2 of his book *The Deep, Hot Biosphere: The Myths of Fossil Fuels*, Thomas Gold discusses the 1977 deep-sea-diving submarine *Alvin*'s exploration of deep-sea vents along the East Pacific Rise, northeast of the Galapagos Islands. The *Alvin* found an "ocean bottom teeming with life" that lived at ranges "far below the deepest possibility for photosynthetic life." A searchlight revealed life was teaming around cracks in the ocean floor that appeared volcanically active in an otherwise barren sea. Here is how Gold described the sea life the *Alvin* found:

> The patch was covered with dense communities of sea animals—some exceptionally large for their kind. Anchored to the rocks, these creatures thrived in the rich borderland where hot fluids from the earth met the marine cold. New to science were species of lemon-yellow mussels and

white-shelled clams that approached a third of a meter in length. Most striking of all were the tube worms, which lurk inside vertical white stalks of their own making, bright red gills protruding from the top. Like the tube worms of shallow waters, these denizens of the deep live clustered together in communities, with tubes oriented outward resembling bristles on a brush. But unlike their more familiar kin, the tube worms of the deep are giants, reaching lengths in excess of two meters.[97]

Further investigations soon found similar hydrothermally active vents in the Atlantic, Pacific, and Indian oceans. In each case, streams of milky fluids and black "smoke" emerged from the seafloor vents. Gold postulated that the hydrocarbon gases venting through volcanically active cracks at the bottom of the sea provided the nutrients these sea bottom microbes and bacteria needed to live at depths where there was no light. "These streams of hydrothermal fluids, heated and enriched in gases and minerals, are now known to be the sources of chemical energy at the base of the vent community's food chain," he wrote.[98]

In 2000, the *Alvin* found a remarkable ecosystem in the mid-Atlantic Ridge at depths of four to five miles below the ocean's surface. Termed the "Lost City," this hydrothermal field was living off deep-Earth hydrocarbon, venting out from calcium carbonate chimneys that reached up almost one hundred yards from the ocean floor. Scientists from the School of Oceanography at the University of Washington in Seattle, joined by scientists from the Woods Hole Oceanographic Institution in Massachusetts, were part of an international team to investigate the Lost City hydrothermal field (LCHF). The international team included Swiss scientists from the Department of Earth Sciences at ETH-Zentrum in Zürich. In March 2005, the oceanographic scientific team published their findings in *Science*.[99] They reported that the serpentinite-hosted Lost City hydrothermal field was "a remarkable submarine ecosystem in which geological, chemical, and biological processes are intimately interlinked." Here is how they summarized their findings:

> Reactions between the seawater and upper mantle peridotite produce methane- and hydrogen-rich fluids, with temperatures ranging from <40° to 90°C at pH 9 to 11, and carbonate chimneys 30 to 60 meters tall. A low diversity of microorganisms related to methane-cycling Archaea thrive in the warm porous interiors of the edifices. Macrofaunal communities show a degree of species diversity at least as high as that of black smoker vent site along the Mid-Atlantic Ridge, but they lack the high

biomasses of chemosynthetic organisms that are typical of volcanically driven systems.[100]

These results appeared to support Gold's hypothesis of sea-bottom life deriving nourishment not from photosynthesis. These sea-bottom creatures, including microorganisms, lived off the abiotic hydrocarbons venting from deep within Earth onto the seafloor through the tall, white carbonate chimney structures of the Lost City.

Then, in February 2008, this same team of ocean scientists published the conclusions of their continued research. Giora Proskurowski of the School of Oceanography at the University of Washington in Seattle was the lead author of the article entitled, "Abiogenic Hydrocarbon Production at Lost City Hydrothermal Field."[101] Proskurowski reported on research led by the University of Washington and the Woods Hole Oceanographic Institution that sampled the hydrogen-rich fluids venting at the bottom of the Atlantic Ocean in the Lost City hydrothermal field. Remarkably, Proskurowski and his team concluded that the hydrogen-rich fluids involved an abiotic synthesis of hydrocarbons generated by seawater chemical reactions with the serpentinite rocks under the Lost City hydrothermal vent field in the Atlantic Ocean. The article rattled the conventional wisdom of petroleum geologists who insisted that all hydrocarbon fuels found in the biosphere had to be organic in origin. Even more impressive, Proskurowski and his team saw the link between the Fischer-Tropsch process and the deep-Earth hydrocarbons exuding from the Lost City hydrothermal chimneys. The article began cautiously introducing the idea that the Lost City chimneys were venting abiogenic hydrocarbons:

> Fischer-Tropsch type (FTT) reactions involve the surface-catalyzed reduction of oxidized carbon to CH_4 and low-molecular-weight hydrocarbons under conditions of excess H_2. This set of reactions has been commonly invoked to explain elevated hydrocarbon concentrations in hydrothermal fluids venting from submarine ultramafic-hosted systems and in springs issuing from ophiolites; however, whether naturally occurring FTT reactions are an important source of hydrocarbons to the biosphere remains unclear.[102]

The oceanographic scientists left no doubt of their conclusion. They had found a source of abiogenic hydrocarbons in the biosphere. They wrote:

> Although CH_4 and higher hydrocarbons have been synthesized by FTT in the gas phase from CO for more than 100 years, only recently were FTT

reactions shown to proceed, albeit with low yields, under aqueous hydrothermal conditions, with dissolved CO_2 as the carbon source.[103]

They concluded:

Here, we show that low-molecular-weight hydrocarbons in high-pH vent fluids from the ultramafic-hosted Lost City Hydrothermal Field (LCHF) at 30°N on the Mid-Atlantic Ridge (MAR) are likely produced abiotically through FTT reactions.[104]

This article in *Science* was the first scientific publication where legitimate scientists were willing to put their reputations on the line by declaring they had found a biosphere case where the interaction of seawater with rocks originating in the mantle produced abiotic hydrocarbons. "Our findings illustrate that the abiotic synthesis of hydrocarbons in nature may occur in the presence of ultramafic rocks, water, and moderate amounts of heat," the oceanographic scientists concluded in print.[105]

The scientists published the FTT equations describing how serpentinization creates a reducing chemical environment characterized by high hydrogen concentrations suited to the production of abiotic hydrocarbons. Serpentinization occurs when this percolation of water transforms ultramafic rocks into the crystal structure of the minerals found in the rock. Ultramafic rocks are high in magnesium and silica, such as igneous olivine and peridotite. These rocks are igneous and meta-igneous rocks typically found in Earth's mantle. In the deep-sea FTT reactions, CO_2 was the carbon source used to combine with the hydrogen produced by serpentinization to form the abiotic hydrocarbons. Proskurowski and his team ruled out seawater bicarbonate as the carbon source for the observed FTT reactions, insisting that "a mantle-derived inorganic carbon source is leached from the host rocks."

Proskurowski and the oceanographic team knew their results would startle traditionally trained petroleum geologists because the hydrocarbons at the Lost City hydrothermal fields were carbon-13 (^{13}C). In both his books, Thomas Gold explained that carbon-13 and carbon-12 (^{12}C) are associated with organic hydrocarbons and abiotic hydrocarbons. But conventional petroleum geologists have always insisted that carbon-12 is the organic isotope of carbon and that all Earth biosphere hydrocarbons are carbon-12. That oil, coal, and natural gas found in Earth's biosphere are carbon-12 is one of the principal arguments conventional petroleum geologists make to argue oil, coal, and natural gas are all organic in origin.

But when the scientists published in *Science* that the carbon they found in the Lost City hydrothermal fields was carbon-13, there was no denying that at least one case of deep-Earth abiogenic hydrocarbons existed.

The Lost Sea scientists had found simple C_1 to C_4 carbon chains coming out of the Lost City's chimneys. The hydrocarbon gases emitted from the Lost City's chimneys were mostly single-bond, straight-chain alkanes, i.e., saturated hydrocarbons, with the chemical formula C_nH_{2n+2}, and ethene (C_2H_4), and other hydrocarbons that the Germans had produced during World War II with the Fischer-Tropsch process. Several Lost City samples found acetylene (C_2H_2), propane (C_3H_8), and propyne (C_3H_4) present. The methane (CH_4) concentrations were a carbon-13 value characteristic of methane produced from serpentinite-hosted vent fields.

Proskurowski and his team also had to rule out that the methane exuding from the LCHF was created organically by organic bicarbonates on the seabed floor. The scientists used accelerator mass spectrometric measurements to apply carbon-14 dating techniques to six methane (CH_4) samples from the LCHF. They wanted to determine if the methane contained any living material, like organic bicarbonates, that could have formed the methane organically. The scientists applied carbon-14 testing to the methane not to date the methane but to see if the methane contained any organic material. Carbon-14 is present in all living and recently expiring things. The half-life of carbon-14 is 5,700 years, meaning every 5,700 years, the amount of carbon-14 in a fossil is only half what it was when the organism died. But if a carbon-14 sample is not radioactive, the carbon is not organic, i.e., not from living things.[106] Thus, the finding that the methane samples contained no radioactive carbon-14 led to the conclusion that whatever produced the methane was not organic. The scientists got carbon-14 (^{14}C) readings from all six methane samples. But the ^{14}C was radioactively dead.

The analysis proved that the carbon source of the methane could not be organic seawater bicarbonate (^{14}C-seawater$_{DIC}$) that had been microbially or abiogenically reduced in the formation of the methane. The absence of organic seawater bicarbonate in the methane samples also suggested to the scientists that organic bicarbonate on the seabed floor had been removed before the production of the hydrocarbons in the vent fluids. The scientists reasoned that precipitation as calcium carbonate ($CaCO_3$) removed the seawater bicarbonate from the seabed floor. After ruling out all possibilities that the methane was organic, the scientists concluded that the serpentinizing basement rock was the source of the carbon in the methane, reasoning that increasing pH leads to carbonate precipitation within the serpentinites. The scientists summarized their findings, saying the "^{14}C content of short-chain hydrocarbons suggests that the requisite carbon for abiotic synthesis is derived by leaching of primordial radiocarbon-dead carbon from mantle host rocks."

Moreover, there was no sediment source rich in organic matter in the Lost City environment. The absence of sea-bottom rich in organic matter alone strongly suggested the methane was abiotic. Thus, the scientists went to great lengths ruling out all possibilities that the methane and other hydrocarbons in the LCHF vents were organic. They took pains to reinforce their conclusion that a thermogenic source had produced the Lost City hydrocarbons. A thermogenic source for the CH_4 would have meant the heat-generating metabolic processes of the deep-sea life forms feeding on the Lost City hydrothermal chimneys had formed the methane. The scientists finally concluded that the high carbon-13 content of the methane from the LCHF vents, as well as the lack of a sediment source rich in organic matter in the path creating the methane, meant the CH_4 was "not appreciably derived from a thermogenic source."

The oceanographic scientists ended their article by stressing that the marine organisms fed on the abiotic hydrocarbons in the Lost City hydrothermal fields. "Lost City may be just one of the many, as yet undiscovered, off-axis hydrothermal systems. Hydrocarbon production by FTT could be a common means for producing precursors of life-essential building blocks in ocean-floor environments or wherever warm ultramafic rocks are in contact with water."

In September 2007, German scientists joined the LCHF team from the Woods Hole Oceanographic Institution to publish in *Nature*.[107] The article showed that marine anaerobic bacteria thrived in consuming short-chain hydrocarbons, including ethane (C_2H_4), propane (C_3H_8), and butane (C_4H_{10}), in a marine environment that lacked sunlight and oxygen. Anaerobic bacteria, by nature, do not live or grow when oxygen is present. In humans, anaerobic bacteria are most common in the gastrointestinal tract.[108] The short-chain hydrocarbons ethane, propane, and butane are all constituents of natural gas. Woods Hole Oceanographic Institution biologist Stefan Sievert, a coauthor of the study, explained that seafloor bacteria could "eat" natural gases like ethane, propane, and butane in a previously unknown way, namely, without oxygen. The bacteria use sulfate instead of oxygen to metabolize natural gases into energy and organic matter. Sievert further explained these marine anaerobic bacteria may have played a role in the evolution of life on early Earth when oxygen was sparse in the oceans and the seafloor hydrothermal vents were spewing hydrocarbons much more common than today. He further detailed these microbes have yet-unknown enzymes that can break down hydrocarbons without heat and oxygen, offering potentially useful catalysts to synthesize compounds.[109]

In 2021, a group of Woods Hole scientists published a study of the Gorda Ridge in the northeast Pacific Ocean off the coast of southern Oregon.[110] This study reported that the consumption of microorganisms that feed off hydrocarbons exuded by hydrothermal vents is essential for Earth's carbon cycles. The scientists found that eukaryotes (or protists) feed off the marine microorganisms, thereby forming a vital link for the food web where carbon transfers to higher tropic levels. The tropic level is the position an organism holds in the food chain. Protists are primarily single-cell microorganisms. Eukaryotes are highly organized protists with a nucleus and specialized cellular machinery called organelles. Algae and amoebas are protists. "The simplest definition is that protists are all the eukaryotic organisms that are not animals, plants, or fungi," explains Alastair Simpson, a professor in the biology department at Dalhousie University.[111] In the paper, the Woods Hole scientists concluded their findings provided the "first estimate of protistan grazing pressure within hydrothermal vent food webs, highlighting the important role that diverse deep-sea protistan communities play in deep-sea carbon cycling."[112]

In his book *The Deep Hot Biosphere*, Thomas Gold noted that mantle-generated methane surging to Earth's surface through deep-sea hydrothermal vents ultimately rose to the surface if not consumed by marine microorganisms. "Any methane that reaches the atmosphere without being oxidized along the way would quickly be oxidized to carbon dioxide in the oxygen-rich atmosphere and there join the pool of atmospheric-oceanic CO_2," he wrote. "What fraction of all the upwelling carbon volatiles would be delivered to the atmosphere as methane, and what fraction as carbon dioxide?"[113] he asked. He pointed out that carbon dioxide coming from volcanoes is well studied. But the large quantities of methane that emerges from abiotic hydrothermal vents in the deep sea go largely unnoticed.

Gold's point is essential. How much of the CO_2 in the atmosphere is abiotic, rising from deep-sea hydrothermal vents that have nothing to do with human beings burning hydrocarbon fuels? If hydrocarbons played a seminal role in creating life on Earth, how could hydrocarbons be organic in origin? The findings of marine microorganisms feeding on the ocean hydrothermal hydrocarbons validated Gold's assumption that life on Earth began with deep-sea microorganisms and was dependent on deep-Earth hydrocarbons as a source of food. Thus, the sequence of life on Earth, the carbon cycle, and the food cycle begin within Earth's mantle.

Gold had an additional deduction essential to this discussion. The exponential growth rates of microbes (and all other forms of life on Earth) require the energy source that supports the life form must arrive in "a metered flow." The energy that sustains life "must be available, but it must not be available all at once." Thus, the flow of life-supporting energy, e.g., the flow of hydrocarbons out of the deep-sea hydrothermal vents, must be metered.[114] If the flow of life-supporting energy stops, the life supported by that energy stops. The inevitable conclusion is that the mantle of Earth continues to make abiotic hydrocarbons on an ongoing basis to continue the life-supporting energy flow at the bottom of the oceans. If Earth's mantle constantly manufactures abiotic hydrocarbons, then abiotic oil, gas, and coal are much more abundant than the organic theory would presuppose. As noted earlier, the IPCC and other fossil fuel believers, presuming dead organic life produces hydrocarbons, must conclude there is a finite amount of organic life in our history. Hence, the amount of oil, coal, and natural gas on Earth must also be limited. If deep-Earth manufactures hydrocarbons continuously, the continuous formation of oil, coal, and natural gas is an ongoing essential part of Earth's carbon cycle, not only in geologic time but also today.

Resolving the "Petroleum Paradox"

In resolving what Gold called "The Petroleum Paradox," i.e., that hydrocarbon fuels contain biological material, Gold came to realize that the deep, hot biosphere was not manifested just by deep-sea hydrothermal vents. The deep, hot biosphere extended miles into Earth. An August 1984 paper published in *Scientific American* by a group of scientists led by Guy Ourisson at the University of Strasbourg found that the quantity of biological debris in petroleum was astonishingly large.[115] The Strasbourg scientific team "expressed the conventional view that biology was essential for the production of hydrocarbons." What Gold came to understand was that petroleum "could be food for a prolific microbial life and thereby create the association between petroleum and biology."[116] In November 1984, Gold published a reply to Ourisson's research in *Scientific American*. Gold's response letter read, in part, as follows:

Abiotic Oil

A widespread early bacterial flora may have arisen when hydrocarbon outgassing of the earth provided a source of chemical energy in the surface layers of the crust where oxygen was abundant owing to the photodissociation of water and the loss of the hydrogen to space. Methane-oxidating bacteria (and possibly oxidizers of hydrogen, carbon monoxide, and hydrogen sulfide) may have been able to thrive in the crustal rocks. In the course of evolution, photosynthesis, with all its complexity, may well have been preceded as a source of energy by hydrocarbon outgassing. The flora the outgassing sustained gave oil and coal its distinctive biological imprint.[117]

One of the molecular signatures of life that Ourisson's team found in oil was hopanoids that Gold described as "slightly oxygenated and enriched versions of the hydrocarbon molecules known as hopanes, which contain anywhere from about 27 to 36 atoms of carbon arranged in contiguous rings in a single molecule."[118] Gold realized that hopanoids are prominent in oil. The Ourisson study also noted the amount of hopanoids was huge. They projected the global stock of hopanoids would be 10^{13} or 10^{14} tons, more than the estimated 10^{12} tons of organic carbon in all living organisms. Gold recognized that the presence of the hopanoids confirmed his theory of the deep, hot biosphere. He wrote:

> Hopanoids are prominent in all of the numerous samples of petroleum that have been tested for them. This includes samples drawn from sediments of widely ranging ages and from all over the world. And there is no dispute that these molecules are derived from the membranes of once-living cells.[119]

Gold commented that "Ourisson and his colleagues were puzzled, however, by the fact that whereas living trees and ferns and algae are known to contain hopanoids at the lower end of the carbon-number spectrum only bacteria contain the higher-carbon molecules, such as C_{35} and C_{36}."[120] The Ourisson team found another interesting molecule (a terpenoid), common in hydrocarbons, "is also present in bacteria known to make their living by oxidizing methane."[121]

All this made sense to Gold. He concluded the following:

> The biogenic molecules discovered in natural hydrocarbons throughout the world can all be linked to constituents of bacteria or archaea, and none is linked exclusively to macroflora or fauna. There is thus no evidence in these observations that anything other than a substantial microbiological

contamination of oils is required to explain all the molecules observed. And this means, in turn, that there is no evidence that any surface life must be invoked to explain the presence of these biological molecules in subsurface hydrocarbons.[122]

Equipped with the theory of the deep, hot biosphere as the solution to explain the petroleum paradox of why organic material is found in petroleum, Gold began estimation on how deep into Earth the biosphere extended and how much biomass the deep, hot biosphere might support. He estimated the biosphere might extend five to ten kilometers (approximately three to six miles) below the surface. He also calculated that the "biomass originating and contained within the deep hot biosphere would be equivalent to a layer of living material that would be approximately 1.5 meters thick if it were spread out over all of the land surface."[123]

Gold noted that deep biomass of this magnitude would be "somewhat more than the existing flora and fauna of the surface biosphere and it comports with the worldwide estimate of biological debris—hopanoids—calculated by the Ourisson team to be present in all crude oils."[124] With these realizations, Gold explained biological material was in crude oil not because organic material produced the oil and not because the petroleum gathered biological debris on its way to the surface. The biological debris was present in crude oil because bacterial microorganisms fed off the hydrocarbons.

In March 1992, Gold published his findings in the *Proceedings of the National Academy of Sciences* in the first publication that Gold entitled "The deep, hot biosphere." He wrote the following about the startling and highly controversial findings:

> There are strong indications that microbial life is widespread at depth in the crust of the Earth, just as life has been identified in numerous ocean vents. This life is not dependent on solar energy and photosynthesis for its primary energy supply, and it is essentially independent of the surface circumstances. Its energy supply comes from chemical sources, due to fluids that migrate upward from deeper levels in the Earth. In mass and volume it may be comparable with all surface life. Such microbial life may account for the presence of biological molecules in all carbonaceous materials in the outer crust, and the inference that these materials must have derived from biological deposits accumulated at the surface is therefore not necessarily valid.[125]

Gold's theory of deep-Earth bacteria does more than simply offer a solution to the petroleum paradox. His approach to understanding deep-Earth bacteria also supports the proposition that deep-Earth hydrocarbons play an essential role in Earth's food cycle, as well as Earth's energy and climate cycles.

Fossil Fuel, Kerogen, and Biogenic Oil Chemistry

Petroleum geologists have developed an idiosyncratic language to describe what they believe is how deceased organic life transforms into hydrocarbons. Let's begin with the concept of source rocks. Traditional petroleum geology identifies periods in geologic time when life on Earth, where there is an "optimum, though not necessarily slow" rate of sedimentation, and anoxic conditions, i.e., a deficiency of oxygen.[126] Source rock is another term used by petroleum geologists. "Many students of individual oil fields have attempted to designate the source rock or rocks that provided the oil," noted Kenneth K. Landes, a professor of geology at the University of Michigan and the author of the 1951 textbook *Petroleum Geology*. "In almost no instances, however, has it been possible to prove definitely that the oil actually came from a certain rock unit."[127] Yet, in 1991, geologists H.D. Klemme and G.F. Ulmishek declared that source rocks of six stratigraphic intervals had created more than 90 percent of all recoverable oil and gas reserves in the world.[128] Sedimentary rock from the Upper Jurassic strata to the Late Jurassic Period (145 to 161 million years ago) accounted for 25 percent of all recoverable hydrocarbon resources; the Middle Cretaceous Period (some 66 to 145 million years ago) for 29 percent; and the Oligocene-Miocene Periods (some 23–34 million years ago) for 25 percent. With this information, practicing petroleum geologists then know to look for recoverable hydrocarbons in the sedimentary rock strata of these three particular periods of geologic time.

Another essential term in traditional petroleum geology is "kerogen." Geologists advising oil company exploration are the primary group whose professionals use terms like "source rock" and "kerogen" to guide them in finding hydrocarbon fuels in sedimentary rock. The term "kerogen" appears mostly in petroleum geology textbooks. Textbooks on organic chemistry rarely if ever discuss kerogen to explain the chemistry of hydrocarbons, even when describing petroleum products as "fossil fuel." Even *Hydrocarbon Chemistry*, a definitive textbook treatment of hydrocarbon chemistry, now in a two-volume 2018 third edition, does not discuss source rock or kerogen. Instead, *Hydrocarbon Chemistry* devotes one or two obligatory paragraphs agreeing oil is organic. The two volumes mainly focus on the chemistry of hydrocarbons after the hydrocarbons are out of the ground. The *Hydrocarbon Chemistry* handbooks present legitimate chemical formulas that professional chemists can understand with reasonable study and effort. The topics of traditional hydrocarbon chemistry, as covered by the two-volume set, include the following issues: cracking processes, dehydrogenation with olefin production, hydrocarbons from methane derivatives, hydrocarbons from methanol, isomerization, alkylation, oxidation of alkanes, reduction-hydrogenation, etc. But there is no discussion whatsoever of source rocks or kerogen.

Kerogen, it turns out, is not a chemist's term. Kerogen is a loose, geological term deriving from the ancient Greek words κηρός, "keros," meaning *wax*, and γένεση "genesis," meaning *birth*. Oil industry glossaries typically define kerogen as a naturally occurring, solid, insoluble organic material that appears in source rocks and can yield oil upon heating. Source rock and kerogen are not terms typically found in chemistry textbooks or specifically used by professional chemists. But the terms generally appear in textbooks or academic discussions devoted to finding and developing hydrocarbon resources in the field. Use of the term "kerogen" is generally a signal the person is a petroleum geologist or engineer focused on hydrocarbon exploration, not a chemist specializing in the chemistry of hydrocarbons.

Kerogen appears to be a term that petroleum geologists use to describe an in-between substance in sedimentary rock containing decaying organic material, including plant and animal tissue as minuscule as phytoplankton (i.e., ocean microalgae). Kerogen is the substance that results from the dead organic material in sedimentary rock as it cooks into the oil. Seppo Korpela of the Ohio State University Department of Mechanical Engineering gives us a fairly typical description.[129] Korpela argues that fossil fuels form when "the early sedimentary layers" at the bottom of a basin are deprived of oxygen such that the organic matter in them did not decay "as it does in the common setting of a kitchen compost." Then, "anaerobic bacteria" can "work and turn the organic material into the substance kerogen. Kerogen can be thought of as immature oil." Again, the term anaerobic refers to a process occurring in the absence of free oxygen. When kerogen is found at depths of between 6,000 and 13,000 feet, and the temperature and pressure are right, the kerogen in the source rock becomes oil. This zone is called the oil window. At depths greater than 13,000 feet, temperatures are so high that oil becomes natural gas. In an article published in 2006, Korpela got the description of the term kerogen down to the following:

As the organic matter is buried deeper by overlying sediment, the pressure consolidates the source rock and prevents oxygen from entering the strata. Anaerobic bacteria convert the existing organic material into kerogen, which upon further burial is converted into oil.[130]

Ker Than, a staff writer for LiveScience.com, provides a commonsense explanation for how kerogen is supposed to transform into fossil fuel.

In the leading theory, dead organic material accumulates on the bottom of oceans, riverbeds, or swamps, mixing with mud and sand. Over time, more sediment piles on top, and the resulting heat and pressure transforms the organic layer into a dark and waxy substance known as kerogen.

Left alone, the kerogen molecules eventually crack, breaking up into shorter and lighter molecules composed almost solely of carbon and hydrogen atoms. Depending on how liquid or gaseous this mixture is, it will turn into either petroleum or natural gas.[131]

The 2015 third edition of Richard C. Selley and Stephen A. Sonnenberg's textbook *Elements of Petroleum Geology* describes three significant phases in the evolution of organic material in response to burial. We will quote the description of these three phases from the textbook as follows:

1. *Diagenesis*: This phase occurs in the shallow subsurface at near normal temperatures and pressures. It includes both biogenic decay, aided by bacteria, and abiogenic reactions. Methane, carbon dioxide and water are given off by the organic matter, leaving a complex hydrocarbon termed kerogen. The net result of the diagenesis of organic material is the reduction of its oxygen content, leaving the hydrocarbon: carbon ratio largely unchanged.

2. *Catagenesis*: This phase occurs in the deeper subsurface as burial continues and temperature and pressure increase. Petroleum is released from kerogen during catagenesis—first oil and later gas. The hydrocarbon: carbon ratio declines, with no significant change in the oxygen: carbon ratio.

3. *Metagenesis*: This third phase occurs at high temperatures and pressures verging on metamorphism. The last hydrocarbons, generally only methane, are expelled. The hydrogen: carbon ratio declines until only carbon is left in the form of graphite. Porosity and permeability are now negligible.[132]

Selley and Sonnenberg explained that chemically, kerogen consists of various elements, including carbon, hydrogen, and oxygen, with small amounts of nitrogen and sulfur.[133] These are what the organic petroleum chemists consider being "soil organic material," or more precisely, "disseminated organic material in sediments that is insoluble in normal petroleum solvents, such as carbon bisulfide."[134] So, the entire process of kerogen-creating oil comes down to placing in the sedimentary rock these common elements that petroleum geologists presume could only come from decomposing organic tissue.

Gold has perhaps the most succinct criticism of the organic theory. "Nobody has yet synthesized crude oil or coal in the lab from a beaker of algae or ferns," he wrote.[135] He also presented a compelling argument that organic theory has the hydrogen-to-carbon ratios wrong. He explained as follows why a synthesis of crude oil or coal from decomposed organic material is extremely unlikely:

> To begin with, remember that carbohydrates, proteins, and other biomolecules are hydrated carbon chains. These biomolecules are fundamentally hydrocarbons in which oxygen atoms (and sometimes other elements, such as nitrogen) have been substituted for one or two atoms of hydrogen. Biological molecules are therefore *not* saturated with hydrogen. Biological debris buried in the earth would be quite unlikely to lose oxygen atoms and to acquire hydrogen atoms in their stead. If anything, slow chemical processing in geological settings should lead to further oxygen gain and thus further hydrogen *loss*. And yet a hydrogen "gain" is precisely what we see in crude oils and their hydrocarbon volatiles. The hydrogen-to-carbon ratio is vastly higher in these materials than it is in undegraded biological molecules. How, then, could biological molecules somehow acquire hydrogen atoms while, presumably, degrading into petroleum?[136]

Note that in the second phase above of catagenesis, the expectation is that the hydrogen-to-carbon ratio declines, the exact opposite of what Gold explained happens.

Gold also objected to petroleum geologists using carbon isotopes to argue that petroleum and natural gas usually appear of biological origin. He noted that carbon has two isotopes: carbon-12 (six protons, six neutrons) and carbon-13 (six protons, seven neutrons). The natural carbon on Earth is predominately carbon-12, with carbon-13 at about 1 percent of the total. A process of fractionation can enrich a product in

Abiotic Oil

one or the other isotope. A fractionated material is isotopically light if the ratio favors carbon-12 and isotopically heavy if the ratio favors carbon-13. Measurements of the carbon isotope ratio's slight variations in samples are not typically done in absolute terms but compared to a norm. The norm is a marine carbonate rock called Pee Dee Belemnite, or PDB, because its carbon isotope value lies in the middle of the distribution of all marine carbonites. The measurements express departures of the carbon-13 content from that norm, described in parts per thousand ("per mill," from the Latin "per mille" that translates "per thousand"). Measurements are symbolically represented as the $\delta^{13}C$ value of the sample. Thus, if the norm is precisely 1 percent carbon-13, then the content of 1.001 percent is a $\delta^{13}C$ value of + 1 per mill, or -1 per mill if the carbon-13 content is 0.999 percent.

Scientists generally assume that a carbon isotope ratio of -30 mill (favoring carbon-12) is of biological origin and above -30 mill (favoring carbon-13) is not organic. The unoxidized carbon in plants, i.e., organic carbon, in contrast to oxidized or inorganic carbon, comes from the atmospheric CO_2 plants absorb in photosynthesis. Concerning petroleum, Gold demonstrated why the carbon-13 ratio in hydrocarbon products is light. Gold pointed to Galimov's Rule, showing that methane rising from the deep Earth tends to be isotopically lighter the shallower at which it is sampled. Thus, as we saw in the Lost City hydrothermal fields, the oceanographic scientists found the methane, when exuded from the vents, was carbon-13 heavy. But according to Galimov's Rule, the abiotic methane would be expected to lose a neutron to be carbon light at the ocean surface. Thus, hydrocarbons measured in sedimentary rock with a low ratio of carbon-13 may yet be abiotic (i.e., carbon-13 heavy) if the hydrocarbons rose from the deep Earth to settle in the sedimentary rock near the surface.[137] There is little dispute that hydrocarbons with heavy ratios of carbon-13 content are abiotic. Gold's point is that just because hydrocarbons with ratios favoring carbon-12 are detected in surface strata, those hydrocarbons may have started with carbon-13 ratios at their deep-Earth point of origin.

Gold pointed out that any methane rising to the surface is oxidized in the atmosphere to CO_2, so "we could no longer distinguish it from gas that had entered the atmosphere already in fully oxidized form."[138] As we noted earlier, oxidized methane is abiogenic. In another technical chemical analysis of surface carbonates, Gold felt confident he was right. He summed up as follows:

In sum, the technical information and arguments in this section lead, in my view, to a straightforward general conclusion: The volumes, ages, and isotope ratios of crustal carbonates represent important evidence in favor of the view that hydrocarbons were primordial constituents of the earth, that they remain still, and that they continuously upwell into the outer crust, finally emerging, oxidizing, and mixing in the atmosphere.[139]

In the final analysis, the organic theory of the origin of oil is yet another example of confusing cause and effect. Petroleum geologists found oil, coal, and natural gas in sedimentary rock at Earth's surface. Organic material is rich in the chemical elements needed to make oil, including both hydrogen and carbon. When conventionally trained petroleum geologists found organic compounds in petroleum, they just assumed decaying organic material created the hydrocarbon fuels in Earth's surface strata.

A fundamental problem remains for the organic theory. To assume that decaying organic material advances to generate hydrocarbon fuels would violate the second law of thermodynamics. Dead organic material decomposes into constituent chemicals. This material does not rise through decomposition into yet a higher form of hydrocarbon energy. Even ancient biological material trapped in swamps or buried in peat decomposes. Fossils found in rock strata are not the original organic material. The chances that an organism will become a fossil are low. Fossils are typically produced by a process in which the original organic material is altered. Several alteration methods, including petrification, carbonization, and silica replacement, preserve the original organism's living structure.[140] The only body fossils likely to be preserved are the parts of the original organism that were not soft tissue, such as bones, shells, teeth, and eggs.[141] The only complete organisms we see preserved from geologic time involve an unusual set of circumstances. The organism has to undergo rapid burial in an environment that lacks oxygen and therefore limits decay. Examples are mammoths frozen in ice or flies caught in ancient amber.[142] But in general, dead organisms decay relatively rapidly into constituent chemicals. Even as a biblical reference, we are instructed that we all face the fate "dust into dust," not "dust into oil."

As Gold pointed out, where are the laboratory experiments demonstrating the chemical equations hypothecated by the organic theory work? We have no laboratory proofs that the temperature and pressures within sedimentary layers are sufficient to synthesize hydrocarbons. Without verification through scientific experimentation, organic chemists' chemical reactions to elaborate their theory are thus theoretical musings, not proven fact. In direct contrast, the Lost City hydrothermal field studies have scientifically proven that methane and other hydrocarbons exuded by the deep-sea vent are abiotic. So, while the organic theory of the origin of oil lacks independent laboratory proof, the scientific validation that Gold is right about the deep, hot biosphere continues to grow.

Fractures in Basement Rock

Now-deceased Houston investment banker Matthew R. Simmons was a lifelong proponent of peak oil who was confident oil depletion was at hand. In his then widely acclaimed 2005 book entitled *Twilight in the Desert: The Coming Saudi Oil Shock and the World Economy*, Simmons advanced his running-out-of-oil fears. Simmons wrote that Saudi Arabia achieved "a remarkable string of exploration successes, from 1940 through 1968, relying largely on technology that seems primitive by today's standards."[143] But then, Simmons argued, Saudi Arabia's success dried up. "As with exploration elsewhere around the world, the effort became a high-stakes game requiring substantial risk for elusive rewards." For Saudi Arabia, Simmons concluded, exploration for new oil reserves since 1968 produced "very meager payoffs."[144]

As a believer in the fossil fuel theory, Simmons concluded Saudi Arabia faced an inevitable dimming future of its oil industry, playing out a script that "was written in the geology eons ago."[145] M. King Hubbert came to see Saudi Arabia as living off production in aging super fields, unable to find additional giants or super giants. Simmons concluded twilight was descending, not only over the oil fields of Saudi Arabia but also over oil fields worldwide. Simmons painted a grim picture of Saudi Arabian oil prospects, arguing that even the giant oil field of Ghawar is depleting and is increasingly cut by water to increase production. In 2005, Simmons felt the Saudi Arabian oil company Aramco was going after the "last of the easily produced, free-flowing oil in the most prolific parts of Ghawar."[146]

In April 2004, the Saudi Minister of Petroleum and Mineral Resources Ali al-Naimi disagreed. He told a conference on Saudi oil held in Washington, D.C., that Matt Simmons had dramatically underestimated Saudi oil reserves:

> Saudi Arabia now has 1.2 trillion barrels of estimated reserves. This estimate is very conservative. Our analysis gives us reason to be very optimistic. We are continuing to discover new resources, and we are using new technologies to extract even more oil from existing reserves.[147]

Simmons acknowledged how difficult it is to obtain accurate data on Ghawar, Saudi's largest field, or on any specific details of Saudi production.

> Ghawar is well known as the world's largest oilfield within the petroleum industry and among analysts and energy journalists. But few people, even among the world's more knowledgeable energy experts, know anything more about Ghawar beyond its colossal size. Rarely has any data been published that provided details about the performance and parameters of this greatest of all oilfields.[148]

But today, the U.S. Energy Information Administration (EIA) agrees with Saudi Arabia. In an update posted on the EIA website in 2019, the EIA credited Saudi Arabia with having 16 percent of the world's proven oil reserves. The EIA noted that Saudi Arabia is the largest exporter of petroleum globally, maintaining oil production capacity at roughly twelve million barrels per day. The EIA stated that eight fields hold more than half of Saudi Arabia's oil reserves. Saudi Arabia is home to the largest onshore and offshore fields in the world. The giant Ghawar field, the world's largest oil field, had an estimated remaining reserve of seventy-five billion barrels. Safaniyah, the world's largest offshore oil field, was estimated to have a remaining reserve of thirty-five billion barrels.[149]

An important but largely neglected study of the Saudi bedrock structure provided strong evidence that the Saudi oil fields resulted from fractures and faults in the basement rock. H.S. Edgell, a geologist at the King Fahd University of Petroleum and Minerals in Dhahran, Saudi Arabia, published the study in 1992. In the paper entitled, "Basement Tectonics of Saudi Arabia as Related to Oil Field Structures," Edgell argued that the Saudi oil fields, including the giant field at Ghawar, were produced by bedrock fractures lying beneath the oil fields.[150] "All the oil fields of Saudi Arabia are of the structural type, and they all lie in the northeastern part of the country, including the Saudi offshore portion of the Persian Gulf," Edgell wrote. Edgell made clear he was referring both to the onshore and offshore fields. "These oil field structures are mostly produced by extensional block faulting in the crystalline Precambrian basement along the predominantly N-S Arabian Trend which constitutes the 'old grain' of Arabia." [151] Precambrian rock dates back geologically some 570 million years ago, back to the origin of Earth 4.6 billion years ago.

Edgell's study argued that oil in Saudi Arabia is abundant because the fault patterns in the underlying bedrock permit oil from Earth's mantle to seep upward into the many porous sedimentary strata lying above. Edgell left no doubt advancing this conclusion: "All the known oil fields of Saudi Arabia and its offshore are thus related to four major directions of basement faulting, namely N-S, NE-SW, NW-SE, and E-W."[152] And again:

> Anticlinal or domal structures in the sedimentary sequence of the northeastern Arabian Platform and its offshore extension contain all the known oil and gas fields of Saudi Arabia. These currently comprise some fifty-six oil fields, all of which owe their origin to deep-seated tectonic movements in the Precambrian crystalline basement.[153]

Translated into simple terms, Edgell is telling us to forget about dinosaurs, ancient forests, plankton, and algae. Saudi Arabia has abundant oil because the fault pattern in Saudi Arabia's fractured bedrock permits oil from the mantle to flow upward. Suppose the Saudis have benefited from basement tectonics that allows deep-Earth oil formed in the mantle to flow upward freely. How can anyone, including Matt Simmons, estimate the amount of oil Saudi Arabia might have at levels far below the surface?

Offshore Drilling and Deep-Earth Hydrocarbons

In 1972, a fisherman named Rudesindo Cantarell reported oil seeping into the Campeche Bay in the Gulf of Mexico about one hundred km (sixty miles) off the Yucatán Peninsula. From that oil slick, Mexico discovered the first oil field in the Cantarell complex in 1976. Production from the Cantarell oil fields began in 1979. The Cantarell oil field complex is one of the largest oil-producing complexes in the world, second only to the Ghawar field in Saudi Arabia. Geoscientists discovered that the giant Chicxulub meteor that impacted the Yucatán at the end of the Mesozoic Era some sixty-six million years ago created the Cantarell oil field by severely fracturing the bedrock structure of the Gulf of Mexico.[154]

The impact crater is massive, estimated to be 100 to 150 miles (160 to 240 kilometers) wide. The seismic shock of the meteor fractured the bedrock below the Gulf and set off a series of tsunami activities that caused a massive section of land to break off and fall back into the crater underwater.[155] The geology of the field suggests that up to 300 meters (approximately 985 feet) of coarse-grained carbonate breccia settled in the impact crater at the bottom of the Gulf, as a result of "a single giant, debris flow generated by the platform-margin collapse due to seismic shaking resulting from the meteorite impact."[156] In other words, the shock of the meteor's impact broke enough soft soil off the mainland to fill in the hole some 300 meters deep with settling sediment dislodged from the shore.

The frequency with which asteroids hit Earth is inversely proportional to asteroid size. Several pebble-sized meteorites hit Earth every year. But asteroids that cause craters twenty to fifty km (twelve to thirty-one miles) in diameter, asteroids large enough to cause widespread catastrophe, hit Earth every one hundred million years or so. Impact craters disappear relatively quickly on Earth. Earth's regenerative processes hide impact craters beneath sediments pulled under subduction zones, folded into mountain ranges, or otherwise buried. An asteroid can hit Earth with a velocity between ten to seventy km/second (i.e., 250,000 km/hour, 160,000 miles/hour), releasing a tremendous amount of kinetic energy. A giant asteroid can produce shock pressures in excess of one hundred gigapascals (GPa), i.e., 14.5 million pounds per square inch (psi) at temperatures greater than 3,000°C (5,400°F).[157]

An impact of that magnitude and speed can fracture bedrock rock, send out massive shockwaves, and throw debris into the atmosphere over thousands of miles away. The duration of the contact and compression stage lasts only a few seconds, even for large asteroids. The following excavation stage lasts up to a few minutes, generating an expanding shock wave. After the shock wave passes, the high pressure creates a pressure wave that ejects material from the impact crater, throwing the ejected material at great distances. Ejected material follows ballistic trajectories upward. Some ejected debris may reach beyond the atmosphere, reentering it thousands of kilometers away. Geoscientists now agree that dozens of impact craters have produced oil and gas at sites around the globe. Geologists working for Petróleos Mexicanos (Pemex), the Mexican petroleum company that developed the Cantarell oil field complex, have determined that the Chicxulub meteor created the Cantarell oil fields, deeply fracturing the bedrock of the Gulf of Mexico.[158]

In September 2006, Chevron Corporation and two oil exploration companies announced the discovery of a giant deep-oil reserve in the Gulf of Mexico. Known as the Jack field, this oil field located some 270 miles southwest of New Orleans was estimated to hold as much as fifteen billion barrels' worth of oil reserves. The *Wall Street Journal* reported that this find alone could boost the nation's current reserves of 29.3 billion barrels by as much as 50 percent.[159]

A few months earlier, in March 2006, Mexico announced the discovery of a huge, new oil find, the Noxal field in the Gulf some sixty miles from the port of Coatzacoalcos on the coast of the Veracruz state. Geologists estimated the Noxal oil fields contain as much as ten billion barrels of oil, making the site a rival to the Cantarell oil field complex, still Mexico's largest oil field in the Gulf.[160] The Noxal field is a deepwater find, relying on new deepwater drilling technology, as does the Jack oil field find. Chevron is drilling the Jack field under some 7,000 feet of water in a 28,175-foot well, in total, nearly seven miles under the surface of the Gulf. The Noxal find was at under 930 meters (0.6 miles) of water and a further 4,000 meters (2.5 miles) underground.

In November 2007, Brazil announced the discovery of a huge offshore oil field, called Tupi, that could contain as much as eight billion barrels of oil, enough to expand Brazil's proven reserves by 40 to 50 percent.[161] The ultra-deep Tupi field lies under 7,060 feet of water (1.34 miles down), 10,000 feet of sand and rocks (another 1.89 miles down), and another 6,600 feet of salt (1.25 miles), for a total of 4.48 miles below the surface of the Atlantic Ocean. Sergio Gabrielli, the CEO of the state-run oil firm Petróleo Brasileiro SA (PBR), told Brazil's President Luiz Inácio Lula da Silva that oil reserves off Brazil's coast contain possibly as much as eighty billion barrels. By specializing in advanced ultra-deep offshore oil exploration, Brazil has moved from being dependent on ethanol for its gasoline consumption to becoming a net oil exporter.

One of the deepest oil wells in the world is the Sakhalin-I in Russia, reaching more than 40,000 feet into the earth, at a depth of 7.7 miles, fifteen times the height of the world's tallest building, the Burj Khalifa in Dubai.[162] Jack Kenney would take delight in this fact, given that Stalin began the Russian-Ukrainian, deep-Earth, Manhattan Project–like efforts to find hydrocarbon fuels in Russia. Organically oriented petroleum geologists still like to identify the geologic age of the strata in deep-Earth offshore oil. Their goal is to claim the strata is source rock that formed the deep-sea oil organically from kerogen. But oil wells were an average of 3,635 feet in the mid-1950s, some sixty-five years ago, at the start of the peak oil hysteria.[163] Then, classically trained petroleum geologists expected to find organically produced oil in sedimentary rock structures near Earth's surface. Over the next few years, as offshore drilling technology continues to advance, we will begin finding oil and natural gas at such deep-Earth levels that the organic theory of the origin of oil will strain credibility.

Extraterrestrial Hydrocarbons

In 2005, four years after Thomas Gold published his book *The Deep Hot Biosphere*[164] in a paperback edition, astronomers confirmed his suspicion that abundant abiogenic hydrocarbons exist in the solar system outside

Earth. On January 14, 2005, NASA scientists, in conjunction with the European Space Agency and the Italian Space Agency, determined from the Cassini-Huygens probe that first landed on Titan, the giant moon of Saturn, that Titan contains abundant methane. "We have determined that Titan's methane is not of biological origin, so it must be replenished by geological processes on Titan, perhaps venting from a supply in the interior that could have been trapped there as the moon formed," Dr. Hasso Niemann of the Goddard Space Flight Center told reporters on November 30, 2005.[165] NASA's announcement confirmed Gold's realization that Earth's hydrocarbon fuels could be abiogenic. Gold got this hint from the spectrographic analysis he had done as an astronomer, showing him abiogenic hydrocarbons were abundant in our solar system. In a study published in *Nature* in 2005, Niemann demonstrated that the instrumentation data from the Cassini-Huygens probe made clear the value of $^{12}C/^{13}C$ in the methane on Titan "provides no support for suggestions of an active biota on Titan."[166]

The Gas Chromatograph Mass Spectrometer (GCMS) took measurements that identified different atmospheric constituents by their mass. The GCMS findings determined that the methane on Titan was composed of carbon-13, the isotope of carbon we have discussed as being abiotic in origin. As noted above, each carbon-13 atom has an extra neutron in its nucleus, making carbon-13 atoms slightly heavier than carbon-12 atoms, permitting the GCMS to distinguish between methane isotopes with light carbon-12 ratios and methane with high carbon-13 ratios. In contrast, living organisms produce hydrocarbons with a lighter ratio favoring carbon-12. The NASA scientists who examined the percentage of carbon-13 heavy atoms to carbon-12 light atoms in the methane on Titan did not observe the carbon-12 enrichment in the methane of Titan, which is associated with organic carbon on Earth.[167]

Titan has hundreds of times more liquid hydrocarbons than all the known oil and natural gas reserves on Earth, according to a team of Johns Hopkins scientists reporting in January 2008 on their new findings from data collected from Cassini-Huygens probe radar data.[168] "Several hundred lakes or seas have been observed, of which dozens are each estimated to contain more hydrocarbon liquid than the entire known oil and gas reserves on Earth," wrote lead scientist Ralph Lorenz of the Johns Hopkins University Applied Physics Laboratory, Laurel, Maryland, in the January 29, 2008, issue of the *Geophysical Research Letters*.[169] Lorenz also reported dark dunes running along the equator that cover 20 percent of Titan's surface, comprising a volume of hydrocarbon material several hundred times larger than Earth's coal reserves. "Titan is just covered in carbon-bearing material—it's a giant factory of organic chemicals," Lorenz wrote. In 2015, scientists reported the methane in Titan's atmosphere originated in Titan's deep, rocky core.[170]

CO_2 Emissions from Deadwood Decomposition

In September 2021, an international team of scientists reported on a field experiment of deadwood decomposition across fifty-five forest sites and six continents. They concluded that decaying wood releases 10.9 gigatons of carbon worldwide every year, 115 percent more than all fossil fuel emissions.[171] One of the coauthors of the scientific report, Professor David Lindenmayer from the Australian National University (ANU), said this study was the first time researchers had been able to quantify the contribution of deadwood to the global carbon cycle. "Until now, little has been known about the role of dead trees," he said. "We know living trees play a vital role in absorbing carbon dioxide from the atmosphere. But up until now, we didn't know what happens when those trees decompose. It turns out, it has a massive impact."[172] The study found that decomposition rates increased with temperature, and the most substantial temperature effect occurred at high precipitation levels. The scientists noted warm temperatures accelerate tree decomposition. Decomposition also depended on the action of wood-boring insects such as longicorn beetles, termites, insects, and microorganisms. "We found both the rate of decomposition and the contribution of insects are highly dependent on the climate, and will increase as temperatures rise," Professor Lindenmayer said. "Higher levels of precipitation accelerate the decomposition in warmer regions and slow it down in lower temperature regions."[173]

The study of deadwood decomposition emphasizes that CO_2 released into the atmosphere is an effect of global warming, not a cause. Warmer temperatures convert atmospheric CO_2 into forest proliferation, which in turn increases the amount of deadwood decomposition. The discovery that methane released from deep-sea vents also gets into the atmosphere raises additional questions regarding the IPCC assumption that all increases in atmospheric greenhouse gases are due to humans burning hydrocarbon fuels. Landfills, wastewater treatment, and the flatulence of cows also emit methane. Methane is the second most abundant greenhouse gas after carbon dioxide. Methane accounts for about 20 percent of global greenhouse gas emissions. But methane is more than twenty-five times more potent than carbon dioxide at trapping heat in the atmosphere.[174] IPCC adherents argue anthropogenic methane emitted to the atmosphere follows a hockey stick-like curve, with the amount of methane in the atmosphere more than doubling since the dawning of the industrial age.[175]

The IPCC assumption is that the increases of CO_2 and CH_4 in the atmosphere are entirely attributable to human action. This assumption fails to consider what we now know about deep-sea CH_4 and deadwood composition releases of CO_2.

Curiously, IPCC activists discount the CO_2 exhaled by human beings. Why? An explanation provided by the McGill University Office for Science and Society argues that every atom of carbon in exhaled carbon dioxide comes from food produced by photosynthesis. The animals we eat also consume plant products. The premise of the McGill University argument then was the following:

> How is it then that we don't worry about the massive amounts of carbon dioxide that are released with every breath taken by billions and billions of people and animals that inhabit the world? Because every atom of carbon in the exhaled carbon dioxide comes from food that was recently produced by photosynthesis. Everything we eat, save for a few inorganic components like salt, was in some way produced by photosynthesis. This is obvious when we eat plant products such as grains, fruits and vegetables, but of course it is also the case for meat. The animals we eat were raised on plant products, Indeed, a growing animal is basically a machine that converts plants into flesh. So, since all the carbon dioxide we exhale originated in carbon dioxide captured by plants during photosynthesis, we are not distributing the carbon dioxide content of the atmosphere by breathing.[176]

The McGill University analysis next distinguished exhaling CO_2 into the atmosphere from the burning of fossil fuels according to the following logic:

> On the other hand, when we burn fossil fuels such as gasoline, we are releasing carbon dioxide that forms from carbon atoms that had been removed from the atmosphere millions and millions of years ago by photosynthesis and had been sequestered in the coal, petroleum and natural gas that forms when plants and animals die and decay. By burning these commodities we are increasing the current level of carbon dioxide. Clearly then, by living and breathing we are not contributing to global warming through the release of carbon dioxide.[177]

This type of reasoning should clarify how tightly the theory of the organic origin of hydrocarbon fuels is intrinsic to the global warming logic. The logic fails to consider that the decomposition of organic matter, as was illustrated by the above-referenced study of decomposing deadwood, adds CO_2 into the atmosphere. A careful analysis of how petroleum geologists visualize the concept of kerogen makes clear the organic theory assumes that most deceased organic material decomposes into constituent chemicals. The idea of kerogen depends upon just how much ancient organic material died in bogs, marshes, or peat, i.e., in conditions where decomposition is arrested long enough for the whole rotting mess to get buried as a sedimentary rock layer accumulating soil. The abiotic origin of hydrocarbon fuels directly threatens this logic. Thomas Gold now has scientific confirmation provided by the study of the Lost City hydrothermal fields that abiotic hydrocarbons originated from the mantle of Earth.

The IPCC reasoning about the accumulation of CO_2 in the atmosphere assumes a linear process. Humans emitting *xyz* amount of CO_2 by burning hydrocarbon fuels adds *xyz* amount of CO_2 to the accumulated total in the atmosphere. Chapter 3 discussed how the COVID-19 lockdowns in 2020 reduced global CO_2 emissions because of the restrictions placed on economic activity. Yet, a recent analysis of CO_2 readings from the Mauna Loa Observatory demonstrated there was no measurable decline in the seasonal Northern Hemisphere atmospheric CO_2 peak, or even in the growth rate, during the 2020 pandemic decline in anthropogenic CO_2 emissions.[178] As noted earlier, Earth's climate system operates to distribute heat around the planet. For instance, the higher the CO_2 concentration is in the atmosphere, the more CO_2 the oceans absorb.[179] This negative feedback mechanism operates to stabilize the concentration of CO_2 in the atmosphere, thereby reducing the warming effect of the greenhouse gas. This nonlinear process suggests that the atmospheric CO_2 concentration is a function of Earth's thermometer, not the quantity of hydrocarbon fuels we burn. The main driving effect of CO_2 is to cause plant life to flourish. Plants absorb CO_2 through photosynthesis, another negative feedback mechanism reducing global warming.

Three Stanford University climate scientists, Carolyn W. Snyder, Michael D. Mastrandrea, and Stephen H. Schneider, published a 2011 article on the complexity of Earth's climate system.[180] While these three authors agree with the IPCC's concern about anthropogenic CO_2, they caution their enthusiasm for solutions by realizing just how complex climate cycles are on Earth. "Complexity is key to understanding the climate system," the three Stanford climate scientists advised. "Complexity is the fundamental nature of the system, not the rare exception."[181] The Stanford scientists reinforced the importance of Edward Lorenz's contributions with chaos theory and René Thom with catastrophe theory in our effort to comprehend how Earth's climate works. They added that complexity theory and systems thinking are needed to understand Earth's climate. One of their major points was that the IPCC's assumption, that a given increase in anthropogenic CO_2 atmospheric concentrations will produce a specified temperature increase with the next few years, is unlikely to be correct. They caution "research has found that there will always be more uncertainty in predictions of abrupt climate change than of gradual climate change."[182] The "cascade of uncertainties" begins multiplying and compounding when human beings start making policy decisions that involve human intervention to alter climate results.[183]

The point is that our policy decisions might accomplish nothing, or worse, might backfire into unintended consequences. The IPCC advisories insist we must decarbonize rapidly now, or else. Given the complexities of the climate and human decision making, why do we blindly follow the IPCC's advice? Truthfully, the IPCC's insistence that we decarbonize is unfounded. We might as well bet the ranch, sure that one of John Holdren's geoengineering schemes would save the planet.

Conclusion

In July 2017, on the twenty-fifth anniversary of Thomas Gold's hardcover publication of his book *The Deep Hot Biosphere*, the National Academy of Sciences published a highly respectful retrospection. The retrospection included the following acknowledgment that many of the controversial theories that Gold proposed in that book had subsequently been proven correct:

> Overwhelming evidence now supports the presence of a deep biosphere ubiquitously distributed on Earth in both terrestrial and marine settings. Furthermore, it has become apparent that much of this life is dependent on lithogenically sourced high-energy compounds to sustain productivity. A vast diversity of uncultivated microorganisms has been detected in subsurface environments, and we show that H_2, CH_4, and CO feature prominently in many of their predicted metabolisms.[184]

The lithosphere is Earth's crust and upper mantle. The retrospection included the realization that Gold was also right on how various cycles work on Earth, including the food and carbon cycles. It continued to acknowledge that "to better understand the subsurface is critical to further understanding the Earth, life, the evolution of life, and the potential for life elsewhere."[185]

If the main driving force of increased atmospheric CO_2 concentration is the greening of the planet, why is that a bad outcome? If decarbonizing results in fewer plants, will eliminating a negative climate feedback mechanism backfire if Earth's current warming turns out to be from causes other than anthropogenic CO_2? We are learning that the carbon cycle on Earth begins in the core and has a lot to do with how the deep oceans work. Do we want to overrule Earth's natural carbon cycle with our decision to reduce CO_2 emissions? What if Earth's thermometer has a rising concentration of atmospheric CO_2 exactly where it decides is best to distribute the heat of the continued interglacial period around the planet? If we disrupt Earth's carbon cycle, we also disrupt the food cycle. Again, Earth is indifferent to human existence on the planet. Earth's climate cycle will adjust to whatever human interventions we plan. But in trying to change Earth's atmospheric CO_2 concentration and temperature to be perfect for us, the climate system response may not be as favorable to human life as we anticipate.

Yet, the IPCC is asking for human intervention into the planet's natural climate system to make Earth more comfortable for us. Earth's climate systems are a temperature-regulating mechanism that works perfectly well whether Earth is in an ice age or a warming period. But Earth's climate systems do not operate to make Earth comfortable for us. While agreeing with the IPCC in principle, the Stanford climate scientists still cautioned that human climate interventions are fraught with uncertainty, unpredictability, and unintended consequences.

The economic system is also a nonlinear system fraught with uncertainty, unpredictability, and unintended consequences. Among these concerns are the economic implications of reducing or eliminating the use of hydrocarbon fuels. The IPCC wants us to discontinue using hydrocarbon fuels, assuming the transfer to "renewable" wind and solar energy is a policy choice sure to succeed. In the next and final chapter, we will examine how likely the IPCC energy policy demands will work as planned.

CHAPTER 10

Renewable Energy Sad Realities

Why Fischer-Tropsch Plants Failed in the U.S., the Limits of Renewable Energy, Renewable Fuel Follies, the Global Economic Retreat Back to Hydrocarbon Energy, and China Ramps Up CO_2 Emissions

> *The Green New Deal is the ultimate wish list of the progressive environmental agenda. And it has almost nothing to do with science or "saving the planet."*
> —Marc Morano, *Green Fraud*, 2021[1]

> *Green energy remains an inconsequential source of energy in America despite more than $80 billion in direct federal taxpayer subsidies under Presidents George W. Bush and Barack Obama.*
> —Stephen Moore and Kathleen Hartnett White, *Fueling Freedom*, 2016[2]

> *My hope is that, amid the often chaotic and confusing debates about climate change and other environmental problems, there exists a hunger to separate scientific facts from science fiction, as well as to understand humankind's positive potential.*
> —Michael Shellenberger, *Apocalypse Never*, 2020[3]

JUST AS EARTH OPERATES ON Earth's rules, not the whims and wishes of humankind, the economy operates on economic laws that are equally unconcerned about human hopes and desires. The decision to move away from hydrocarbon fuels in a switch to renewable energy involves a public policy decision loaded with unproven assumptions. We can acknowledge, for instance, that the sun generates solar power, giant turbine engines can harness the wind, and corn can transform into ethanol. But if these and other forms of renewable fuels were commercially successful compared to hydrocarbon fuels, we would not be having this discussion.

According to the U.S. Energy Information Administration (EIA), the statistical and analytical division of the U.S. Department of Energy, in 2020, the primary energy consumption by energy source was as follows: petroleum, 35 percent; natural gas, 34 percent; coal, 10 percent; nuclear electric power, 9 percent; and renewable energy, 12 percent. Thus, hydrocarbon fuels in 2020 still accounted for nearly 80 percent of all primary energy consumption in the United States.[4]

For all the hoopla, renewable fuels in 2020 were only 12 percent of all U.S. energy consumption. When we break down the renewable energy category, solar energy and wind energy together were 37 percent of the renewable energy used, but only 4.44 percent of all energy consumed in the United States in that year. These numbers should be surprising for a person who only follows the mainstream media hysteria on the need to drop hydrocarbon fuels in favor of renewable energy.

Just for emphasis, consider the contribution of each type of renewable energy source to the total mix. Solar energy was only 11 percent of all renewable energy in 2020, 1.32 percent of all energy consumed in the US. Wind energy was 26 percent of all renewable energy, 3.12 percent of all energy consumed. Biomass waste and biofuels were 21 percent of all renewable energy, 2.5 percent of all energy consumed. Hydroelectric was 22 percent of all renewables, 2.6 percent of all energy consumed. The IPCC has been pushing the climate change fear button for more than two decades. Despite the IPCC declaring a scientific consensus that we must stop using hydrocarbon fuels to save the planet, the EIA energy consumption data for the United States in 2020 shows the American public is not convinced.

Today, petroleum companies would readily embrace, develop, and implement cheaper renewable power if the economics worked. If the cost and energy output per unit of renewable fuels were cost-competitive with hydrocarbon fuels, private enterprise petroleum companies would switch to renewable fuels willingly. Petroleum companies have no political axe to grind. The point of this chapter is that we are not yet there. Today, renewable fuels remain more expensive and less efficient than hydrocarbon fuels. Given the economic realities of renewable energy versus hydrocarbon fuels, there is no contest. Hydrocarbon fuels remain dominant in the United

States. If governments at all levels in the United States stopped subsidies and tax breaks for renewable energies, the political push to decarbonize would collapse, despite the IPCC hysteria over anthropogenic CO_2 and global warming.

Why Fischer-Tropsch Plants Failed in the USA

After World War II, U.S. Army intelligence officers had the first opportunity to confiscate Nazi scientific documents and interview Nazi scientists. By 1948, British intelligence, Canadian intelligence, and Russian intelligence all joined in, focusing their efforts on understanding how the Nazis had produced synthetic petroleum products so successfully.

Ultimately, under the auspices of Operation Paperclip, the Office of Strategic Services (OSS), the predecessor agency to the CIA, hundreds of Nazi scientists and engineers were secretly brought to the United States. This was despite their complicity in some of the Nazi's most horrific war crimes, including using political prisoners from the Holocaust as their guinea pigs in terrifying "scientific experiments" and in employing Jews and other political prisoners as slave labor in Nazi war-machine factories.[5]

In 1949, the U.S. Bureau of Mines opened a synthetic fuels demonstration plant in Louisiana, Missouri, on 390 acres of a former War Department ammonia plant located seventy-five miles north of St. Louis. Bechtel Corporation operated this $10 million coal hydration plant, with some 400 employees, including seven Nazi synthetic fuel scientists brought to the United States after World War II as part of Operation Paperclip. Among the Nazi scientists working at the Missouri synthetic fuel plant was Helmut Pichler, who had worked as Franz Fischer's assistant at the Kaiser Wilhelm Institute in Berlin. After World War II, U.S. intelligence agents interviewed Pichler in Germany and agreed to bring him to the United States. Pichler had possession of his extensive body of lab notes and unpublished working papers detailing his work developing the Fischer-Tropsch process.

Once in the United States, Pichler joined Hydrocarbon Research Inc., where he helped construct a commercial Fischer-Tropsch plant in Brownsville, Texas. In his later years, Pichler confessed that the German scientists and engineers interviewed by U.S. intelligence operatives at the end of World War II did not divulge all they knew. The truth is that up until 1940, German scientists and engineers, with the consent of the Nazi government, had been transferring a considerable amount of accurate Fischer-Tropsch technical information to a consortium of six companies that had been members of the old Standard Oil Company. Beginning in 1938 and 1939, Standard Oil also began purchasing common stock of Hydrocarbon Research, Inc.[6]

While the U.S. government's postwar efforts to develop synthetic fuel plants were successful, the project never took root in a global economy where the production of petroleum fossil fuels was both abundant and commercially profitable. While interesting to U.S. oil companies and government officials, synthetic oil production was too costly to pursue when oil reserves in the United States were still relatively abundant and reasonably cheap to discover, develop, and bring to market. Simply put, U.S. oil companies had no reason to create a synthetic oil industry when they were making billions of dollars in profits bringing naturally produced oil and natural gas products to market.

By the 1960s, the U.S. government's interest in synthetic fuels was largely academic. The taxpayer funding for the Fischer-Tropsch process dried up. In the 1960s, the U.S. Bureau of Mines transferred all records on synthetic oil to the Office of Coal Research in the Department of the Interior. Then, in the 1970s, the Energy Research and Development Administration took possession of the Fischer-Tropsch records. In 1977, Congress created the U.S. Department of Energy, and the public policy emphasis shifted to the fossil fuel program. On June 30, 1980, the Energy Security Act was signed into law, creating the U.S. Synthetic Fuels Corporation, which provided financial assistance to the private sector to stimulate the production of synthetic fuels. Still, the one plant Pichler supervised was the only Fischer-Tropsch plant ever built in the United States.[7]

As a result of the public policy emphasis on utilizing abundant fossil fuel resources, the Nazi petroleum secrets languished. Hundreds of thousands of pages of confiscated German scientific papers on the Fischer-Tropsch process remained classified until the late 1970s. In October 1975, Texas A&M University's Center for Energy and Mineral Resources initiated a project to locate, retrieve, abstract, and index the German World War II industrial records to publicize the Fischer-Tropsch processes Nazi Germany had used to produce synthetic fuel.[8] By 1977, the project's staff of twelve full- and part-time members brought to Texas A&M 310,000 pages of documents, consisting primarily of the 305 Technical Oil Mission microfilm reels and 25 microfilm reels collected by Air Force intelligence at the end of World War II.

But, even today, countless thousands of pages of Fischer-Tropsch scientific studies lie deteriorating, never translated, in aging and neglected paper and microfilm archives. Remarkably, despite the efforts of Texas A&M and the National Archives, the process of locating confiscated Nazi synthetic petroleum documents for scientific study remains difficult, if not virtually impossible. When found, most records stayed in the original condition when first seized in 1945, never summarized or abstracted in English, let alone translated in full.

Over time, the U.S. energy industry forgot about liquefying coal and creating synthetic fuels. The Fischer-Tropsch records became the largely forgotten property of the government archives. Why bother liquefying coal when the U.S. still had abundant oil and natural gas reserves available domestically or on international markets at a relatively reasonable price? Even in oil crises, such as the 1975 OPEC oil embargo under President Jimmy Carter, the Fischer-Tropsch process remained forgotten. Few serious politicians or scientists thought seriously about reviving interest in the Fischer-Tropsch process to supplement politically restricted oil and natural gas supplies with synthetic liquid fuel.

Today, few Americans know anything about the World War II achievements of the Nazis in developing synthetic fuel. How different this was from the enthusiasm of the U.S. military's Technical Oil Mission, which at the end of World War II had defined the following as targets of opportunity: (1) all Nazi synthetic fuel plants, including refineries and chemical plants; (2) all research laboratories, including the Kaiser Wilhelm Institute; and (3) corporate headquarters, including IG Farben. Even today, rather than study the Fischer-Tropsch equations to unravel the code of producing synthetic oil, U.S. petroleum geologists remain happy to designate the Nazi documents to obscurity because they consider synthetic oil production a waste of time.

The Limits of Renewable Energy

Suppose we had a battery the size of a flashlight battery that could store enough solar energy to light a city. If solar technology that powerful existed, the entire world would drop hydrocarbon fuels in an instant to switch to these new, powerful solar batteries. Suppose we had a wind energy storage device equally as powerful. Suppose ten wind turbines placed a few miles outside a metropolitan area could provide all the electricity that the city and suburbs needed for a week, whether or not the wind blew. Again, a mighty wind energy capacity would dislodge hydrocarbon fuels from their current position as the world's preferred energy choice. The physics of solar and wind power limit their usefulness. The sun does not shine in the sky at night, and the equinox, two times each year, marks the point where day lasts for twelve hours and night occurs for twelve hours in both the Northern and the Southern Hemispheres. Like every other component of Earth's weather and climate systems, the wind is also variable. Wind and solar devices capture no energy when the wind does not blow and the sun does not shine. Current technology for generating and storing solar and wind energy is limited. Installations of solar and wind energy collection devices require enormous facilities that use up vast amounts of territory. The cost of providing sufficient backup or storage to run a stable electric grid from wind or solar power could multiply the cost of generating electricity by a factor of five or more, given the problem of intermittency. Intermittency is an inherent problem of wind and solar power given that wind and solar power generation fluctuates wildly not only between day and night, but also because of a score of factors including seasonal variations and weather events.[9] While wind and solar technologies are advancing, the ability to create a mighty wind or solar battery the size of a flashlight battery is nowhere on the horizon.

The U.S. Energy Information Administration (EIA), in addition to reporting on the total energy consumption by source in the United States each year, also reports separately on the sources of energy used to generate utility-scale energy. The EIA reported in 2021 that natural gas provided 60 percent of all utility-scale electricity generation in the United States and in 2020 the energy came from hydrocarbon fuels, including natural gas (40.3 percent), coal (19.3 percent), and petroleum (0.4 percent). Nuclear power accounted for 19.7 percent. Today, renewable fuels still provide only 19.8 percent of all utility-scale electricity generation, with wind power at 8.4 percent, hydropower at 7.3 percent, solar power at 2.3 percent, and biomass at 1.4 percent.[10] The EIA also clarified in its *Annual Energy Outlook for 2021* that the growth of wind and solar utility-scale electricity generation depends on federal tax subsidies. The EIA reported the following regarding the outlook for wind and solar power in 2021:

> The projection now assumes the production tax credit (PTC) for wind runs for an extra year, or through 2024, following a one-year extension under the Taxpayer Certainty and Disaster Tax Relief Act of 2019, Division Q of the Further Consolidated Appropriations Act of 2020 passed in December 2019 and under the Internal Revenue Service's Notice 2020-41 issued in May 2020. Although capital costs for both wind and solar continue to decline throughout the projection period, without additional

policy intervention, wind is not as cost-competitive as solar. More than two-thirds of cumulative wind capacity additions from 2020 to 2050 occur before the PTC expires at the end of 2024. The steadier pace of solar additions in part reflects the continued availability of a 10% investment tax credit (ITC), which continues in perpetuity after 2023 when the current 30% phases out.[11]

The costs for generating wind and solar energy have decreased dramatically over the past decade, primarily due to technological advances. The International Renewable Energy Agency (IRENA) reported in 2020 that the global weighted-average leveraged costs of electricity for newly commissioned utility-scale solar photovoltaic (PV) projects fell by 85 percent between 2010 and 2020, from $0.381/kWh (i.e., per kilowatt-hour) to $0.057/kWh. Over the same period, costs for onshore wind projects declined by 56 percent, from $0.089/kWh to $0.039/kWh. Francesco La Camera, IRENA's director-general, was enthusiastic about these results. "The last decade has seen [concentrating solar power], offshore wind and utility-scale solar PV all join onshore wind in the cost range for new capacity fired by fossil fuels, when calculated without the benefit of financial support," he said. "Indeed, the trend is not only one of renewables competing with fossil fuels, but significantly undercutting them."[12]

While technological advances have reduced the costs of using solar and wind power to generate utility-scale electricity, the cost figures do not tell the whole story. Renewable fuels are gaining in cost-affordability in a unit-by-unit comparison with hydrocarbon fuels. Yet, today, solar and wind power lack the scalability to provide reliably operating, utility-scale electricity generation able to supply energy 24/7 (i.e., 24 hours a day for 7 days a week) for large metropolitan areas on a global basis.

An international study that the International Conference on Innovation, Modern Applied Science and Environment Studies published in 2020 (ICIES2020) examined the effect of regulatory policies and fiscal incentives on achieving renewable energy targets in twenty-eight European countries over the 1990–2008 period. The study warned that renewable energy still faces severe operational problems in large-scale applications despite technological advances. The report concluded the following politically correct but realistically cautious evaluation:

> The development of renewable energy innovation systems is one of the most critical aspects of socio-economic growth and energy security as well as tools for achieving inclusive and sustainable goals. Renewable energy

innovation is cost-competitive in many countries, but their large-scale distribution can lead to operational challenges.[13]

The authors of the ICIES2020 report also cautioned that renewable energy use in utility-scale electricity generation remained dependent upon government tax credits and fiscal support to maintain their economic viability. The massive up-front investments needed to put scalable renewable energy generation in place limited the private investment capital available to be put at risk in developing large-scale solar or wind energy generation. The ICIES2020 report continued as follows:

> The authors also focus on the need to implement additional fiscal incentives and public financing policies for achieving renewable energy targets. In addition to payment for capacity, it is necessary to distribute such financial instruments as investment subsidies and grants, fiscal (tax) discounts, biofuel blend obligation, net metering/billing and subsidizing the cost of credit. Without these support measures, renewable energy innovations may be problematic for some countries, especially in the current context of limited access to credit. All stakeholders (policymakers, regulators, technology providers, utilities, etc.) should work in collaboration to achieve cost-effective energy sector.[14]

Earlier, in chapter 3, we reviewed the blackout and brownout problems in utility-scale electricity generation when government legislators and regulators mandate a percentage of electricity generated by utilities to be from solar or wind sources. An honest evaluation of renewable energy applications in utility-scale electricity generation compels the conclusion that IPCC-oriented public policy generation is supporting the movement to switch from hydrocarbon fuels to renewable fuels to generate electricity for modern metropolitan areas around the globe. Despite the technological advances making unit-for-unit costs of renewable energy diminish significantly, the increased cost competitiveness of renewable energy is necessary but insufficient to drive a purely economic decision to abandon hydrocarbon fuels in a large-scale, utility-energy generation. Without government subsidies, renewable energies remain a failure in this context.

Renewable Fuel Follies[15]

The green agenda to advance renewable fuels is promoted by IPCC true believers with such confidence and enthusiasm that it is appropriate for us to ask a few critical questions. How exactly do renewable energies work?

THE TRUTH ABOUT ENERGY

Will renewable energies reduce CO_2 emissions? How much will implementing renewable energies in our lives cost the typical person?

Let's start by expanding the discussion in chapter 3 about Texas's problems implementing legislative mandates to increase the use of wind and solar power.

Texas Renewable Follies

Under Texas state law, the Texas Public Utility Commission since 1999 has implemented a renewable portfolio standard (RPS) regulatory mandate that requires a specified year-to-year increase in the amount of electricity generated in the state from renewable solar and wind sources. According to the U.S. Energy Information Administration (EIA), Texas leads the nation in wind power, producing about 28 percent of all wind-powered electricity generated in the United States in 2020. The EIA further noted that wind power surpassed the state's nuclear generation for the first time in 2014. By 2020, wind power generated twice as much electricity as the state's two nuclear power plants combined. Wind power constitutes approximately 90 percent of the renewable energy Texas produces.[16]

The severely cold winter of 2021 was a disaster for the Electric Reliability Council of Texas (ERCOT), a state supervisory council that the Texas Public Utility Commission regulates. The state's electric grid collapsed at the height of the bitterly cold 2021 winter, leaving close to 4.5 million Texas homes and businesses without power. Texans spent long stretches freezing as blackouts spread and persisted. More than one hundred Texans died from the cold, and homes suffered severe property damage, including burst pipes. Overall, the electric grid failure caused Texans more than an estimated $295 billion in economic damages.

The failure of the Texas power grid happened at the worst time. In the winter of 2021, Texas, like most states, was under COVID-19 lockdown conditions. Before that winter, Texans had spent an estimated $66 billion implementing wind and solar before the deadly February 2021 Winter Storm "Uri" hit. Texas had also collected an additional $21.7 billion in local, state, and federal subsidies and incentives to implement renewable energy. According to ERCOT's data, the state faced another crisis from June 12–20, 2021, when the summer peak power demand reached 70,000 megawatts. The wind energy dropped first to 3,000 megawatts and then to zero.[17] After these power grid failures, the Texas legislature passed new laws requiring power generation facilities to weatherize so they could withstand extreme weather better. But the lawmakers did nothing to change the state's mandates to use an ever-increasing percentage of renewable sources to generate electricity for Texans.[18]

Electric Vehicles (EVs) Limitations

How long does it take to charge an electric car? A typical electric vehicle (EV) with a sixty kWh (i.e., sixty kilowatt hour) battery takes approximately eight hours to charge from empty to full using a seven kWh charging device. Most electric car drivers "top off" their electric charge before hitting empty. A fifty kWh rapid charging device will take about thirty-five minutes to add one hundred miles of range in a "top-off" refill. The rule is the larger the battery and the slower the charger, the longer the refill takes. Drivers of electric cars must learn how to calculate the Pod Point Confidence Range, the maximum "feel safe" distance to go before recharging. The average gasoline-powered vehicle takes less than five minutes to fill at the typical gas station, not counting waiting time.[19]

The bottom line is that electric cars mean waiting longer to refill. According to one analysis, a vehicle occupies a spot at the gas pump for about five minutes. So, in fifteen minutes, a single gas pump can refuel three cars. But even if the time to charge an electric car is reduced to fifteen minutes per charge, an electric car recharging station would need three times the charging plugs to reach a similar throughput. But what if you are third or fourth in line?[20]

The Biden administration's 2021 infrastructure bill allocated $7.5 billion to build a national network of 250,000 fast-charging stations for electric cars. In 2021, the nation only has some 216,000 fast-charging stations. In 2020, only about 2 percent of all cars sold in the United States were electric, with half that total in California. The Biden administration's goal is to have 50 percent of all vehicles sold be zero emission by 2030. To accomplish this goal, the United States will need 2.4 million public and workplace chargers to handle going from electric cars being 2 percent of all vehicles sold to 50 percent.[21]

EV electric batteries add excess weight. An analysis of EVs on the road in 2021 showed the average EV battery added 30.7 kilograms of weight (about 68 pounds) per kilowatt-hour (kWh). So, the Ford Mustang Mach-E that is illustrated in a video analysis weighs 2,300 kg (about 5,070 pounds) with a battery capacity of almost 98.9 kWh. That translates into 23.2 kilograms (kg), or about 51 pounds, per kWh.[22] The GMC Hummer EV Edition 1 truck adds on extra batteries for the additional driving range and power. As a result, the GMC Hummer EV weighs over 9,000 pounds, roughly three times the weight of a Honda Civic.[23] Few people will want to drive a Hummer as a family vehicle. But the principle is that EVs weigh more than gas-powered cars because the electric batteries add weight. The Ford F-150 Lightning weighs about 1,600 pounds more than a similar gas-powered F-150 truck. As another example, the electric Volvo XC40 Recharge weighs about 1,000 pounds more than a gas-powered Volvo XC40.[24] Heavier vehicles use up more energy per mile, which translates to more time spent at charging stations.

The manufacturing of EVs demands more CO_2 emissions mainly because the BEV (i.e., battery electric vehicle) takes more energy to produce, mainly due to the cost of the materials and the fabrication process to make a BEV lithium-ion battery. A study conducted by the Union of Concerned Scientists in 2015 found that producing a midsize, midrange BEV capable of delivering eighty-four miles per charge adds more than one ton of CO_2 emissions to the total manufacturing emissions. The emissions needed to manufacture the BEV result in 15 percent greater emissions than manufacturing a similar gasoline vehicle. The tradeoff expected was that the EV would reduce overall emissions by 51 percent over the vehicle's life.[25] Yet, the point is the EV is nowhere near "zero emissions" given the CO_2 emissions required to make the BEVs.[26]

Moreover, as we pointed out earlier in this chapter, slightly less than 20 percent of all electricity produced in the United States in 2021 comes from renewable sources. Thus, approximately 80 percent of the energy used in recharging stations comes from hydrocarbon fuels. So, while an EV may not have a tailpipe emitting CO_2, the CO_2 emissions are transferred to the power plant that emits CO_2 because the power plant uses hydrocarbon fuels 80 percent of the time to generate the electricity that supplies the charging station. A contributor to Forbes on energy topics, Jude Clemente, pointed out that more electric vehicles mean more hydrocarbon fuels. "The anti-fossil-fuel business tends to forget and/or ignore the fact that electric cars are, obviously, just that…powered by electricity, a secondary energy source that is mostly generated by the combustion of coal and natural gas both here in the U.S. and around the world."[27]

In 2021, EVs will still cost considerably more than conventional ICE (internal combustion engine) vehicles. A September 2021 article in the Washington Post that was promoting EVs pointed out that the average transaction price for a new car was nearly $40,000, according to Kelley Blue Book. Then the article switched gears, pointing out that the starting price of several excellent EVs, including the Chevy Bolt, the Nissan Leaf, and the Volkswagen ID.4, "slide under that average." But the starting price is not the average price. The Washington Post article did not want to admit that EVs are still considerably more expensive than ICEs. "But electric cars aren't necessarily pricier than the equivalent gasoline vehicles," the Washington Post article stressed. "You don't have to buy a $100,000 Tesla Model S to enjoy the benefits of switching." You know that if EVs were cheaper than ICEs, the Washington Post would not have had to work so hard to continue burying the truth.[28] EV lithium batteries remain expensive, and lithium prices skyrocketed 254 percent from January to September 2021 as more auto manufacturers jumped on the EV bandwagon.[29] Most homes lack the various types of 240-volt plugs required by EV home charging units.[30] Besides Tesla, finding a public EV charging station compatible with your vehicle may take some doing.[31]

Unreliable Solar Energy and Wind Energy

In chapter 8, we saw that obliquity, Earth's tilt measured from its axis, is why the planet has seasons. Two significant limitations make solar power only intermittently available. First, the sun does not shine at night, and second, the sun shines considerably fewer hours per day in each hemisphere during winter. Solar energy requires at least four peak sunlight hours a day to work. A "peak sunlight hour" is an hour of sunlight (i.e., sun irradiance) that delivers 1,000 watts of photovoltaic power per square meter (roughly 10.5 feet) per hour. The equation for that is one peak sun hour = 1,000 W/m^2 of sunlight per hour. Solar panels need to face the sun directly at midday, when the sun is at its strongest, to have a chance of receiving one peak sunlight hour. The number of peak sunlight hours increases the closer a given geographic location is to the equator and summer.[32] In the United States, Arizona and California are top solar states with higher average peak sunlight hours, and New York and Massachusetts are at the low end for peak sunlight hours. Clouds and other weather conditions can limit peak sunlight hours even in top solar states.

Another solar energy limitation is that commercial solar cells are only about 15 percent efficient in capturing sunlight on average, with a few top-tier panels reaching 20 percent. Even the best laboratory solar cells are only 40 percent efficient. In a home solar energy installation, one square meter covered with solar panels of average efficiency should absorb about 150 joules of energy every second, or about 150 watts of power. This amount of electricity could charge home solar storage batteries at night to light the house the next day.[33] Powering New York City with solar energy would require solar panels to be installed on approximately 26.5 percent of the city's landmass. Powering Paris by solar would require solar panels on 44.2 percent of the city's total area.[34]

Utility-scale solar energy installations require industrial-scale, lithium-ion battery storage systems to function. The average duration of these storage systems is only 1.7 hours, with a maximum capacity of 4 hours.[35] So, the sun does not shine all the time. Even when the sun shines, it may provide few hours of the peak sunlight required. And the battery storage capacity of utility-scale, solar energy power systems is limited.

The story with wind energy is much like the story with solar energy. Wind availability varies dramatically by state, with Florida getting the slightest wind and Alaska getting the most wind.[36] Utility-scale wind turbine farms, just like utility-scale solar panel installations, require lithium-ion storage batteries. Once again, utility-scale wind storage lithium-ion batteries have a limited, four-hour maximum capacity.[37] The factors that limit utility-scale wind energy applications are the same factors that limit utility-scale solar energy applications. Wind speeds vary considerably even when the wind blows, requiring utility companies to install hydrocarbon fuel-powered backup systems. Lithium-ion battery storage capacity for both wind and solar is expensive and limited in maximum duration. As a result of the undependability of wind and solar energy, utility systems implementing these renewable power sources experience a decrease in electric grid reliability and an increase in operating costs.[38] Fundamental principles of physical science place severe limitations on the ability of wind and solar energy to reduce CO_2 emissions significantly. Political goals to mandate increased wind and solar energy use in utility-scale power grids create operational demands because power plants constantly must switch back to hydrocarbon fuels. The increased operating costs and greater likelihood of grid blackout failures make justifying wind and solar energy questionable on purely economic grounds. The increased implementation of utility-scale wind and solar power in the United States, the U.K., and Europe reflects the political power of IPCC-driven political correctness, not sound decision-making on a strictly economic basis.

Combustion releases the energy in hydrocarbon fuels. In the previous chapter, we saw that octane is a saturated, straight-chain hydrocarbon that is "one of the most important and well-known molecules in petroleum."[39] Octane chemically is C_8H_{18} consisting of a chain of eight carbon atoms and eighteen hydrogen atoms. Another way to express the formula chemically for octane is $CH_3(CH_2)_6CH_3$. When octane burns, the bonds between the atoms break, releasing the energy stored in the bonds that hold the molecule together. Burning hydrocarbon fuels generates electricity through a steam process that rotates a turbine that drives a generator to produce electricity.[40] The energy density of hydrocarbon fuels is higher than the energy density of wood, for instance. The energy density of wood is measured at 15,000 Joules/gram (J/g), compared to coal at 24,000 J/g, propane at 46,000 J/g, and gasoline-petrol at 47,000 J/g.[41]

Solar and wind generate electricity differently:

- *Here is how solar power works.* Photovoltaic (PV) cells absorb photons from the sun. An electronic process inside the PV cells converts the photons into electrons, creating electricity. Solar-generated electricity is online immediately, without needing to go through the process of a turbine or a generator.[42]
- *Here is how wind power works.* Wind energy turns the rotors of a wind turbine. A wind turbine turns wind into energy by using the aerodynamic force from the rotor blades to turn. The rotor connects to a generator. The connection can be through a direct-drive turbine. Or the connection can be indirect, operating through a shaft and a series of gears that force the rotation of a generator that creates the electricity.[43]

There is no combustion in the conversion of solar or wind energy into electricity. Solar and wind energy require storage in a battery to pick up slack periods when the sun is not shining or the wind is not blowing. Hydrocarbon-fueled power plants need no battery storage facilities. When there is a need for more electricity, the power plant simply burns more hydrocarbon fuels. Switching back and forth between wind and solar energy to hydrocarbon fuels is operationally tricky and costly. Today utility-scale wind- and solar-driven power plants are not zero-emission operations if only because hydrocarbon fuel backups must always be in place, ready to use.

Why Bother with Biofuel?

Ethanol was the rage during the OPEC energy crisis under President Jimmy Carter in the 1970s. With President Carter fully aboard with the peak oil crowd, his administration championed ethanol as the way America would achieve energy independence. Ethanol was a biofuel produced from corn through a fermentation process.[44] Farmers in corn-producing states like Iowa embraced the ethanol movement. Since the 1970s, government mandates have set a specified percentage requirement to blend ethanol into gasoline. Ethanol has enjoyed decades of favorable treatment under IRS tax laws and generous government subsidies. Taxpayers spent tens of billions of dollars over the past forty years to prop up a biomass fuel that the Biden administration appears ready to abandon.[45]

In September 2021, the Biden administration began circulating suggestions of significant cuts to the Environmental Protection Agency–administered Renewable Fuel Standard (RFS) biofuels mandate. On September 22, 2021, Reuters reported that one of the largest ethanol manufacturers in the United States, Archer-Daniels-Midland Co., and the nation's Corn Belt farmers would oppose the cuts. Reuters noted that ethanol is by far the nation's most widely used biofuel.[46]

On May 25, 2016, in the early months of the Trump administration, C. Ford Runge, the McKnight University professor of applied economics and law at the University of Minnesota, penned a strong case that corn-based ethanol was an unsound economic idea from the beginning. Professor Runge urged the EPA not to ramp up to a higher percentage of the 30 percent ethanol to gasoline blend President Obama's fuel economy standards established in 2012. "Higher-ethanol blends still produce significant levels of air pollution, reduce fuel efficiency, jack up corn and other food prices, and have been treated with skepticism by some car manufacturers for the damage they do engines," Professor Runge wrote. "Growing corn to run our cars was a bad idea 10 years ago. Increasing our reliance on corn ethanol in the coming decades is doubling down on a poor bet."[47]

On August 27, 2009, an influential front-page article in the *Wall Street Journal* declared "the biofuels revolution that promised to reduce America's dependence on foreign oil is fizzling out."[48] The article reported that ethanol is in financial trouble as two-thirds of U.S. biodiesel production capacity stood unused. Moreover, ethanol proved to be an inefficient fuel. A 10 percent ethanol blend in gasoline reduces gas mileage by 3 to 5 percent; a 20 percent blend reduces it 6 to 10 percent.

But the death blow to the ethanol craze of the Jimmy Carter years was the realization that the production of ethanol consumed more hydrocarbon fuel than ethanol saved. Studies conducted by Cornell University ecologist David Pimentel demonstrated that ethanol production from corn requires 29 percent more hydrocarbons than it saves. Even worse, switchgrass requires 45 percent more hydrocarbons and wood biomass requires 57 percent more. "There is just no energy benefit using plant biomass for liquid fuel," Pimentel said. "These strategies are not sustainable."[49]

The Global Economic Retreat Back to Hydrocarbon Energy

The massive expenditure of public funds needed to put a utility-scale wind and solar capacity in place is only the beginning of the economic catastrophe that switching to renewable fuels will cause.

In September 2021, as economies were restarting after the COVID-19 economic shutdowns, a worldwide energy shortage caused surging coal, oil, and natural gas prices. A frigid winter in Europe and wind speeds in the North Sea that were among the slowest in the past twenty years during August to September 2021 compounded the problem for the European Union.[50] In the U.K., surging natural gas prices have caused energy companies serving more than 1.7 million customers to collapse, forcing the government to intervene.[51] On October 8, 2021, Alexander Smith, a senior reporter for NBC News Digital, summed up the developing world energy crisis as follows:

> Homes and factories across China are shrouded in darkness. India's coal-fired power stations are running on scraps. Dozens of British utilities firms have gone bust. Spain announced emergency legislation after household bills shot up more than a third in one year. And there are fears that a harsh winter in the United States could deliver Americans' most expensive heating costs in years. Energy shortages are sweeping the world even before winter's cruelest months freeze the Northern Hemisphere, and officials and experts point out that the multiple issues behind the crunch will make solutions harder to come by.[52]

Still, NBC News preferred to attribute the 2021 energy crisis to the COVID-19 lockdown and the economic problems occasioned by restarting the global economy. "It's like a car that's been taken off the road for a while and now we want to restart it quickly—it takes time," said Jianzhong Wu, a professor specializing in energy infrastructure at Wales' Cardiff University.[53] NBC explained that global energy consumption shrank by 4.5 percent in 2020, the most significant drop since World War II. But with vaccines encouraging governments to reopen economies, energy producers are struggling to ramp up production quickly. "All of this is happening at a time when most of the world is trying to wean itself off fossil fuels and onto renewable energy," Smith wrote, attempting an explanation that blamed something other than the inherent limitations of renewable energy power and the increased costs that are the inevitable consequences of relying on renewable fuels to power the global economy. "But during this transition period, countries still need to rely on oil, gas, and coal—particularly when the weather doesn't cooperate," he concluded, trying to maintain that the energy crisis involved in moving away from hydrocarbon fuels was only temporary.[54] The global warming lobby does not want to admit there are adverse economic consequences that necessarily result from political decisions like shutting down the Keystone XL Pipeline and refusing to use clean coal technologies because they involve fossil fuels.

In September 2021, the Chinese government ordered the top state-owned energy firms to secure coal and oil reserves "at all costs."[55] As the power supply shock of fall 2021 gripped Europe, three Chinese northeast provinces experienced unannounced power cuts as the electricity shortage that first hit businesses extended to homes. As blackouts hit millions of Chinese, unexpected outages became the new normal. China experienced a warm 2021 summer that led to extreme air condition consumption, while an ongoing trade dispute with Australia hampered China's coal supply and sent coal prices skyrocketing.[56] Amid the 2021 energy crisis, Beijing gave up all pretenses of decarbonizing. Instead, China invested billions of U.S. dollars into fracking technology. The state-owned PetroChina announced it planned to start producing 20,000 barrels of shale oil per day from the shale formations in the Gulong area in its major Daqing oil field complex.[57] In September 2021, Goldman Sachs estimated that power shortages had hit as much as 44 percent of China's industrial activity. China is the largest coal producer and consumer in the world. Nearly 60 percent of China's energy comes from coal, with China still operating about one thousand coal-burning power stations.[58] In the global energy crisis of 2021, China predictably put out a call to suppliers urging more coal imports.[59] As natural gas prices in the European Union reach record high levels equivalent to an oil shock of paying $190 for a barrel of oil, panicked EU "green" governments have begun putting out calls for coal.[60]

Legendary gas trader John Arnold, the billionaire founder of the energy-focused Centaurus Capital hedge fund, issued a warning at the end of September 2021. He said that the United States could experience colossal energy price increases, causing chaos in Europe if the country tries to move to renewable fuels too fast. "The energy crisis in Europe is a wake-up call that [the] U.S. must ensure a smooth transition to decarbonization," he tweeted. Arnold warned that a sharp shift in the United States away from oil and gas could be damaging when the renewable energy infrastructure might not make up the shortfall. "While some advocate trying to destroy the fossil fuel industry as quickly as possible, the reality is oil and gas consumption will be high this decade in any decarbonization scenario," he said. "Very high oil/gas prices risk a voter backlash against decarbonization policies, which are vital to a cleaner future." Despite this tip of the hat to global warming political correctness, Arnold understood that the natural gas crunch will force users to turn to oil. The WTI crude index, the U.S. benchmark price, neared eighty dollars per barrel at the end of September 2021, a surge of 57 percent over the previous year. [61]

The U.S. Energy Information Administration (EIA) releases its forecast for future international energy use every two years in the *International Energy Outlook*. The 2013 issue of the report was realistic about the future of renewable fuels. That EIA report projected that global energy consumption would grow by 56 percent by 2040, with oil, coal, and natural gas the dominant energy sources. Strong economic growth in developing countries would be the dominant force driving world energy markets until 2040. "Rising prosperity in China and India is a major factor in the outlook for energy demand," said EIA administrator Adam Sieminski in a press release covering the 2013 report. The EIA report said energy use in developing countries would increase by 90 percent by 2040, while industrialized nations would experience only a 17 percent increase. By 2040, China's energy demand should be twice that of the United States.[62]

In September 2019, the EIA released *International Energy Outlook 2019*, the most recent EIA report with projections at the time of this writing, from 2019 to 2050. The EIA again noted that while renewables are "the world's fastest growing form of energy," hydrocarbon fuels "continue to meet much of the world's energy demand." The report projected that by 2050, world energy consumption would grow nearly 50 percent. But almost 70 percent of global energy consumption would remain using oil, natural gas, and coal, with the oil price possibly reaching as high as $175/barrel. Despite the worldwide push for renewable fuels, the EIA projected that by 2050, renewable fuels would still only be 28 percent of global energy consumption.[63]

September 2021 proved that in a global energy squeeze, world governments would scramble to get more hydrocarbon fuels. In that month, with the need for heat during the winter season rapidly advancing on the Northern Hemisphere, even the globally oriented European Union went on the lookout for more natural gas resources. In September 2021, options traders began buying call options expiring in December 2022 for Brent crude oil at $200/barrel. A call option gives the options trader the right to buy at that price. In other words, options trades buying Brent crude call options for December 2022 were betting that crude oil would be selling for more than $200/barrel at that time in the future.[64]

Neo-Marxists may aim to cripple capitalism by eliminating hydrocarbon fuels in capitalist countries, especially the United States. Elected government officials, including GOP Texas Governor Greg Abbott, have been willing to get on the bandwagon supporting renewable fuels. Government officials who must seek reelection are typically happy to tout politically correct themes popular with voting constituents. But the moment the consumer faces energy shortages, energy blackouts, and massive energy costs, these same government officials drop renewable fuels in their scramble back to the safe reliability of hydrocarbon fuels. Suddenly, the need to save the planet by reducing CO_2 emissions gets less compelling, even in China, where a freezing, starving, and unemployed population—the biggest of any country on Earth—may prefer rebellion to submission when the alternative is living without sufficient and affordable energy.

China Ramps Up CO_2 Emissions

What good will the Green New Deal decarbonization hysteria do if China continues to rely upon hydrocarbon energy to fuel an economy that China intends to make number one in the world?

China is the world's leading country in CO_2 emissions.[65] In 2020, China's coal-intensive economy emitted more CO_2 than the United States, the European Union, and other developed nations combined. In that year, China emitted 27 percent of all greenhouse gas emissions worldwide. The country's draconian lockdowns during the COVID-19 pandemic allowed its economy to bounce back quickly. As the Chinese economy grew over the last thirty years, China's CO_2 emissions more than tripled. Given China's huge but relatively stable population, its per capita emissions have also grown. Thus, their per capita CO_2 emissions at 10.1 tons per person in 2020 were just below the 10.5 ton per capita average of the thirty-eight-nation Organisation for Economic Co-operation and Development (OECD).[66] Each year, the country burns half of the coal burned worldwide.[67] As we saw earlier in this chapter, China is poised in 2021 to combat the global energy crisis by importing and burning even more coal.

China also operated the world's largest Fischer-Tropsch plant, converting coal to petroleum products. The state-owned Shenhua Ningxia Coal Industry Group operates the Ningxia coal-to-liquid (CTL) plant that produces 100,000 barrels of petroleum products per day.[68] China's Shenhua Group, currently restructured as part of China Energy, had not allowed a Western journalist to visit the Ningxia plant until 2017. That year Chinese researcher Xing Zhang toured the plant accompanied by the firm's vice chairman, Dr. Yao Min. She wrote her findings in a blog for the International Energy Agency (IEA). She found that by 2017 Shenhua had invested some fifty-five billion yuan (£6.2 billion, or $8.4 billion) in the Ningxia plant that then was turning out some twenty million tons of coal into 2.7 million tons of diesel fuel, a million tons of naphtha petroleum, and 340,000 tons of liquid gas. On September 24, 2021, reporting on the Ningxia plant for the *Daily Mail*, David Rose observed that making oil from coal doubles the amount of CO_2 emissions per unit of energy produced.[69]

So, with China giving only lip service to the IPCC's mandates to decarbonize, the world's CO_2 emissions will only increase regardless of what the United States, the U.K., or the European Union might do. This realization that China has no intent to decarbonize is perhaps the ultimate realization of "renewable fuel follies." Let's revisit the intensity with which Green New Deal advocates have pursued their cause in the United States, the U.K., and the European Union. We can see more clearly the extent to which the IPCC warnings on global warming and climate change have transformed into an attack on capitalism. The hypocrisy is astounding.[70] IPCC adherents go out of their way to ignore China's wanton use of coal. Instead, IPCC adherents believe China's promise while ignoring the reality that China has no intention of complying with IPCC decarbonization targets. A study published in *Science* in April 2021 shockingly found that China's pledge to achieve carbon neutrality before 2060 is largely consistent with the Paris Accord's aim of limiting global warming to 1.5°C. Dr. Hongbo Duan, an associate professor at the University of Chinese Academy of Sciences and the paper's lead author, explained the study's conclusion as follows: "If China keeps to the 1.5°C pathways, it will be able to become carbon neutral by 2060." One of the paper's principal authors, Detlef P. van Vuuren, a professor at the PBL Netherlands Environment Assessment Agency and the Copernicus Institute of Sustainable Development at Utrecht University, said China's professed goal to decarbonize by 2060 was "an ambitious step in the right direction."[71]

The adherents of the Paris Accord warnings and the IPCC decarbonization targets turn a blind eye to China. They accept China's politically correct but imaginary decarbonization targets. At the same time, these global warming true believers continue to punish the Western industrial world despite massive efforts the United States, the U.K., and the European Union have made to replace hydrocarbon fuels with less efficient and less reliable renewable energy sources.

Conclusion

On November 9, 2020, the Environmental Protection Agency released the 2019 greenhouse gas (GHG) data collected under the EPA's Greenhouse Gas Reporting Program (GHGRP). These data showed that between 2018 and 2019, the total GHG emissions from large facilities in the United States fell 5 percent. "President Trump was right to leave the Paris Climate Accords," said EPA Administrator Andrew Wheeler when announcing these findings. "We have done more to reduce our GHG emissions over the past four years than our international competitors who cling to the ceremonial and arbitrary agreement."[72]

Despite U.S. reductions in CO_2 emissions under the Trump administration, President Biden has refused to acknowledge any progress the Trump administration made reducing CO_2 emissions.[73] On April 22, 2021, President Biden announced a new target for the United States to achieve a 50 to 52 percent reduction from the 2005 levels of CO_2 emissions to be achieved in an economy-wide effort by 2030. It was announced during the U.S. State Department–sponsored 2021 two-day Leaders Summit on Climate event. The White House press release said Biden's announcement challenged the world to increase the ambition to fight climate change. It stressed that setting these 2030 goals as part of Biden's focus "on building back better in a way that will create millions of good-paying, union jobs, ensure economic competitiveness, advance environmental justice, and improve the health and security of communities across America."[74] That the Biden administration has entered a new area of politicized science could not be any clearer.

CONCLUSION

Quo Vadimus? (Where Are We Going?)

There is no longer any debate that global warming is real, and that it is happening now at an alarming rate. It is transforming the global climate system before our eyes.
—John Brooke, Michael Bevis, and Steve Rissing,
Ohio State University, in *Time*, 2019[1]

The reality is that climate change has become a secular religion with an orthodoxy that brooks no dissent.
—Gary Anderson, George Washington University,
in *The American Spectator*, 2021[2]

ON MAY 24, 2015, POPE FRANCIS issued an encyclical entitled *Laudato Si'*—a phrase borrowed from a hymn by his namesake, Saint Francis of Assisi, "*Laudato si', mi' Signore*" (Praise be to you, my Lord.)[3] With this encyclical, Pope Francis issued not an infallible papal statement but a matter of faith and morals delivered as the supreme pastor of all Catholics worldwide.[4] Pope Francis's encyclical embraced the IPCC's concerns about human beings destroying Planet Earth. In the encyclical, the pope addressed the IPCC's central climate issue as follows:

> It is true that there are other factors (such as volcanic activity, variations in the earth's orbit and axis, the solar cycle), yet a number of scientific studies indicate that most global warming in recent decades is due to the great concentration of greenhouse gases (carbon dioxide, methane, nitrogen oxides and others) released mainly as a result of human activity. As these gases

build up in the atmosphere, they hamper the escape of heat produced by sunlight at the earth's surface. The problem is aggravated by a model of development based on the intensive use of fossil fuels, which is at the heart of the worldwide energy system.[5]

Many Catholics were appalled that Pope Francis, a pope with decidedly socialist views, should take as a matter of faith an unqualified endorsement of the secular arguments blaming for global warming the burning of hydrocarbon fuels and the resulting anthropogenic CO_2. On September 18, 2015, conservative columnist George F. Will attacked Pope Francis's encyclical for embracing the IPCC's position on global warming with the following language:

> Pope Francis embodies sanctity but comes trailing clouds of sanctimony. With a convert's indiscriminate zeal, he embraces ideas impeccably fashionable, demonstrably false and deeply reactionary. They would devastate the poor on whose behalf he purports to speak—if his policy prescriptions were not as implausible as his social diagnoses are shrill.[6]

Will argued Pope Francis's endorsement of global warming and climate change theories was consistent with his leftist views on divorce and same-sex marriage. Will continued as follows:

> Francis's fact-free flamboyance reduces him to a shepherd whose selectively reverent flock, genuflecting only at green altars, is tiny relative to the publicity it receives from media otherwise disdainful of his church. Secular people with anti-Catholic agendas drain his prestige, a dwindling asset, into promotion of policies inimical to the most vulnerable people and unrelated to what once was the papacy's very different salvific mission.[7]

Somewhat surprisingly, Paul Ehrlich joined the attack on Pope Francis's encyclical. On September 24, 2015, Ehrlich coauthored an article in *Nature Climate Change* with John Harte from the Energy and Resources Group, University of California, Berkeley. The paper entitled "Biophysical limits, women's rights and the climate encyclical" attacked Pope Francis not for his position on global warming but for his failure to see that the global warming problem is a problem of overpopulation. "Pope Francis needs to heed his own comments on the Church's 'obsession' with contraception and abortion, and assume a leadership position in support of women's rights and family planning," Ehrlich and Harte insisted.[8] In an interview with *The Guardian* in London, Ehrlich made clear that he dismissed as "raving nonsense" Pope Francis's call for action on climate change so long as the leader of the world's one billion Catholics rejects the need for population control. Ehrlich told *The Guardian* that the pope was right about global warming, but the pope was wrong not to address the rising population's strain on Earth's natural resources.[9]

Let's return to the themes of chapter 2. In their 1977 textbook *Ecoscience: Population, Resources, Environments*,[10] Paul R. Ehrlich, Anne H. Ehrlich, and John P. Holdren argued a fetus has no right to live under the U.S. Constitution because a "green abortion" might be required to save Earth from overpopulation. In the textbook, the authors stated the following:

> The common law and drafters of the U.S. Constitution did not consider a fetus a human being. Feticide was not murder in common law because the fetus was not considered to be a human being, and for purposes of the Constitution a fetus is probably not a "person" within the meaning of the Fourteenth Amendment. Thus, under the Constitution, abortion is apparently not unlawful, although infanticide obviously is. This is a very important distinction, particularly since most rights, privileges, and duties in our society are dated from birth and not from some earlier point in time.[11]

Earlier in the text, the authors categorically stated that from the point of view of biology, "a fetus is only a *potential* human being, with no particular rights."[12] And, again: "To most biologists, an embryo or a fetus is no more a complete human being than a blueprint is a complete building."[13] The authors further insist that because "the moment of birth is easier to ascertain than the moment of conception, implantation, or quickening," Constitutional rights should begin only at birth.[14] "Such an easily ascertainable point in time [the moment of birth] is a sensible point from which to date Constitutional rights, which should not depend upon imprecisions," they insisted.[15] Paul R. Ehrlich, Anne H. Ehrlich, and John P. Holdren articulated a central thesis of the textbook: "The maximum size the human population can attain is determined by the physical capacity of Earth to support people."[16]

In the nearly four decades spanning the 1977 publication of the textbook with Holdren to his 2015 criticism of Pope Francis, Ehrlich had not changed his views on overpopulation. Yes, he agreed in 2015 that anthropogenic CO_2 caused global warming. But even in 2015, he still felt anthropogenic CO_2 would be much less of a problem if only the world did not have so many people. In the final analysis, Ehrlich has always believed the global warming/climate change hysteria today is nothing more than an ecological/environmental version of the persisting Malthusian overpopulation fear.

Today, as Earth approaches eight billion people, neither overpopulation nor climate change has destroyed us all. Julian Simon was right. The doomsday predictions Ehrlich made in his 1990 book *The Population Bomb* failed to materialize. We conclude as we started, with the realization that global warming and climate change hysteria derive not from scientifically valid arguments but from subliminal fears locked deep within the subconsciousness of human psychology.

The Future of Nuclear Energy

Three Mile Island in 1979, Chernobyl in 1986, and the Fukushima Daiichi nuclear disaster in 2011 have frightened people worldwide that atomic power is not safe. In his 2013 book *How Big Is Big and How Small Is Small*, Timothy Paul Smith, the physics and environmental studies research professor at Dartmouth College we cited before, pointed out an essential connection between hydrocarbon fuels and nuclear fission. What makes both forms of energy powerful is that combustion breaks apart the energy bonds holding together the atoms that form the molecules of the exploding energy source. This principle applies whether the energy source in question is octane (C_8H_{18}) or uranium-235 (^{235}U). Smith pointed out that in both cases, "a fuel is characterized by the fact that it starts out stable, needs a 'spark' to ignite it, but then is reconfigured into something with weaker or fewer bonds, releasing the excess energy." He explained that the excess energy must be greater than the "spark," or in the case of nuclear fission with ^{235}U, such that the number of neutrons released in the chain reaction is sufficient to sustain the burning. He summarized the point as follows:

> What we are seeing is that potential energy that is stored in a material, a fuel, is closely related to the forces that bind that material. The amount of energy in nuclear fuel is related to the binding of neutrons and protons by the nuclear force. The amount of energy in other fuels, such as oil or food, is related to the binding of the atoms by chemical forces or bonds.[17]

Biomass fuels like ethanol resemble the Fischer-Tropsch process. A chemical fermentation process replaces the chemical reaction in the case of the Fischer-Tropsch process to turn corn into gasoline or diesel-like petrol.

Quo Vadimus? (Where Are We Going?)

Ethanol, like octane and uranium-235, needs to be combusted to release energy. But, as we pointed out earlier in this chapter, renewable energies, including solar and wind, do not release energy by combustion. The force of the wind is a mechanical one that makes turbines and generators turn. The force of the sun is a photovoltaic force that transforms photons into electrons. Moving to solar and wind power involves abandoning a more powerful energy source, namely hydrocarbons, to utilize a less powerful energy source. The problem with ethanol is the same problem with the Fischer-Tropsch process converting coal to saturated straight-chain hydrocarbons. Like the Fischer-Tropsch process, producing gasoline and diesel-like petrol from corn is too costly. The bottom-line conclusion is inescapable. The Green movement wants to force the U.S., the U.K., and the EU to use renewable energies that are less powerful, more challenging to operate, and more costly to produce so it would dampen economic growth in the developed world, thereby reducing the appeal of capitalism.

Two developing scientific advances promise to fulfill the 1950s' expectation that atomic energy would become the limitless, safe, and reliable future energy source. The first advance involves the development of Generation IV nuclear reactors.[18] Generation IV nuclear plants promise to be "walk-away safe, running down instead of running out of control if left unattended." Moreover, Generation IV nuclear reactors can be small modular reactors placed in underground containment units, enhancing safety and reducing the cost of atomic energy. Finally, Generation IV nuclear reactors use thorium, not plutonium, thereby producing no by-products to build nuclear weapons.[19]

The second development involves scientific breakthroughs in nuclear fusion technology. While nuclear fission, the energy used to create atomic weapons, consists in breaking apart atoms, nuclear fusion involves merging two light nuclei to form a single heavier nucleus. As the U.S. Department of Energy explains, the process releases energy because the mass of the resulting single nucleus is less than the mass of the original two nuclei. The leftover mass becomes energy. As DOE explained, Albert Einstein's famous equation ($E = mc^2$) describes how fusion works. Energy is a function of mass. Most research on nuclear fusion has focused on the deuterium-tritium (DT) fusion reaction that produces a neutron and a helium nucleus. In the process, the DT fusion reaction releases much more energy than in other fusion reactions.[20] Let's turn again to Timothy Paul Smith to understand more clearly how nuclear fusion works:

Our present models of the universe start out with the Big Bang. A few hundred million years later the universe was made primarily of hydrogen and a bit of helium. Stars were formed and started to burn that hydrogen. Through nuclear fusion these hydrogen atoms combined into helium and released energy, in the same way our Sun works today. This process of creating heavier elements is called nucleosynthesis: helium burns to form beryllium, and then carbon, nitrogen, oxygen and other elements. But the process of fusion can only go so far up the periodic table and it stops at iron. You cannot fuse iron with another atom and get energy. Iron is the most stable element.[21]

In August 2021, scientists at the U.S. Department of Energy's flagship laser facility, the U.S. National Ignition Facility, housed at the Lawrence Livermore National Laboratory in California, generated more than ten quadrillion watts of fusion power for a fraction of a second. This amount of energy was roughly 700 times the generating capacity of the entire U.S. electric grid at any given moment.[22]

While this research has implications for nuclear weapons, nuclear fusion holds promise also as the potential energy of the future, more potent even than hydrocarbon energy. James E. Hanley, a Ph.D. political scientist who writes for the American Institute for Economic Research, described the potential for fusion energy as follows:

> Fusion has the potential to cost only one-fourth the current price of nuclear energy, half the cost of natural gas, and be cheaper than onshore wind. All this is in addition to its safety, reliability, and lack of pollution. And by using the most abundant element on earth, hydrogen, it provides a limitless future.[23]

Truthfully, we may be driving vehicles in the future with Generation IV atomic batteries before we ever run out of hydrocarbon fuels. That would be a lesson Julian Simon would understand as certainly as he would anticipate the possibility of fusion energy superseding, in our perhaps not-too-distant future, both hydrocarbon energy and energy generated by nuclear fission.

Renewable Energy Economic Suicide

On September 9, 2021, President Biden announced that the U.S. would aim to replace conventional hydrocarbon-based aviation fuel with sustainable renewable aviation fuel by 2050.[24] Biden's executive action mandated

Quo Vadimus? (Where Are We Going?)

the Departments of Energy, Transportation, Agriculture, and Defense, the National Aeronautics and Space Administration, the General Services Administration, and the Environmental Protection Agency to fund the development of aviation fuel from renewable energy sources to cut aviation emissions 20 percent by 2030. Noting that total nonmilitary aviation represents 11 percent of all U.S. transportation-related greenhouse gas emissions, Biden's executive order warned that aviation's share of emissions "is likely to increase as more people and goods fly." The executive order stressed the following:

> In the future, electric and hydrogen-powered aviation may unlock affordable and convenient local and regional travel. But for today's long-distance travel, we need bold partnerships to spur the development of billions of gallons of sustainable aviation fuels quickly.[25]

The Biden administration did not explain how commercial aircraft could carry enough lithium-ion batteries to power commercial aircraft on regional flights. Nor did the Biden administration explain how battery-type devices could hold enough difficult-to-compact hydrogen molecules to fit within the design limits of commercially viable regional aircraft. Current technology will not allow wind turbines or solar photovoltaic panels to be attached to jet airplanes. So, the import of the executive order was that the renewable aviation fuels would have to be produced synthetically through waste biomass or through some synthetic Fischer-Tropsch–like process fermenting corn, plants, or other organic materials.

What is the difference between aviation fuel and gasoline? Gasoline consists of hydrocarbons that contain from seven to eleven carbon atoms, with hydrogen atoms attached. Jet aviation fuel contains hydrocarbons with a larger number of carbon atoms, typically twelve to fifteen atoms. Kerosene is added to jet fuel because it has a lower freezing point, and temperatures in commercial jet aircraft flights can reach heights where temperatures drop to under -40°C. Gasoline would freeze at these heights. Finally, jet aviation fuel has additives that gasoline does not have, including anti-static chemicals, deicing agents, anticorrosive agents, and antibacterial agents.[26] Gasoline and diesel fuel are compression fuels designed to power piston engines. Jet aviation fuel is an ignition fuel designed to power turbine engines.[27]

In September 2020, the U.S. Department of Energy released a report describing the technical paths to producing sustainable aviation fuel (SAF).[28] The report acknowledged that "the low energy density of even the best batteries severely limits opportunities for electrification" of commercial jet aircraft.[29] The report also acknowledged that SAF is currently limited to the renewable energy portion of sustainable jet fuel, representing only 10 percent of the total blend today. The report specified that the SAF blend is still predominately hydrocarbon energy. The renewable SAF blend portion consists of synthetic gas (syngas) produced through Fischer-Tropsch-like processes, and fats, oils, greases, sugars, and alcohols typically transformed into fuel through synthetic fermentation processes.[30] Today hydrocarbon-based aviation fuel is less costly than gasoline or diesel fuel, mainly because aviation fuel is easier to refine. The 2020 DOE report admitted that today SAF is more expensive to produce than petroleum-based Jet-A fuel. The report acknowledged that the additional cost of producing SAF creates a hurdle for the commercial aviation industry because fuel is typically 20 to 30 percent of an airline's operating cost.[31]

On March 15, 2020, conservative public policy author Rael Jean Isaac published an article in the *American Spectator* warning that the Biden administration's war on hydrocarbon fuels amounts to economic suicide. She wrote the following:

> With the wave of executive orders and legislation coming from the Biden administration, and the cultural antics of his woke supporters, Biden's war on fossil fuels has received insufficient attention. Yet energy is the lifeblood of our economy, and making traditional energy sources vastly more expensive is the single most destructive aspect of Biden's policies. If this country does not successfully mobilize against these policies, the vast majority will experience a dramatic drop in their standard of living.[32]

On his first day as president, Biden canceled the Keystone XL pipeline. But on May 19, 2021, the Biden administration waived sanctions on Russia's Nord Stream 2 pipeline to Germany. Reuters reported the State Department had sent to Congress a report concluding that Matthias Warnig, the CEO of Nord Stream 2 and an ally of Russian President Vladimir Putin, had engaged in sanctionable activity. But Biden's secretary of state, Antony Blinken, waived those sanctions, saying it was in the United States' national interest. Reuters explained the decision came as the Biden administration sought to rebuild ties with Germany after relations had deteriorated under former President Donald Trump.[33] Russia, like China, has moved aggressively forward with hydrocarbons, without regard for Biden's wishes. In October 2021, Russia's energy giant Gazprom angered Washington and Kyiv by signing a deal with Hungary to supply Hungary

Quo Vadimus? (Where Are We Going?)

with natural gas through a Black Sea pipeline that bypassed the previously used pipeline through Ukraine.[34]

The Biden administration appears determined to commit economic suicide with an aggressive green energy policy designed to achieve net-zero CO_2 emissions. Russia and China have taken the exact opposite path. With the United States embracing an energy policy to abandon hydrocarbon fuels in a switch to more costly renewables, Russia and China see an opportunity for economic advancement achieved by increasing their reliance on hydrocarbon fuels. But economic suicide may be precisely what the neo-Marxists wish for the United States, given their hatred of capitalism and the economic prosperity the use of hydrocarbon fuels stimulates.

Final Thoughts

Suppose the global energy crisis of 2021 teaches us something. In that case, we learn that in an energy crunch, even "progressive" governments like the U.K. and the EU will move to avoid severe economic downturns by making greater use of hydrocarbon fuels, with natural gas being the preferred option.

The "woke" leftists embracing critical race theory have now mobilized the Modern Monetary Theory to fund an authoritarian government with potentially limitless draconian power. Incoming President Joe Biden promised an "all-of-government" approach to fight climate change, requiring virtually all federal agencies, from the Defense Department to the Treasury, to help the administration achieve its goal of sharply slashing greenhouse gas emissions.[35] The Biden administration appears to be on a mission to implement their green plan, whether or not the energy benefits promised from renewable fuels ever materialize into economic reality. With John Holdren joining Paul Ehrlich, Thomas Malthus's overpopulation fears morphed into ecological and environmental worries. With the Green New Deal, these ecological fears have morphed into a form of neo-Marxist, anti-capitalism that aims to deprive nations like the United States of the economic prosperity hydrocarbon energy has fueled since the dark days of the 1930s Depression and the end of World War II. Even with a global population estimated at ten billion people by 2050, economic prosperity can continue if we persist with the intelligent utilization of the available, nonfinite supply of hydrocarbon fuels on Earth. More likely than not, we will reach the ten billion population mark by 2050 because Americans will decide to continue the dominant use of hydrocarbon fuels. That seems a much better bet today than thinking we will create a brighter and more prosperous economic future through Green New Deal policies.

In the 1970s, Jimmy Carter pushed ethanol to ensure continued American energy independence. Despite 1970s Malthusian predictions, we did not run out of oil. Instead, under the leadership of President Donald Trump, the United States reclaimed its position as a world leader in the production of hydrocarbon fuels. What failed was ethanol. As we noted earlier, we are dropping the legal mandates and government subsidies for ethanol today because ethanol did not work. Ethanol proved to be too expensive to produce and too challenging to use, expending more hydrocarbon fuels than CO_2 emissions saved.

The climate science reviewed in previous chapters should question any assumption we may have that the dramatic Earth changes experienced in geologic time have anything to do with human causation. Earth is a planet where climate chaos is only interrupted by abrupt catastrophic disruptions. At any instant, Earth's temperature and climate can change through variances of the Milankovitch Cycles that may take tens of thousands or millions of years to materialize, through the impact of another Chicxulub-like asteroid, through continued shifting continental plates, or a dozen other unforeseen events. We should contemplate the possibility of volcanism around the Pacific's Ring of Fire erupting for tens of thousands or even a million years in the future. But why do humans persist first in worrying that we will run out of coal or oil, only to be followed by worrying next that we should never have used coal or oil in the first place? As we have seen from the chemistry of abiogenic oil, hydrocarbons formed in Earth's mantle during pre-Cambrian time may have been responsible for the oxygen the planet needed to sustain life on the surface. Instead of being an unfortunate accident on Earth, hydrocarbons are fundamental to life here.

Timothy Paul Smith, in his 2013 book, contemplated Earth as a changing environment. He wrote:

> The Earth seems so solid and rocks seem so rigid and permanent. But given enough time—and we are talking about hundreds of millions of years—and the heat that rises up from the core of the Earth, the continents are plastic and malleable. We should think of the Earth as being in a slow—very slow—boil, with rocks rising in one place and being drawn down in another. Giant convection currents exist, with continental plates instead of air or water.[36]

Quo Vadimus? (Where Are We Going?)

He also contemplated Earth's history, understanding just how insignificant we humans are. Again, he wrote:

> All of these techniques: radiometric dating of the stars, luminosity of stars in globular clusters, nucleosynthesis and the abundance of elements and finally the expansion of the universe all agree 13.7 billion years as the age of that universe. I actually find it amazing that Earth has been here for a lot of that time, about a third of the life of the cosmos.[37]

Let's consider that Earth is some 4.6 billion years old and that the first humanoid creatures on Earth date back to only about three million years. We see how insignificant we are to geologic time. Yet, as we observed earlier, we humans like to think we measure all things. The expanding Earth theory is complex for humans because we have no experience measuring time in tens of thousands or millions of years. All recorded human history only stretches back some 5,000 years, with the development of the Sumerian cuneiform script. Yet, today's climate alarmists like Michael Mann want to start pushing alarms because he considered a few years in the 1990s to have been unusually warm.

Humans will most likely persist on Earth for the foreseeable future. But if humans ever disappear from this planet, the cause is unlikely to be anthropogenic CO_2 emitted into the atmosphere by burning coal, oil, and natural gas.

Endnotes

Introduction

1. Marc Morano, *Green Fraud: Why the Green New Deal Is Even Worse Than You Think* (Washington, D.C.: Regnery Publishing, 2021), p. 244.
2. Stephen Moore and Kathleen Hartnett White, *Fueling Freedom: Exposing the Mad War on Energy* (Washington, D.C.: Regnery Publishing, 2016), "Prologue," p. ix.
3. Michael Shellenberger, *Apocalypse Never: Why Environmental Alarmism Hurts Us All* (New York: Harper, 2020), p. 264.
4. Gustave Le Bon, *The Crowd: A Study of the Popular Mind* (Lexington, Kentucky: Schastlivye Books, 2013), "Introduction," pp. i-ii.
5. Ibid., Chapter IV, p. 41.
6. Charles Mackay, *Extraordinary Popular Delusions and the Madness of Crowds* (London: L.C. Page & Company, 1932).
7. Ibid., "Preface to the Edition of 1852," p. xix.
8. Christopher Booker and Dr. Richard North, *Scared to Death: From BSE to Global Warming: Why Scares Are Costing Us the Earth* (London and New York: Continuum, 2007).
9. Ibid., "Introduction," p. viii.
10. Ibid., p. ix.
11. S. Warren Carey, *Theories of the Earth and Universe: A History of Dogma in the Earth Sciences* (Stanford, California: Stanford University Press, 1988), "Epilogue," p. 365.
12. Ibid.

Chapter 1

1. Julian Simon, *Hoodwinking the Nation* (New Brunswick, New Jersey: Transactions Publishers, 1999), p. 1. Emphasis in original.
2. Ibid., p. viii.
3. Ibid.
4. Ben Joseph Wattenberg, "Foreword" to Julian Simon's book, *Hoodwinking the Nation,* p viii.
5. Julian L. Simon, *A Life Against the Grain: The Autobiography of an Unconventional Economist* (New Brunswick, New Jersey: Transaction Publishers, 2002).
6. Ibid., pp. 7-8.
7. Julian L. Simon, *A Life Against the Grain: The Autobiography of an Unconventional Economist,* "Preface," p. xi.
8. Julian L. Simon, editor, *The Ultimate Resource* (Princeton, New Jersey: Princeton University Press, 1981).
9. Chapter 11, "When Will We Run Out of Oil? Never!" *The Ultimate Resource 2* (Princeton, New Jersey: Princeton University Press, 1996, revised edition, 1998), pp. 162-181, at p. 162.

Endnotes

10 Ibid., pp. 162-163.
11 Ibid., p. 163.
12 W. Stanley Jevons, *The Coal Question: An Inquiry Concerning the Progress of the Nation, and the Probable Exhaustion of our Coal-mines* (London: Macmillan and Co., Limited, 1865), http://books.google.com/books?id=cUgPAAAAIAAJ&dq=stanley+-jevons+%2B+the+coal+question&printsec=frontcover&source=bl&ots=FR5jis89Q-J&sig=ObEbEGRBqAs0YrGRQnu_ekOkvOw&hl=en&ei=lYShSZD6PIT6Mqvvi-M8L&sa=X&oi=book_result&resnum=1&ct=result#PPR6,M1.
13 Simon, "When Will We Run Out of Oil? Never!" p. 165.
14 "United Kingdom," assessment of the UK energy and economy conducted by Euracoal.eu, no date, https://euracoal.eu/info/country-profiles/united-kingdom/. The heavy decline in U.K. coal mining since the 1980s is further documented here: Dorothy Musariri, "Here's where the UK's last remaining mines are still being operated—and where others are planned," NS Energy, March 12, 2020, https://www.nsenergybusiness.com/features/coal-mining-uk/.
15 Ibid. For additional confirmation that UK carbon taxes have led to an over 90 percent drop in coal-fired electricity since 2013, when carbon taxes were introduced on carbon dioxide emissions in Great Britain, see: University College London, "British carbon tax leads to 93% drop in coal-fired electricity," Phys.org, January 27, 2020, https://phys.org/news/2020-01-british-carbon-tax-coal-fired-electricity.html.
16 Simon, "When Will We Run Out of Oil? Never!" p. 165.
17 Ibid., p. 177.
18 Ibid., p. 180.
19 The data are taken directly off spreadsheets published by the Energy Information Administration, U.S. Department of Energy. The data represent the official energy statistics from the U.S. government. See: "International: Petroleum and other liquids, Crude oil including lease condensate reserves (billion b) 2020," https://www.eia.gov/international/data/world/petroleum-and-other-liquids/more-petroleum-and-other-liquids-data?pd=5&p=00000000000000000008&u=0&f=A&v=mapbubble&a=-&i=none&vo=value&t=C&g=00000000000000000000000000-00000000000000000001&l=249-ruvvvvvfvtvnvv1vrvvvvvfvvvvvvfvvvou20evvvvvvvvvnvvvs0008&s=315532800000&e=1609459200000&. See also: "Statistical Review of World Energy, 2020," sixty-ninth edition, BP.com, https://www.bp.com/content/dam/bp/business-sites/en/global/corporate/pdfs/energy-economics/statistical-review/bp-stats-review-2020-full-report.pdf. According to the BP Statistical Review of World Energy 2020, the world's total proved oil reserves came to 1.73 trillion barrels. For additional discussion, see: "World's Biggest Oil Reserves: Here Are the Top 10 Countries," *The National Interest,* March 12, 2021, https://nationalinterest.org/blog/buzz/world%E2%80%99s-biggest-oil-reserves-here-are-top-10-countries-180126.
20 For instance, see: A.I. Levorsen, *Geology of Petroleum* (San Francisco: W.H. Freeman and Company, 1954). In his textbook, Levorsen cites two academic papers written by Hubbert: (1) M. King Hubbert, "The Theory of Ground-water Motion," *Journal of Geology,* Volume 48 (November-December 1940), pp. 785-944; and (2) M. King Hubbert, "Entrapment of Petroleum under Hydrodynamic Conditions," *Bulletin of the American Association of Petroleum Geologists,* Volume 37 (August 1953), pp. 1954-2026.
21 M. King Hubbert, "Nuclear Energy and the Fossil Fuels," presented before the Spring Meeting of the Southern District Division of Production, American Petroleum

Institute, Plaza Hotel, San Antonio, Texas, March 7-9, 1956, emphasis in the original. The paper is archived here: https://web.archive.org/web/20080527233843/http://www.hubbertpeak.com/hubbert/1956/1956.pdf.
22. Ibid.
23. Ibid.
24. Ibid.
25. Michael Lynch, president and director of Global Petroleum Service, Strategy Energy & Economic Research Inc. (SEER), 2003, http://www.energyseer.com/MikeLynch.html.
26. Michael C. Lynch, "The New Pessimism about Petroleum Resources: Debunking the Hubbert Model (and Hubbert Modelers)," *Minerals & Energy—Raw Materials Report,* Volume 18, Issue 1 (2003), emphasis in original, published online on November 5, 2010, https://www.tandfonline.com/doi/abs/10.1080/14041040310001966a. The paper in its entirety is archived here: http://www.energyseer.com/NewPessimism.pdf.
27. Kenneth S. Deffeyes, *Hubbert's Peak: The Impending World Oil Shortage* (Princeton, New Jersey: Princeton University Press, revised and updated paperback edition, 2001), p. 135.
28. Kenneth S. Deffeyes, *Beyond Oil: The View from Hubbert's Peak* (New York: Hill and Wang, 2005), Chapter 1, "Why Look Beyond Oil?" p. 3.
29. Colin Campbell and Jean H. Laherrère, "The End of Cheap Oil," *Scientific American,* March 1998, https://dieoff.com/page140.htm.
30. Ugo Bardi, "Peak oil, 20 years later: Failed prediction or useful insight?" *Energy Research & Social Science,* Volume 48 (February 2019), pp. 257-261, https://www.sciencedirect.com/science/article/pii/S2214629618303207. The once prominent peak oil website, The Oil Drum (www.theoildrum.com), was published by the Institute for the Study of Energy and Our Future, a Colorado nonprofit corporation, from 2015 to 2013. The last issue of The Oil Drum website, which commonly demonized anyone who disagreed with their peak oil beliefs, was published on October 15, 2013, appearing with only one article, entitled, "The Oil Drum writers: Where Are they now?" Issues of The Oil Drum website are archived here: http://theoildrum.com/special/archives.
31. "The U.S. leads global petroleum and natural gas production with record growth in 2018," U.S. Energy Information Administration, U.S. Department of Energy, August 20, 2019, https://www.eia.gov/todayinenergy/detail.php?id=40973.
32. "Coal explained: How much coal is left?" U.S. Energy Information Administration, U.S. Department of Energy, October 9, 2020, https://www.eia.gov/energyexplained/coal/how-much-coal-is-left.php.
33. "Oil and petroleum products explained," U.S. Energy Information Administration, U.S. Department of Energy, April 13, 2021.
34. Marc Morano, "Shock Epic Fails: The decade that blew up energy predictions—USA energy boom, CO_2 emissions drop defied predictions! See the charts!" ClimateDepot.com, December 24, 2019, https://www.climatedepot.com/2019/12/23/shock-epic-fails-the-decade-that-blew-up-energy-predictions-usa-energy-boom-defied-every-prediction-see-the-charts/.
35. Emel Akan, "U.S. Reliance on Russian Oil Surges to Record High Amid Tensions," *The Epoch Times,* June 17, 2021, https://www.theepochtimes.com/us-reliance-on-russian-oil-surges-to-record-highs-amid-tensions_3863635.html/?utm_source=partner&utm_campaign=TheLibertyDaily.

Endnotes

36 Marc Morano, "Biden begs OPEC to increase oil as he shuts down U.S. production—U.S. oil imports from Russia hit 11 year high," ClimateDepot.com, July 11, 2021, https://www.climatedepot.com/2021/07/11/biden-begs-opec-to-increase-oil-as-he-shuts-down-u-s-production-u-s-oil-imports-from-russia-hit-11-year-high/.

Chapter 2

1 Associated Press, "First Person: John Holdren on Global Warming," televised interview, April 8, 2009, https://www.youtube.com/watch?v=A5yWUvzi6iM.
2 Francis F. Chen, *Introduction to Plasma Physics (New York: Plenum Press, 1974)*.
3 Harrison Brown, *The Challenge of Man's Future: An Inquiry Concerning the Condition of Man During the Years That Lie Ahead* (New York: Viking Press, 1954).
4 John P. Holdren, "Introduction to Part II: Dimensions of the Human Predicament: Population, Resources, and Environment," in *Earth and the Human Future: Essays in Honor of Harrison Brown,* edited by Kirk R. Smith, Fereidun Fesharaki, and John P. Holdren (Boulder and London: Westview Press, 1986), pp. 73-79, at p. 73.
5 Ibid.
6 Harrison Brown, *The Challenge of Man's Future,* p. 7.
7 Ibid., p. 6.
8 John P. Holdren, "Introduction to Part II: Dimensions of the Human Predicament: Population, Resources, and Environment," p. 75.
9 Ibid.
10 Ibid.
11 John P. Holdren, "Science and Technology for Sustainable Well-Being," Presidential Address, February 15, 2007, American Association for the Advancement of Science (AAAS), published in *Science Magazine,* Volume 319, Issue 5862 (January 25, 2008), pp. 424-434, https://science.sciencemag.org/content/319/5862/424.full.
12 Ibid., quoting from a slide included with the original paper but not published in the *Science* reprint of Holdren's speech.
13 Nobel lecture by John P. Holdren, chair of the Executive Committee of the Pugwash Council, Pugwash Conferences on Science and World Affairs, "Arms Limitation and Peace Building in the Post-Cold-War World," NobelPrize.org, 1995, https://www.nobelprize.org/prizes/peace/1995/pugwash/lecture/. Holdren began that lecture with the following: "It is a special privilege for me, as Chair of the Executive Committee of the Pugwash Council, to present this lecture on behalf of the Pugwash Conferences on the occasion of our organization's sharing the Nobel Peace Prize with our founder and President, Professor Joseph Rotblat."
14 Harrison Brown, *The Challenge of Man's Future,* p. 105.
15 Ibid., p. 106.
16 Ibid., p. 105.
17 Ibid.
18 Ibid., pp. 104-105.
19 Ibid., p. 106.
20 Ibid., p. 103.
21 Ibid., pp. 86-87.
22 Ibid., p. 263.
23 Ibid., pp. 263-264.
24 Ibid., p. 260.
25 Ibid., p. 261.

26 Ibid., p. 221.
27 Ibid., p. 220.
28 Ibid., p. 236.
29 Ibid., pp. 236-237.
30 Ibid., p. 255.
31 Ibid., p. 262.
32 Ibid., p. 261.
33 Ibid., p. 157.
34 Ibid.
35 Ibid., p. 168.
36 Glen T. Seaborg, "Harrison Brown's Role in the Manhattan Project," in *Earth and the Human Future: Essays in Honor of Harrison Brown,* edited by Kirk R. Smith, Fereidun Fesharaki, and John P. Holdren (Boulder and London: Westview Press, 1986), pp. 81-109, at p. 81.
37 Harrison Brown, *Must Destruction Be Our Destiny? A Scientist Speaks as a Citizen* (New York: Simon & Schuster, 1946).
38 Ibid., pp. 90-91.
39 Paul R. Ehrlich and Anne H. Ehrlich, *How to Know the Butterflies: Illustrated Keys for Determining to Species All Butterflies Found in North America* (Dubuque, Iowa: William C. Brown Publishers, 1961).
40 Paul R. Ehrlich, *The Population Bomb* (New York: a Sierra Club/Ballentine book, 1968).
41 For the story of Paul Ehrlich's early career and his meeting John Holdren, see: Paul Sabin, *The Bet: Paul Ehrlich, Julian Simon, and Our Gamble over Earth's Future (*New Haven and London: Yale University Press, 2013), pp. 29-30. Holdren grew up in San Mateo, California, not far from Stanford. After studying aeronautics and physics as an undergraduate at MIT, Holdren enrolled in the doctoral program in physics at Stanford University. Sabin notes that Holdren was inspired by Ehrlich's lectures and writings. On the advice of his wife, Cheri, who studied biology in Ehrlich's laboratory, Holdren sought out to meet him. Despite the twelve-year age difference and the Holdren's status as a graduate student, the two became close friends and collaborators.
42 Paul R. Ehrlich, *The Population Bomb*, p. 17.
43 Ibid., "Prologue," *The Population Bomb,* first paragraph, page unnumbered.
44 Ibid., last paragraph, page unnumbered.
45 "A long fuse: 'The Population Bomb' is still ticking 50 years after its publication," TheConversation.com, July 10, 2018, https://theconversation.com/a-long-fuse-the-population-bomb-is-still-ticking-50-years-after-its-publication-96090.
46 Marc Morano, "Flashback 1980: Paul Ehrlich calls oil 'a resource which we know damn well is going to be gone in 20 or 30 years' (By year 2000 or 2010)," ClimateDepot.com, October 2, 2017, https://www.climatedepot.com/2017/10/02/flashback-1980-paul-ehrlich-calls-oil-a-resource-which-we-know-damn-well-is-going-to-be-gone-in-20-or-30-years-by-year-2000-or-2010/.
47 Ibid.
48 Paul R. Ehrlich and John P. Holdren, "Population and Panaceas: A Technological Perspective," *Bioscience,* Volume 19, Number 12, (December 1969), pp. 1065-1071.
49 Julian L. Simon, *A Life Against the Grain: The Autobiography of an Unconventional Economist* (New Brunswick, New Jersey: Transactions Publishers, 1999), "Preface," p. xiii. Emphasis in original.

Endnotes

50 Paul R. Ehrlich, Anne H. Ehrlich, and John P. Holdren, *Ecoscience: Population, Resources, Environments* (San Francisco: W.H. Freeman and Company, 1977).
51 Paul R. Ehrlich and Anne H. Ehrlich, *Population, Resources, Environment: Issues in Human Ecology* (San Francisco: W.H. Freeman and Company, 1970; reprinted in 1972).
52 Paul R. Ehrlich, Anne H. Ehrlich, and John P. Holdren, *Ecoscience, Population, Resources, Environment* (San Francisco: W.H. Freeman and Company, 1970), p. 1. The quotation can be found here: Harrison Brown, The Challenge of Man's Future, pp. 66-67.
53 Paul R. Ehrlich, Anne H. Ehrlich, and John P. Holdren, *Ecoscience: Population, Resources, Environments*, p. 943. All page references in this subsection to the chapter are from the 1977 paperback edition that is a revision of the textbook originally published by Paul and Anne Ehrlich without John Holdren.
54 Ibid., p. 786.
55 Ibid, p. 838. Emphasis in original.
56 Ibid.
57 Ibid.
58 Ibid.
59 Ibid., pp. 786-789. The quotation is on p. 788.
60 Paul R. Ehrlich and Anne H. Ehrlich, *Population, Resources, Environment: Issues in Human Ecology*, p. 1. All page references in this subsection are to the original 1970 hardcover textbook.
61 Ibid., p. 51.
62 Ibid., p. 1.
63 Ibid., pp. 145-148, at p. 147.
64 The Impact Team, *The Weather Conspiracy: The Coming of the New Ice Age* (New York: Ballantine Books, 1977).
65 Harrison Brown, *The Challenge of Man's Future*, pp. 140-141.
66 Ibid., p. 142.
67 Ibid.
68 Paul R. Ehrlich, *The Machinery of Nature: The Living World Around Us—And How It Works* (New York: Simon & Schuster, 1986), pp. 273-274. Emphasis in original.
69 John P. Holdren and Paul R. Ehrlich, editors, *Global Ecology: Readings Toward a Rational Strategy for Man* (New York: Harcourt Brace Jovanovich, Inc., 1971).
70 Paul R. Ehrlich and John P. Holdren, "Overpopulation and the Potential for Ecocide," in *Global Ecology: Readings Toward a Rational Strategy for Man*, pp. 64-78, at pp. 76-77.
71 Ibid., p. 77.
72 Ibid.
73 Julian L. Simon, "Paul Ehrlich Saying It Is So Doesn't Make It So," *Social Science Quarterly*, Volume 63, Number 2 (June 1982), pp. 381-385, at p. 381.
74 Paul Sabin, *The Bet: Paul Ehrlich, Julian Simon, and Our Gamble Over Earth's Future*, pp. 134-137.
75 "The Highest and Purest Democracy: Rabbi Roland Gittelsohn's Iwo Jima Eulogy to His Fallen Comrades," The National World War II Museum, New Orleans, February 19, 2020, https://www.nationalww2museum.org/war/articles/highest-and-purest-democracy-rabbi-roland-gittelsohns-iwo-jima-eulogy-his-fallen.
76 Julian L. Simon, *The Ultimate Resource* (Princeton, New Jersey: Princeton University Press, 1981), pp. 9-10. Cited in Paul Sabin, The Bet: Paul Ehrlich, Julian Simon, and Our Gamble Over Earth's Future, pp. 78-79. On p. 219, Sabin rightly concluded that

"human history over the past forty years has not conformed to Paul Ehrlich's predictions." Yet, Sabin in his book is far more sympathetic to Ehrlich than he is to Julian Simon. See: Robert Whaples, "Book Review," The Independent Review, Volume 19, Number 1 (Summer 2014), pp. 137-140.

77 Simon gave a bitter account of *Science* refusing to publish his 1970 draft article in which he referenced the University of Illinois speech because in his draft article he had referred to Ehrlich et al. as "coercionists." See: Julian L. Simon, *Population Matters: People, Resources, Environment, and Immigration* (New Brunswick, New Jersey: Transaction Publishers, 1990), pp. 495-497. Simon's 1970 Earth Day speech at the University of Illinois was published in *Population: A Clash of Prophets*, edited by Edward Pohlman (New York: New American Library, 1973), pp. 48-62.

78 "Answer to Malthus? Julian Simon Interviewed by William Buckley," *Population and Development Review*, Volume 8, Number 1 (March 1982), pp. 205-218, at pp. 205-206.

79 Alok Jha, green technology correspondent, "Obama climate adviser open to geoengineering to tackle global warming," *The Guardian*, April 8, 2009, https://www.theguardian.com/environment/2009/apr/08/geoengineering-john-holdren.

80 David G. Victor, M. Granger Morgan, Jay Apt, John Steinbruner, and Katherine Ricke, "The Geoengineering Option: A Last Resort Against Global Warming?" *Foreign Affairs*, March/April 2009, https://www.foreignaffairs.com/articles/arctic-antarctic/2009-03-01/geoengineering-option.

81 Christopher Mims, "'Albedo Yachts' and Marine Clouds: A Cure for Climate Change?" *Scientific American*, October 21, 2009, https://www.scientificamerican.com/article/albedo-yachts-and-marine-clouds/.

82 David G. Victor, M. Granger Morgan, Jay Apt, John Steinbruner, and Katherine Ricke, "The Geoengineering Option: A Last Resort Against Global Warming?"

83 Amy Fleming, "Cloud spraying and hurricane slaying: how ocean geoengineering became the frontier of the climate crisis," *The Guardian*, June 23, 2021, https://www.theguardian.com/environment/2021/jun/23/cloud-spraying-and-hurricane-slaying-could-geoengineering-fix-the-climate-crisis.

84 Statement of Dr. John P. Holdren, director, Office of Science and Technology Policy, Executive Office of the President, before the Committee on Commerce, Science, and Transportation, U.S. Senate, July 30, 2009.

85 Eric Hoffer, *The True Believer: Thoughts on the Nature of Mass Movements* (New York: Harper & Row, 1951), "Preface."

Chapter 3

1 Eric Hoffer, *The True Believer*, Chapter XIV, "Unifying Agents: Hatred," Part 65, p. 91 (paperback edition).

2 John Houghton, *Global Warming: The Complete Briefing* (Cambridge, U.K.: Cambridge University Press, third edition, 2004), p. 199.

3 "History of the IPCC," Intergovernmental Panel on Climate Change, "History of Climate Change," IPCC.ch, no date, https://www.ipcc.ch/about/history.

4 *Global Warming of 1.5°C*, Intergovernmental Panel on Climate Change, 2018. An IPCC Special Report on the impacts of global warming of 1.5°C above preindustrial levels and related global greenhouse gas emission pathways, in the context of strengthening the global response to the threat of climate change, sustainable development, and efforts to eradicate poverty [V. Masson-Delmotte, P. Zhai, H.-O. Pörtner, D.

Endnotes

Roberts, J. Skea, P.R. Shukla, A. Pirani, W. Moufouma-Okia, C. Péan, R. Pidcock, S. Connors, J.B.R. Matthews, Y. Chen, X. Zhou, M.I. Gomis, E. Lonnoy, T. Maycock, M. Tignor, and T. Waterfield (eds.)], https://www.ipcc.ch/sr15/.

5. Quoted in the following: Jeff Tollefson, "IPCC says limiting global warming to 1.5°C will require drastic action," *Nature*, October 8, 2018, https://www.nature.com/articles/d41586-018-06876-2.
6. Madeline Ostrander, "Thirty Years Warmer," *Slate*, July 2, 2018, https://slate.com/technology/2018/07/james-hansens-1988-climate-change-warning-30-years-on.html.
7. Guy Darst, "AP Was There: The age of climate change begins," Associated Press, June 18, 2018, https://apnews.com/article/59db44d726fa4a608987e674e6f13dd8.
8. Daniel Bell, *The End of Ideology: On the Exhaustion of Political Ideas in the Fifth* (Glencoe, Illinois: The Free Press of Glencoe, 1960).
9. Daniel Bell, *The Coming of Post-Industrial Society* (New York: Basic Books, 1973).
10. Daniel Bell, *The Coming of Post-Industrial Society* (New York: Basic Books, 1976).
11. Ann Pettifor, "Top 10 books for a greener economy," *The Guardian*, June 2, 2021, https://www.theguardian.com/books/2021/jun/02/top-10-books-for-a-greener-economy-ann-pettifor-green-new-deal. See also: "The GND (Green New Deal) Group," https://greennewdealgroup.org/about-the-group/.
12. Ann Pettifor, "Top 10 books for a greener economy."
13. Ibid.
14. Ibid.
15. Ann Pettifor, *The Case for the Green New Deal* (London: Verso, 2019), p. 18.
16. Ibid.
17. Ibid, p. 171.
18. "Colin Hines," New Era Network, no date, https://notthembutus.wordpress.com/archives/colin-hines/. See also: Colin Hines, *Localization: A Global Manifesto* (London and Sterling, VA: Earthscan Publications Ltd., 2000).
19. Ann Pettifor, *The Case for the Green New Deal*, back cover copy.
20. Michael Jacobs and Mariana Mazzucato, "Rethinking Capitalism: An Introduction," in *Rethinking Capitalism: Economics and Policy for Sustainable and Inclusive Growth*, edited by Michael Jacobs and Mariana Mazzucato (Malden, Massachusetts, and Oxford, U.K.: Wiley Blackwell, 2016), pp. 1-27, at p. 11.
21. Ibid., p. 2.
22. Ibid., pp. 2-10.
23. Ibid., p. 10.
24. Ibid.
25. Ibid.
26. Ibid.
27. Ibid., p. 11.
28. Mariana Mazzucato, *Mission Economy: A Moonshot Guide to Changing Capitalism* (New York: Harper Business, 2021).
29. Ibid., p. 138.
30. Ibid., p. 140.
31. Sharon Zhang, "Why the Social Policies in the Green New Deal Are Essential to Its Success," *Pacific Standard* magazine, February 27, 2019, https://psmag.com/ideas/why-the-social-policies-in-the-green-new-deal-are-essential-to-its-success.

32 Naomi Klein, *This Changes Everything: Capitalism vs. the Climate* (New York: Simon & Schuster, 2014), p. 3.
33 Ibid., p. 7.
34 Ibid., p. 8.
35 Ibid., p. 7.
36 Ibid., pp. 7-8.
37 Naomi Klein, *On Fire: The (Burning) Case for a Green New Deal* (New York: Simon & Schuster, 2019), "Introduction," pp. 1-53, at p. 1. Capitalization in original.
38 Ibid. p. 3.
39 Ibid.
40 Ibid., p. 4.
41 "Confirmation Bias," American Psychology Association, APA Dictionary of Psychology, no date, https://dictionary.apa.org/confirmation-bias.
42 Ibid., p. 194. Emphasis in original.
43 Miranda Green, "Ocasio-Cortez joins climate change sit-in at Pelosi's office," *The Hill*, November 13, 1918, https://thehill.com/policy/energy-environment/416411-youth-protestors-fill-nancy-pelosis-office-demanding-climate-change.
44 Lefty Coaster, "Sunrise Movement uses chains to block WH gates as Biden guts Green Infrastructure to satisfy GOP," *Daily Kos*, June 30, 2021. Also see: Ewan Palmer, "White House Shutdown as Protesters Chain Themselves Across Entrances," *Daily Kos*, June 30, 2021, https://www.dailykos.com/stories/2021/6/30/2037797/-Sunrise-Movement-use-chains-to-block-WH-gates-as-Biden-guts-Green-Infrastructure-to-satisfy-GOP.
45 Varshini Prakash and Guido Girgenti of the Sunrise Movement (editors), *Winning the Green New Deal: Why We Must, How We Can* (New York: Simon & Schuster Paperback, 2020), "Introduction," pp. vii-xxii, at p. xv. Emphasis in original.
46 Ian Haney López, "Averting Climate Collapse Requires Confronting Racism," in *Winning the Green New Deal: Why We Must, How We Can*, edited by Varshini Prakash and Guido Girgenti of the Sunrise Movement, pp. 38-52, at pp. 38-39.
47 Ibid., p. 39.
48 Ibid., p. 47.
49 Ibid., p. 52.
50 Ibid.
51 Mariana Mazzucato, "Avoiding a Climate Lockdown," Project Syndicate, September 22, 2020, https://www.project-syndicate.org/commentary/radical-green-overhaul-to-avoid-climate-lockdown-by-mariana-mazzucato-2020-09.
52 Jonathan Watts, "Climate crisis: in coronavirus lockdown, nature bounces back—but for how long?" *The Guardian*, April 9, 2020, https://www.theguardian.com/world/2020/apr/09/climate-crisis-amid-coronavirus-lockdown-nature-bounces-back-but-for-how-long.
53 Ibid.
54 Ibid.
55 Corinne Le Quéré, Robert B. Jackson, Matthew W. Jones, et al., "Temporary reduction in daily CO_2 emissions during the COVID-19 forced confinement," *Nature Climate Change*, Volume 10 (July 2020), pp. 647-653, https://www.nature.com/articles/s41558-020-0797-x. See also: Carlie Porterfield, Forbes Staff, Business, "Report: World Needs Equivalent of Pandemic Lockdown Every Two Years to Meet Paris Carbon Emission Goals," *Forbes*, March 3, 2021, https://www.forbes.com/

Endnotes

sites/carlieporterfield/2021/03/03/report-world-needs-equivalent-of-pandemic-lockdown-every-two-years-to-meet-paris-carbon-emission-goals/?sh=7e589db46dee.

56 Corinne Le Quéré, Robert B. Jackson, Matthew W. Jones, et al., "Temporary reduction in daily CO_2 emissions during the COVID-19 forced confinement," *Nature Climate Change*.

57 Corinne Le Quéré, Glen P. Peters, Piere Friedlingstein, et al., "Fossil CO_2 emissions in the post-COVID-19 era," *Nature Climate Change*, Volume 11 (March 2, 2021), pp. 197-199, https://www.nature.com/articles/s41558-021-01001-0.

58 Ibid.

59 Mariana Mazzucato, "Avoiding a Climate Lockdown." Regarding "the disease of the anthropocene," Mazzucato was quoting a journal article published by two Spanish health scientists. See: Cristina O'Callaghan-Gordo and Josep M. Antó, "COVID-19: The disease of the anthropocene," *Environmental Research*, Volume 187 (August 2020), https://www.ncbi.nlm.nih.gov/pmc/articles/PMC7227607/.

60 Mariana Mazzucato, "Avoiding a Climate Lockdown."

61 Ibid.

62 Mariana Mazzucato, "The Covid-19 crisis is a chance to do capitalism differently," *The Guardian*, March 18, 2020, https://www.theguardian.com/commentisfree/2020/mar/18/the-covid-19-crisis-is-a-chance-to-do-capitalism-differently.

63 Ibid.

64 Marguerite Ward, "The founder of the World Economic Forum explains why 'a new mindset' is giving him hope for climate action, and shares which companies are getting it right," Business Insider, April 22, 2021, https://www.businessinsider.com/wef-chief-klaus-schwab-bill-gates-right-about-climate-change-2021-4.

65 Klaus Schwab and Thierry Malleret, *COVID-19: The Great Reset* (Geneva, Switzerland: Forum Publishing, 2020).

66 Klaus Schwab, founder and executive chairman of World Economic Forum, "Now is the time for a 'great reset,'" World Economic Forum, WEForum.org, June 3, 2020, https://www.weforum.org/agenda/2020/06/now-is-the-time-for-a-great-reset/.

67 Ibid.

68 Ibid.

69 Klaus Schwab and Thierry Malleret, *COVID-19: The Great Reset*, "Introduction," pp. 11-20, at p. 11.

70 Ibid.

71 For an explanation of the Overton Window, see: "The Overton Window," Mackinac Center for Public Policy, https://www.mackinac.org/OvertonWindow.

72 Klaus Schwab and Thierry Malleret, *COVID-19: The Great Reset*, p. 145.

73 Ibid., p. 149.

74 Ibid., p. 151.

75 Alex Gilbert and Morgan Bazillan, "What can the Texas electricity crisis tell us about the future of energy markets?" World Economic Forum, WEForum.org, February 23, 2021, https://www.weforum.org/agenda/2021/02/texas-electricity-crisis-energy-transition/.

76 "What's the Difference Between a Blackout and a Brownout?" DirectEnergy.com, no date, https://www.directenergy.com/learning-center/difference-between-blackout-brownout.

77 Seth Hancock, "World Economic Forum: Controlling Our Energy and Destroying Our Future," LibertyLoft.com, July 2, 2021, https://thelibertyloft.com/world-economic-forum-controlling-our-energy-and-destroying-our-future/.

78   Kate Aronoff, Alyssa Battistoni, Daniel Aldana Cohen, and Thea Riofrancos, *A Planet to Win: Why We Need a Green New Deal* (London and New York: Verso, 2019), p. 7.
79   Eric Hoffer, *The True Believer*, Chapter XIII, "Factors Promoting Self-Sacrifice," Part 56, p. 79.
80   Ibid.
81   Ibid.
82   Ibid., p. 79.
83   Ibid., p. 80.

Chapter 4

1   "Climate: 10 Million Clean Energy Jobs," Senator Joe Biden campaign promise running for president in 2020, Biden Harris Democrats, "Battle for the Soul of America," https://joebiden.com/climate-labor-fact-sheet/.
2   Barrasso's entire investigative report can be read here: Senator John Barrasso, M.D., "Solyndra Syndrome & The Green Stimulus Delusion," Special Investigative Report, May 5, 2021, https://www.energy.senate.gov/services/files/424E7FA4-AD66-454B-8174-978411AB0447.
3   Senator John Barrasso, M.D., "Ranking Member Barrasso Releases Investigative Report," The Solyndra Syndrome & the Green Stimulus Delusion, Republican News, U.S. Senate Committee on Energy and Natural Resources," May 5, 2021, https://www.energy.senate.gov/2021/5/ranking-member-barrasso-releases-investigative-report-the-solyndra-syndrome-the-green-stimulus-delusion. Senator Barrasso's full investigative report is available here: Senator John Barrasso, M.D., Ranking Member, U.S. Senate Committee on Energy and Natural Resources, *Solyndra Syndrome & The Green Stimulus Delusion*, Energy.Senate.gov, April 29, 2021, https://www.energy.senate.gov/services/files/424E7FA4-AD66-454B-8174-978411AB0447.
4   Ibid.
5   "Vice President Biden Announces Finalized $535 Million Loan Guarantee for Solyndra," press release, Office of the Vice President, The White House, ObamaWhiteHouse.archives.gov, September 4, 2009, https://obamawhitehouse.archives.gov/the-press-office/vice-president-biden-announces-finalized-535-million-loan-guarantee-solyndra.
6   Ibid.
7   David Rotman, "Can We Build Tomorrow's Breakthroughs?" *MIT Technology Review*, December 19, 2011, https://www.technologyreview.com/2011/12/19/117045/can-we-build-tomorrows-breakthroughs/.
8   Chris Gentilviso, "President Obama Visits Green Energy Start-up Solyndra," *Time*, May 26, 2010, https://newsfeed.time.com/2010/05/26/president-obama-visits-green-energy-start-up-solyndra/.
9   Tom Hals, "U.S. solar firm Solyndra files for bankruptcy," Reuters, September 6, 2011, https://www.reuters.com/article/us-solyndra/u-s-solar-firm-solyndra-files-for-bankruptcy-idUSTRE77U5K420110906.
10   "Key Facts: Solyndra Solar," U.S. Department of Energy, no date, https://www.energy.gov/key-facts-solyndra-solar.
11   Ibid.
12   Ben Sills, "Solar Panel Makers Face Supply-Glut," *Bloomberg*, November 15, 2010, at http://www.bloomberg.com/news/2010-11-16/solar-panel-makers-face-supply-glut-armageddon-chart-of-day.html.

Endnotes

13 "Special Report: The Department of Energy's Loan Guarantee to Solyndra, Inc.," U.S. Department of Energy, Office of Inspector General, Document Number 11-0078-I, August 24, 2015, https://www.energy.gov/sites/prod/files/2015/08/f26/11-0078-I.pdf.
14 Ibid.
15 Joe Stephens and Carol D. Leonnig, "Solyndra: Politics infused Obama programs," *Washington Post*, December 25, 2011, https://www.washingtonpost.com/solyndra-politics-infused-obama-energy-programs/2011/12/14/gIQA4HllHP_story.html.
16 See also: Carol D. Leonnig, Joe Stephens, and Alice Crites, "Obama's focus on visiting clean-tech companies raises questions," *Washington Post*, June 25, 2011, https://www.washingtonpost.com/politics/obamas-focus-on-visiting-clean-tech-companies-raises-questions/2011/06/24/AGSFu9kH_story.html?tid=a_inl_manual.
17 Joe Stephens and Carol D. Leonnig, "Solyndra: Politics infused Obama programs."
18 Ibid.
19 Ibid.
20 Peter Schweizer, "What is China Buying in the Biden Administration?" Gatestone Institute, July 6, 2021, https://www.gatestoneinstitute.org/17518/china-biden-administration.
21 Institute for Energy Research, "Obama Subsidizes U.S. Solar Energy and Promises to Do the Same in India," InstituteForEnergyResearch.org, February 20, 2015, https://www.instituteforenergyresearch.org/renewable/solar/obama-subsidizes-u-s-solar-energy-promises-india/.
22 Stephen Dinan, "Obama clean energy loans leave taxpayers in $2.2 billion hole," *Washington Times*, April 27, 2015, https://www.washingtontimes.com/news/2015/apr/27/obama-backed-green-energy-failures-leave-taxpayers/. See also: Ashe Schow, "President Obama's Taxpayer-Backed Green Energy Failures," The Daily Signal, October 18, 2012, https://www.dailysignal.com/2012/10/18/president-obamas-taxpayer-backed-green-energy-failures/.
23 Senator John Barrasso, M.D., "Solyndra Syndrome & The Green Stimulus Delusion," Special Investigative Report, p. 8.
24 Ben Wolfgang, "Abengoa, Obama green energy project, on verge of bankruptcy; demise recalls Solyndra," *Washington Times*, November 25, 2015, https://www.washingtontimes.com/news/2015/nov/25/abengoa-obama-green-energy-project-on-verge-of-ban/.
25 Marita Noon, "How Democrats Say 'Crony Corruption' in Spanish: Abengoa," *Townhall*, August 4, 2012, https://finance.townhall.com/columnists/maritanoon/2012/08/04/how-democrats-say-crony-corruption-in-spanish-abengoa-n878147.
26 Veronique de Rugy, "Assessing the Department of Energy Loan Guarantee Program," Mercatus Publications, Mercatus Center, George Mason University, June 19, 2012, https://www.mercatus.org/publications/government-spending/assessing-department-energy-loan-guarantee-program.
27 Danielle Ola, "Abengoa continues divestment plan with sale of four PV plants," *PV Tech*, April 11, 2016, https://www.pv-tech.org/abengoa-continues-divestment-plan-with-sale-of-four-pv-plants/.
28 Tom Kenning, "Abengoa sells share in Abu Dhabi CSP plant to Masdar," *PV Tech*, February 5, 2016, https://www.pv-tech.org/abengoa-sells-share-in-abu-dhabi-csp-plant-to-masdar/.

29 Tom Kenning, "Spain's Abengoa begins insolvency proceedings with shares tumbling," *PV Tech*, November 26, 2015, https://www.pv-tech.org/spains-abengoa-begins-insolvency-proceedings-with-shares-tumbling/.

30 Jose Elías Rodríguez and Robert Hetz, "UPDATE 3—Spain's Abengoa starts insolvency proceedings, shares dive," Reuters, November 25, 2015, https://www.reuters.com/article/abengoa-gonvarri-idUSL8N13K16B20151125#4FDlchkxtHltgObo.97.

31 "Court orders seizure of Abengoa Mexico assets," Renewables Now, April 13, 2016, https://renewablesnow.com/news/court-orders-seizure-of-abengoa-mexico-assets-520843/.

32 "How Democrats Say 'Crony Corruption' in Spanish: Abengoa," The Green Corruption Files, updated version, August 8, 2012, https://corruption485.rssing.com/chan-11189954/article6-live.html.

33 Katie Fehrenbacher, "Goldman Sachs to invest $150 billion in clean energy," *Fortune*, November 2, 2015, https://fortune.com/2015/11/02/goldman-sachs-clean-energy/.

34 Marita Noon, "Ex-Im Bank Crucial to Clinton Crony Capitalism Faces Closure," Breitbart.com, June 8, 2015, https://www.breitbart.com/politics/2015/06/08/ex-im-bank-crucial-to-clinton-crony-capitalism-faces-closure/.

35 CJ Ciaramella, "Ex-Im Bank Funds Green Jobs Overseas," *Washington Free Beacon*, January 6, 2013, https://freebeacon.com/politics/ex-im-bank-funds-green-jobs-overseas/.

36 Ibid.

37 Brian Eckhouse, "Abengoa Yield Insulated from Potential Bankruptcy of Founder," *Bloomberg*, March 4, 2016, https://www.bloomberg.com/news/articles/2016-03-04/abengoa-yield-insulated-from-potential-bankruptcy-of-founder.

38 Jennifer Oldham and Shai Oster, "Solar Jobs Join Harry Reid to Chinese Billionaire in Price Drop," *Bloomberg*, April 3, 2012, https://www.bloomberg.com/news/articles/2012-04-03/solar-jobs-join-harry-reid-to-chinese-billionaire-in-price-drop.

39 Ibid.

40 Wynton Hall, "Harry Reid's Son Representing Chinese Solar Panel Plant in $5 Billion Nevada Deal," Breitbart.com, September 4, 2012, https://www.breitbart.com/politics/2012/09/04/Harry-Reid-s-Son-Representing-Chinese-Solar-Panel-Plant-In-5-Billion-Nevada-Deal/.

41 Marcus Stern, "U.S. Senator Reid, son combine for China firm's desert plant," Reuters, August 31, 2012, https://www.reuters.com/article/us-usa-china-reid-solar/u-s-senator-reid-son-combine-for-china-firms-desert-plant-idUSBRE87U06D20120831.

42 Chuck Neubauer and Richard T. Cooper, "In Nevada, the Name to Know Is Reid," *Los Angeles Times*, June 23, 2003, https://www.latimes.com/archives/la-xpm-2003-jun-23-na-sons23-story.html.

43 Chris Clarke, "Nevada Solar Factory Cancelled," KCET, Los Angeles, July 10, 2013, https://www.kcet.org/redefine/nevada-solar-factory-canceled.

44 "Reid calls for solar plant at site of closed station," *Knight Ridder Tribune*, archive, December 2007, https://www.electricityforum.com/news-archive/dec07/Solarplanturgedforclosedstation.

45 "Company drops plans for solar project in Nevada," Yahoo News, June 17, 2013, https://news.yahoo.com/company-drops-plans-solar-project-145357457.html.

46 Conor Shine, "Company dumps big Laughlin solar project, says market won't support it," *Las Vegas Sun*, June 14, 2013, https://lasvegassun.com/news/2013/jun/14/company-dumps-big-laughlin-solar-project-says-mark/.

47 U.S. Patent Office, Patent #6904336, entitled "System and Method for Residential Emissions Trading," applied for by Franklin Raines and his associates on November 8,

Endnotes

2002, and issued on June 7, 2005, https://patft.uspto.gov/netacgi/nph-Parser?Sect1=PTO1&Sect2=HITOFF&d=PALL&p=1&u=%2Fnetahtml%2FPTO%2Fsrchnum.htm&r=1&f=G&l=50&s1=6904336.PN.&OS=PN/6904336&RS=PN/6904336.

48 U.S. Patent Office, Assignments (3) of Patent #6904336, https://assignment.uspto.gov/patent/index.html#/patent/search/resultAbstract?id=6904336&type=patNum.

49 BGC Environmental Brokerage Services, "What we do," EmissionTrading.org, no date, http://www.emissionstrading.com/AboutUs/. See also the "Emissions" subsection of the website at http://www.emissionstrading.com/Emissions/.

50 U.S. Patent Office, Patent #7133750, entitled "System and Method for Residential Emissions Trading," applied for by Franklin Raines and his associates on April 28, 2005, and issued on November 7, 2006, https://patft.uspto.gov/netacgi/nph-Parser?Sect1=PTO1&Sect2=HITOFF&d=PALL&p=1&u=%2Fnetahtml%2FPTO%2Fsrchnum.htm&r=1&f=G&l=50&s1=7133750.PN.&OS=PN/7133750&RS=PN/7133750.

51 Letter sent by Fannie Mae general counsel Alfred M. Pollard, to Reps. Darrell Issa (R-CA) and Jason Chaffetz (R-UT) of the Committee on Oversight and Government Reform, U.S. House of Representatives, dated May 25, 2010.

52 See for instance: "The Politics of Climate Change Legislation," *Harvard Magazine*, February 15, 2013, https://www.harvardmagazine.com/2013/02/environmentalist-failure-to-pass-cap-and-trade.

53 "Franklin Raines: Fannie Mae," *Business Week*, January 10, 2005, at http://www.businessweek.com/magazine/content/05_02/b3915646.htm.

54 Associated Press, "Ex-Fannie Mae Chief Ends Pay Dispute," *New York Times*, November 15, 2006, at http://www.nytimes.com/2006/11/15/business/15fannie.html.

55 Associated Press, "Scandal to Cost Ex-Fannie Mae Officers Millions," *New York Times*, April 19, 2008, http://www.nytimes.com/2008/04/19/business/19fannie.html?_r=2.

56 Andy Sullivan, "Analysis: Obama's 'green jobs' have been slow to sprout," Reuters, April 13, 2012, https://www.reuters.com/article/us-usa-campaign-green-idUSBRE83C08D20120413.

57 Benn Steil, Dinah Walker, and Romil Chouhan, "President Obama's Green Jobs Cost Taxpayers Big Bucks," *Forbes*, November 2, 2012, https://www.forbes.com/sites/realspin/2012/11/02/president-obamas-green-jobs-cost-taxpayers-big-bucks/?sh=37e6506f5699.

58 Gordon Hughes, "The Myth of Green Jobs," The Global Warming Policy Foundation (GWPF), Report 3, 2011, https://www.thegwpf.org/images/stories/gwpf-reports/hughes-green_jobs.pdf.

59 Ibid., "Summary," p. 5.
60 Ibid.
61 Ibid.
62 Ibid., p. 6.
63 Douglas Ernst, "Sen. Cory Booker: Embracing 'Green New Deal' is 'bold,' like fighting Nazis," *Washington Times*, February 8, 2019, https://www.washingtontimes.com/news/2019/feb/8/cory-booker-embracing-green-new-deal-is-bold-like-/.

64 Kamala Harris, "Green New Deal," KamalaHarris.medium.com, February 8, 2019, https://kamalaharris.medium.com/green-new-deal-2699b33ba666.

65 "Transcript, Democratic Debate, Second Round, in Detroit, Michigan, on Night 1, July 30, 2019," Rev.com, https://www.rev.com/blog/transcript-of-july-democratic-debate-night-1-full-transcript-july-30-2019.

66 William Cummings, "'The world is going to end in 12 years if we don't address climate change,' Ocasio-Cortez says," *USA Today*, January 22, 2019, https://www.usatoday.com/story/news/politics/onpolitics/2019/01/22/ocasio-cortez-climate-change-alarm/2642481002/.

67 Matthew Boyle, "Alexandria Ocasio-Cortez: 'The World Is Going to End in 12 Years if We Don't Address Climate Change,'" Breitbart.com, January 21, 2019, https://www.breitbart.com/politics/2019/01/21/alexandria-ocasio-cortez-the-world-is-going-to-end-in-12-years-if-we-dont-address-climate-change/.

68 Robert Kraychik, "Greenpeace Founder: Global Warming Hoax Pushed by Corrupt Scientists 'Hooked on Government Grants,'" Breitbart.com, March 7, 2019, https://www.breitbart.com/radio/2019/03/07/greenpeace-founder-global-warming-hoax-pushed-corrupt-scientists-hooked-government-grants/.

69 "Weather Channel Co-Founder John Coleman slams Federal climate report," ClimateDepot.com, May 7, 2014, https://www.climatedepot.com/2014/05/07/600-page-litany-of-doom-weather-channel-co-founder-john-coleman-slams-obama-climate-report-a-total-distortion-of-the-data-and-agenda-driven-destructive-episode-of-bad-science-gone-berserk/.

70 "John Coleman Letter to UCLA," Real Science, StevenGoddard.wordpress.com, October 19, 2014, https://stevengoddard.wordpress.com/2014/10/19/john-coleman-letter-to-ucla/. See also: Brandon Jones, "John Coleman shocks global warming supporters, 'science not valid,'" *TheGlobalDispatch*.com, October 23, 2014, https://www.theglobaldispatch.com/john-coleman-shocks-global-warming-supporters-science-not-valid-48622/.

71 Text of Green New Deal resolution as posted on the website of Sen. Edward J. Markey (D-MA), February 8, 2019, https://www.markey.senate.gov/imo/media/doc/Green%20New%20Deal%20Resolution%20SIGNED.pdf.

72 Douglas Holtz-Eakin, Dan Bosch, Ben Gitis, Dan Goldbeck, Philip Rossetti, "The Green New Deal: Scope, Scale, and Implications," American Action Forum, February 25, 2019, https://www.americanactionforum.org/research/the-green-new-deal-scope-scale-and-implications/.

73 Salvador Rizzo, "What's actually in the 'Green New Deal' from Democrats?" *Washington Post*, February 11, 2019, https://www.washingtonpost.com/politics/2019/02/11/whats-actually-green-new-deal-democrats/.

74 Jeff Cox, "Ocasio-Cortez's Green New Deal offers 'economic security' for those 'unwilling to work,'" CNBC.com, February 7, 2019, https://www.cnbc.com/2019/02/07/ocasio-cortezs-green-new-deal-offers-economic-security-for-those-unwilling-to-work.html?__source=twitter%7Cmain.

75 Office of Representative Alexandria Ocasio-Cortez, notes on the Green New Deal released on Thursday, February 7, 2019, at 8:30 a.m., https://assets.documentcloud.org/documents/5729035/Green-New-Deal-FAQ.pdf#page=2.

76 Ibid.

77 "Gross Domestic Product, Fourth Quarter and Annual 2018 (Initial Estimate)," news release, U.S. Department of Commerce, Bureau of Economic Analysis, February 28, 2019, https://www.bea.gov/news/2019/initial-gross-domestic-product-4th-quarter-and-annual-2018.

78 Dylan Stableford, senior writer, Yahoo News, "Ocasio-Cortez floats 70 percent tax on rich to pay for 'Green New Deal,'" Yahoo.com, January 4, 2019, https://news.yahoo.com/o/casio-cortez-floats-70-percent-tax-rich-pay-green-new-deal-170848121.html.

Endnotes

79 Ryan Dezember, "China's Footprint in U.S. Oil: A State-by-State List," *Wall Street Journal*, March 6, 2012, https://www.wsj.com/articles/BL-DLB-37024.
80 Ryan Dezember and James T. Areddy, "China Foothold in U.S. Energy," *Wall Street Journal*, updated March 6, 2012, https://www.wsj.com/articles/SB10001424052970204883304577223083067806776.
81 Chen Aizhu, Muyu Xu, Gavin Maguire, "China's oil buying binge to run on in 2021 as tank operators, refiners stock up," Reuters, November 17, 2020, https://www.reuters.com/article/us-china-oil-stockpiles/chinas-oil-buying-binge-to-run-on-in-2021-as-tank-operators-refiners-stock-up-idUSKBN27Y0A2.
82 Sara Schonhardt, "China's Greenhouse Gas Emissions Exceed Those of All Other Developed Countries," E&E News, *Scientific American*, May 6, 2021, https://www.scientificamerican.com/article/chinas-greenhouse-gas-emissions-exceed-those-of-all-other-developed-countries-combined/. See also: Kate Larsen, Hannah Pitt, Mikhail Grant, and Trevor Houser, "China's Greenhouse Gas Emissions Exceeded the Developed World for the First Time in 2019," Rhodium Group, May 6, 2021, https://rhg.com/research/chinas-emissions-surpass-developed-countries/.
83 Bart Sweerts, Stefan Pfenninger, Su Yang, Doris Folini, Bob van der Zwaan, and Martin Wild, "Estimation of losses in solar energy production from air pollution in China since 1960 using surface radiation data," *Nature Energy*, July 8, 2019, pp. 657-663, https://www.nature.com/articles/s41560-019-0412-4.
84 Michael Standaert, "Why China's Renewable Energy Transition Is Losing Momentum," *Yale Environment 360*, published at the Yale School of the Environment, September 26, 2019, https://e360.yale.edu/features/why-chinas-renewable-energy-transition-is-losing-momentum.
85 "China to stop subsidies for new solar power stations, onshore wind projects in 2021," Reuters, June 11, 2021, https://www.reuters.com/business/energy/china-stop-subsidise-new-solar-power-stations-onshore-wind-projects-2021-2021-06-11/.
86 Li Xing, "Population control called key to deal," *China Daily*, December 10, 2009, http://www.chinadaily.com.cn/china/2009-12/10/content_9151129.htm.
87 Thomas Wire, "Fewer Emitters, Lower Emissions, Less Cost. Reducing Future Carbon Emissions by Investing in Family Planning," Technical Report, London School of Economics, 2009.
88 "Fewer feet, smaller footprint: Fewer people would mean lower greenhouse-gas emissions," *The Economist*, September 21, 2009, https://www.economist.com/international/2009/09/21/fewer-feet-smaller-footprint.
89 William Mitchell, L. Randall Wray, and Martin Watts, *Macroeconomics* (London, UK: Red Globe Press, distributed in the U.S. by Macmillan International, 2019), p. 13.
90 See, for instance: Tom Leonard, "Broken down and rusting, is this the future of Britain's 'wind rush'?" *London Daily Mail*, March 18, 2012, https://www.dailymail.co.uk/news/article-2116877/Is-future-Britains-wind-rush.html#ixzz1pbANJuGk. See also: Michael Shellenberger, *Apocalypse Never: Why Environmental Alarmism Hurts Us All* (New York: HarperCollins, 2020).

Chapter 5

1 Henrik Svensmark and Nigel Calder, *The Chilling Stars: A Cosmic View of Climate Change* (Cambridge, U.K.: Icon Books, 2007), updated edition 2008, p. 2.
2 Roy W. Spencer, *The Bad Science and Bad Policy of Obama's Global Warming Agenda* (New York: Encounter Books, 2010), pp. 24 and 26.

3   Nigel Calder, *The Manic Sun: Weather Theories Confounded* (London: Pilkington Press, 1997), p. 41.
4   Roy W. Spencer, *The Great Global Warming Blunder: How Mother Nature Fooled the World's Top Climate Scientists* (New York: Encounter Books, 2010), p. xxiii.
5   Ibid., p. xxxi.
6   Roy W. Spencer, *Climate Confusion: How Global Warming Hysteria Leads to Bad Science, Pandering Politicians and Misguided Policies That Hurt the Poor* (New York: Encounter Books, 2008), pp., 62-64, at p. 63.
7   Ibid., p. 63.
8   Ibid., p. 64.
9   Ibid., p. 90.
10 Ibid. p. 55.
11 Ibid.
12 Ibid.
13 Ibid., p. 57.
14 Roy W. Spencer, *The Great Global Warming Blunder: How Mother Nature Fooled the World's Top Climate Scientists*, pp. 38-43.
15 Ibid., p. 43.
16 Ibid., pp. 48-49, at p. 49.
17 Ibid., p. 41.
18 Ibid., p. 61.
19 Ibid., p. 72.
20 Ibid.
21 "Prominent Geologist Dr. Robert Giegengack dissents—Laments 'hubris' of those who believe that we can 'control' climate—Denounces 'semi-religious campaign,'" ClimateDepot.com, November 11, 2019, https://www.climatedepot.com/2019/11/11/prominent-geologist-dr-robert-giegengack-dissents-laments-the-hubris-that-leads-us-to-believe-that-we-can-control-climate-denounces-semi-religious-campaign/.
22 Thom Nelson, "CO_2 Is Not the Climate Control Knob," December 6, 2011, quoting a Phil Jones email (email #3165) released in the Climategate FOIA documents, http://tomnelson.blogspot.com/2011/12/2000-warmist-phil-jones-goes-to.html.
23 Roy W. Spencer, *The Great Global Warming Blunder: How Mother Nature Fooled the World's Top Climate Scientists*, p. 15.
24 Ibid.
25 Roy W. Spencer and William D. Braswell, "Potential Biases in Feedback Diagnosis from Observational Data: A Simple Model Demonstration," Earth System Science Center, University of Alabama in Huntsville, *Journal of Climate*, Volume 21 (November 1, 2008), pp. 5624-5628, https://journals.ametsoc.org/view/journals/clim/21/21/2008jcli2253.1.xml?tab_body=pdf.
26 Roy W. Spencer, *Climate Confusion: How Global Warming Hysteria Leads to Bad Science, Pandering Politicians and Misguided Policies That Hurt the Poor*, pp. 14-21, at p. 16.
27 Roy W. Spencer, *The Great Global Warming Blunder: How Mother Nature Fooled the World's Top Climate Scientists*, pp. 19-20.
28 Ibid. p. 21.
29 Ibid., p. 111.
30 Ibid., p. 119.
31 Ibid., p. 140.
32 Ibid., p. 156.

Endnotes

33 Ibid., pp. 106-107.
34 Ibid., p. 153.
35 Ibid.
36 Ibid., p. 155.
37 Ibid.
38 Ibid., p. 111.
39 Ibid., p. 155.
40 Christopher Booker and Richard North, *Scared to Death: From BSE to Global Warming: Why Scares Are Costing Us the Earth* (New York: Continuum US, 2007), p. 391.
41 Habibullo Abdussamatov, "Energy Imbalance Between the Earth and Space Controls the Climate," *Earth Sciences*, Volume 9, Number 4 (July 2020), pp. 117-125, https://pdfs.semanticscholar.org/9052/342acb91d4eaf9fad2625ce8fc499fe900ec.pdf.
42 Ibid., p. 119.
43 Christopher Booker and Richard North, *Scared to Death*, p. 392.
44 Habibullo I. Abdussamatov, Pulkovo Observatory of the RAS, St. Petersburg, Russia, "Bicentennial Decrease of the Total Solar Irradiance Leads to Unbalanced Thermal Budget of the Earth and the Little Ice Age," *Applied Physics Research*, Volume 4, Number 1 (February 2012), pp. 178-184, http://citeseerx.ist.psu.edu/viewdoc/download?doi=10.1.1.353.4136&rep=rep1&type=pdf.
45 Habibullo Abdussamatov, "Current Long-Term Negative Average Annual Energy Balance of the Earth Leads to the New Little Ice Age," *Thermal Science*, Volume 19, Supplement 2 (2015), pp. S279-S288, http://www.doiserbia.nb.rs/img/doi/0354-9836/2015/0354-98361500018A.pdf.
46 Ibid., p. S284.
47 "Solar Cycle 25 Is Here. NASA, NOAA Scientists Explain What It Means," NASA.gov, September 15, 2020, https://www.nasa.gov/press-release/solar-cycle-25-is-here-nasa-noaa-scientists-explain-what-that-means.
48 "NOAA Confirms a 'Full-Blown' Grand Solar Minimum," ElectroVerse.net, September 2, 2020, https://electroverse.net/noaa-confirms-a-full-blown-grand-solar-minimum/.
49 News Wire, "Global Cooling? NOAA Confirms 'Full-blown' Grand Solar Minimum," 21stCenturyWire.com, September 5, 2020, https://21stcenturywire.com/2020/09/05/global-cooling-noaa-confirms-full-blown-grand-solar-minimum/. Also picked up by: Cap Allon, "NOAA confirms a 'full-blown' Grand Solar Minimum," Electroverse.net, September 2, 2020, https://electroverse.net/noaa-confirms-a-full-blown-grand-solar-minimum/. For NOAA sunspot projection data projecting Cycle 25 sunspot activity as diminishing, see: NOAA, "Solar Cycle Progression," predicting Solar Cycle Sunspot Number Production through 2034," https://www.swpc.noaa.gov/products/solar-cycle-progression. And this: NOAA, "Predicted Sunspot Number and Radio Flux, 2021-2040," swpc.noaa.gov, https://www.swpc.noaa.gov/products/predicted-sunspot-number-and-radio-flux.
50 Valentina Zharkova, "Modern Grand Solar Minimum will lead to terrestrial cooling," *Temperature*, Volume 7, Number 3 (2020), pp. 217-222, https://www.ncbi.nlm.nih.gov/pmc/articles/PMC7575229/.
51 Jamie Carter, "A 'Termination Event' on the Sun May Be Imminent as Solar Activity Escalates Say Scientists," *Forbes*, June 16, 2020, https://www.forbes.com/sites/jamiecartereurope/2021/06/16/a-termination-event-may-be-imminent-on-the-sun-as-solar-activity-escalates-say-scientists/?sh=3ed477c0611f.

52 "Is the Sun causing global warming?" NASA Global Climate Change, no date, Climate. NASA.gov, https://climate.nasa.gov/faq/14/is-the-sun-causing-global-warming/.
53 National Centers for Environmental Information of the National Oceanic and Atmospheric Administration (NOAA), "Glacial-Interglacial Cycles," no date, https://www.ncdc.noaa.gov/abrupt-climate-change/Glacial-Interglacial%20Cycles.
54 "Victor Hess discovers cosmic rays: April 7, 1912," no date, CERN Accelerating Science, https://timeline.web.cern.ch/victor-hess-discovers-cosmic-rays-0.
55 Elizabeth Howell, "What Are Cosmic Rays?" Space.com, May 10, 2018, https://www.space.com/32644-cosmic-rays.html.
56 Ibid.
57 U.S. Department of Energy, Office of Science, "DOE Explains …Muons," Energy.gov, no date, https://www.energy.gov/science/doe-explainsmuons.
58 "Cosmic rays: particles from outer space," CERN, no date, https://home.cern/science/physics/cosmic-rays-particles-outer-space. For an easily comprehensible discussion of ionization, see: Brad Cole, "What is Ionization," with pictures, InfoBloom.com, no date, https://www.infobloom.com/what-is-ionization.htm.
59 E. Friis-Christensen and K. Lassen, "Length of the Solar Cycle: An Indicator of Solar Activity Closely Associated with Climate," *Science*, New Series, Volume 254, Number 5032 (November 1, 1991), pp. 698-700, https://science.sciencemag.org/content/254/5032/698. See also: Christopher Booker and Richard North, *Scared to Death*, p. 393.
60 E. Friis-Christensen and K. Lassen, "Length of the Solar Cycle," p. 700.
61 Christopher Booker and Richard North, *Scared to Death*, p. 393.
62 Henrik Svensmark and Eigil Friis-Christensen, "Variation of cosmic ray flux and global cloud coverage—a missing link in solar-climate relationships," *Journal of Atmospheric and Solar-Terrestrial Physics*, Volume 59, Number 11 (1997), pp. 1225-1232, ftp://ftp.spacecenter.dk/pub/Henrik/FB/Svensmark1997%28GCR-clouds%29.pdf.
63 Henrik Svensmark and Nigel Calder, *The Chilling Stars: A Cosmic View of Climate Change* (Cambridge, U.K.: Icon Books Ltd., 2007, updated edition published 2008), "Overview," p. 1.
64 Henrik Svensmark et al., "The connection between cosmic rays, clouds, and climate," address to EIKE (https://eike-klima-energie.eu), 2018, https://www.eike-klima-energie.eu/wp-content/uploads/2018/12/SvensmarkMunic2018.pdf. For a video of Svensmark's presentation, see https://www.youtube.com/watch?v=Rg3MqdBX0_k, posted on YouTube.com, January 24, 2019, by "Rathnakumar S."
65 "CERN's CLOUD experiment provides unprecedented insight into cloud formation," CERN, press release, August 25, 2011, https://home.cern/news/press-release/cern/cerns-cloud-experiment-provides-unprecedented-insight-cloud-formation.
66 Ibid.
67 Jasper Kirby, Joachim Curtius, Markku Kulmala, et al., "Role of sulphuric acid, ammonia, and galactic cosmic rays in atmospheric aerosol nucleation," *Nature* 476 (August 24, 2011), pp. 429-433, https://www.nature.com/articles/nature10343.
68 Christopher Booker and Richard North, *Scared to Death*, p. 393.
69 Nir J. Shaviv, "The Spiral Structure of the Milky Way, Cosmic Rays, and Ice Age Epochs on Earth," *New Astronomy*, Volume 8 (September 12, 2002), pp. 39-77., https://arxiv.org/pdf/astro-ph/0209252.pdf.
70 "Milky Way Spiral Structure," Messier.Seds.org, no date, https://www.messier.seds.org/more/mw_arms.html.

Endnotes

71 "How Long to Orbit Milky Way's Center," EarthSky.org, November 28, 2016, https://earthsky.org/astronomy-essentials/milky-way-rotation/.

72 Kevin Jardine, "Basic plan of the Milky Way," GalaxyMap.org, no date, http://ymap.org/drupal/node/171.

73 Karen Masters, "How often does the Sun pass through a spiral arm in the Milky Way?" Cornell University, "Ask an Astronomer," last updated on April 18, 2016, http://curious.astro.cornell.edu/about-us/55-our-solar-system/the-sun/the-sun-in-the-milky-way/207-how-often-does-the-sun-pass-through-a-spiral-arm-in-the-milky-way-intermediate.

74 Nir J. Shaviv and Ján Veizer, "Celestial driver of Phanerozoic climate?" *GSA Today* (July 2003), pp. 4-10, https://www.researchgate.net/publication/235451724_Celestial_driver_of_Phanerozoic_climate_GSA_Today.

75 Ibid.

76 Jean-François Ghienne, André Desrochers, Thijs R. A. Vandenbroucke, et al., "A Cenozoic-style scenario for the end-Ordovician glaciation," *Nature Communications*, Volume 5, Number 4485 (September 1, 2014), https://www.nature.com/articles/ncomms5485.

77 Nir J. Shaviv and Ján Veizer, "Celestial driver of Phanerozoic climate?" See also: Nir Shaviv, "How Might Climate be Influenced by Cosmic Rays?" Institute for Advanced Study, Princeton, New Jersey, 2015, https://www.ias.edu/ideas/2015/shaviv-milkyway. For Shaviv and Veizer's rebuttal of various attempts by climate change true believers to disprove his CRF theory and the conclusion CO_2 has little effect on Earth's climate, see: Nir J. Shaviv and Ján Veizer, "On the Role of Cosmic Ray Flux Variations as a Climate Driver: The Debate," Nir Shaviv's blog at ScienceBits.com, no date, http://www.sciencebits.com/ClimateDebate.

78 S. Fred Singer and Dennis T. Avery, *Unstoppable Global Warming Every 1,500 Years* (Lanham, Maryland: Rowman & Littlefield Publishers, Inc., 2007), pp. 195-196.

79 Ibid.

80 "Five fascinating Gaia revelations about the Milky Way," European Space Agency, September 21, 2020, https://www.esa.int/About_Us/ESAC/Five_fascinating_Gaia_revelations_about_the_Milky_Way.

81 Data on carbon dioxide concentrations in Earth's atmosphere over geological time are summarized here: C.D. Idso, R.M. Carter, and S.F. Singer, editors, *Climate Change Reconsidered II: Physical Science* (Chicago, Illinois: The Heartland Institute, 2013), p. 151. Contemporary readings of CO_2 concentrations are drawn from NOAA research at the Mauna Loa Observatory. See: NOAA Research News, "Carbon dioxide peaks near 420 parts per million at Mauna Loa Observatory," Research.NOAA.gov, June 7, 2021, https://research.noaa.gov/article/ArtMID/587/ArticleID/2764/Coronavirus-response-barely-slows-rising-carbon-dioxide.

82 Daniel H. Rothman, "Atmospheric carbon dioxide levels for the past 500 million years," *Proceedings of the National Academy of Sciences of the United States of America (PNAS)*, Volume 99, Number 7 (April 2, 2020), pp. 4167-4171, https://www.pnas.org/content/99/7/4167.

83 Daniel H. Rothman, "Atmospheric carbon dioxide levels for the past 500 million years," *Proceedings of the National Academy of Sciences of the United States of America*.

84 C.D. Idso, R.M. Carter, and S.F. Singer, editors, *Climate Change Reconsidered II: Physical Science*, p. 156.

85 Nicolas Caillon, Jeffrey P. Severinghaus, Jean Jouzel, et al., "Timing of Atmospheric CO_2 and Antarctic Temperature Changes Across Termination III," *Science*, Volume 299 (April 2003), pp. 1728-1731, https://www.researchgate.net/publication/10855143_Timing_of_Atmospheric CO_2 and_Antarctic_Temperature_Changes_Across_Termination_III.
86 Ibid., p. 1730.
87 Marshall Shepherd, "Why Solar Activity and Cosmic Rays Can't Explain Global Warming," *Forbes*, August 10, 2019, https://www.forbes.com/sites/marshallshepherd/2019/08/10/why-solar-activity-and-cosmic-rays-cant-explain-global-warming/.
88 S.K. Solanki and M. Fligge, "A reconstruction of total solar irradiance since 1700," *Geophysical Research Letters*, Volume 26, Number 16 (August 15, 1999), pp. 2465-2468, https://agupubs.onlinelibrary.wiley.com/doi/pdfdirect/10.1029/1999GL900370.
89 Robin Crockett, University of Northampton, U.K., *A Primer on Fourier Analysis for the Geosciences* (Cambridge, U.K.: Cambridge University Press, 2019).
90 Blanca Mendoza, "Total solar irradiance and climate," *Advances in Space Research*, Volume 35 (2005), pp. 882-890, https://connaissance-innovation.typepad.fr/index/files/2005_mendoza_tsi_and_climate.pdf.
91 Ibid., p. 889.
92 J. Beer, M. Vonmoos, and R. Muscheler, "Solar Variability Over the Past Several Millenia," *Space Science Reviews*, Volume 125, August 2006, pp. 67-79, https://link.springer.com/chapter/10.1007/978-0-387-48341-2_6.
93 Ibid., Abstract, p. 67.
94 Ibid. All remaining quotes from Beer and his colleagues are drawn from this same article.
95 Ibid., p. 76.

Chapter 6

1 Stephen McIntyre and Ross McKitrick, "Corrections to the Mann et al. (1998) Proxy Data Base and Northern Hemisphere Average Temperature Series," *Energy & Environment*, Volume 14, Number 6 (2003), pp. 751-771, at p. 751, http://www.multi-science.co.uk/mcintyre-mckitrick.pdf.
2 A.W. Montford, *The Hockey Stick Illusion, Climategate and the Corruption of Science* (London, U.K.: Stacey International, 2010), p. 390.
3 Don Easterbrook, *Evidence-Based Climate Science: Data Opposing CO_2 Emissions as the Primary Source of Global Warming* (Amsterdam, Netherlands: Elsevier, second edition, 2016), pp. 28-29.
4 Jonathan Watts, "Climatologist Michael E. Mann: 'Good people fall victim to doomism. I do too sometimes,'" interview, *The Guardian*, February 27, 2021, https://www.theguardian.com/environment/2021/feb/27/climatologist-michael-e-mann-doomism-climate-crisis-interview.
5 Michael E. Mann, curriculum vitae, Pennsylvania State University, Department of Meteorology, University Park, PA, no date, http://www.meteo.psu.edu/holocene/public_html/Mann/about/cv.php.
6 Michael E. Mann, Raymond S. Bradley, and Malcolm K. Hughes, "Global-scale temperature patterns and climate forcing over the past six centuries," *Nature*, Volume 392 (April 23, 1998), https://meteor.geol.iastate.edu/classes/ge515/papers/Mann_et_al_Nature1998.pdf.

Endnotes

7 Malcolm K. Hughes and Henry F. Diaz, "Was There a 'Medieval Warm Period, and If So, When and Where?" *Climatic Change*, Volume 26 (March 1994), pp. 109-142, http://www.climateknowledge.org/figures/Rood_Climate_Change_AOSS480_Documents/Hughes_Medieval_Warm_ClimaticChange_1994.pdf.
8 Ibid., pp. 136-137.
9 A.W. Montford, *The Hockey Stick Illusion: Climategate and the Corruption of Science*, p. 27.
10 Ibid., pp. 26-27, at p. 26, in the title of the chapter subsection.
11 Ibid., p. 27.
12 Ibid.
13 Ibid.
14 Ibid., pp. 32-33.
15 Ibid., pp. 30-36.
16 Michael E. Mann, Raymond S. Bradley, and Malcolm K. Hughes, "Northern Hemisphere Temperatures During the Past Millennium: Inferences, Uncertainties, and Limitations," *Geophysical Research Letters*, Volume 26, Number 6 (March 15, 1999), pp. 759-762, https://agupubs.onlinelibrary.wiley.com/doi/pdfdirect/10.1029/1999GL900070.
17 A.W. Montford, *The Hockey Stick Illusion: Climategate and the Corruption of Science*, p. 40.
18 C.K. Folland, T.R. Karl, and K.Y.A. Vinnikov, "Observed Climate Variations and Change," Section 7, in J.T. Houghton, G.J. Jennings, and J.J. Ephraums, *Climate Change: The IPCC Scientific Assessment* (Cambridge, U.K.: Cambridge University Press, 1990), pp. 194-238, Figure 7.1 at p. 202, https://archive.ipcc.ch/ipccreports/far/wg_I/ipcc_far_wg_I_full_report.pdf.
19 Ibid.
20 C.K. Folland, T.R. Karl, and K.Y.A. Vinnikov, "Observed Climate Variations and Change," "Executive Summary," p. 199.
21 John L. Daly, "The 'Hockey Stick': A New Low in Climate Science," *John-Daly.com*, no date, http://www.john-daly.com/hockey/hockey.htm.
22 Dr. David Deming, Testimony before a U.S. Senate Committee on Environment and Public Works, Full Committee Hearing on Climate Change and the Media, December 6, 2006, 9:30 a.m., EPW.Senate.gov, https://www.epw.senate.gov/public/index.cfm/hearings?Id=BFE4D91D-802A-23AD-4306-B4121BF7ECED&Statement_id=361256C4-11DC-4E5D-8D1D-9FEDF082D081. For Deming's academic paper referenced in his testimony, see: David Deming, "Climatic Warming in North America: Analysis of Borehole Temperatures," *Science*, Volume 268 (June 16, 1995), pp. 1576-1577.
23 Ibid.
24 Ibid.
25 "Summary for Policymakers: A Report of Working Group 1 of the Intergovernmental Panel on Climate Change," in *Climate Change 2001: The Scientific Basis,* edited by J.T. Houghton, Y. Ding, D.J. Greggs, et al., IPCC (Cambridge, U.K.: Cambridge University Press, 2001), pp. 1-20, at p. 3, https://www.ipcc.ch/site/assets/uploads/2018/03/WGI_TAR_full_report.pdf.
26 NOAA National Centers for Environmental Information, *State of the Climate: Global Climate Report for Annual 1998*, published online January 1999, https://www.ncdc.noaa.gov/sotc/global/199813.

27 Donald A. Graybill and Sherwood B. Idso, "Detecting the Aerial Fertilization Effect of Atmospheric CO_2 Enrichment in Tree-Ring Chronologies," *Global Biogeochemical Cycles*, Volume 7, Number 1 (March 1993), pp. 81-95, http://www.climateaudit.info/pdf/others/graybill.idso.1993.pdf.
28 Ibid.
29 Ibid., "Conclusions," p. 92.
30 Ibid., "Abstract," p. 81.
31 Ibid.
32 Ibid.
33 Stephen McIntyre and Ross McKitrick, "Corrections to the Mann et al. (1998) Proxy Data Base and Northern Hemisphere Average Temperature Series," *Energy & Environment*, pp. 766-767.
34 Stephen McIntyre and Ross McKitrick, "Hockey sticks, principal components, and spurious significance," *Geophysical Research Letters*, Volume 32, Issue 3 (February 12, 2005), https://agupubs.onlinelibrary.wiley.com/doi/epdf/10.1029/2004GL021750.
35 Ibid.
36 Ross McKitrick, "Memorandum by Ross McKitrick, Associate Professor of Economics, University of Guelph," published by the Select Committee on Economic Affairs, House of Lords, Parliament U.K., February 24, 2005, https://publications.parliament.uk/pa/ld200506/ldselect/ldeconaf/12/12we16.htm#note82.
37 Ross McKitrick, Department of Economics, University of Guelph, "What Is the 'Hockey Stick' Debate About?" Invited Special Presentation to the Conference "Managing Climate Change—Practicalities and Realities in a Post-Kyoto Future," Parliament House, Canberra, Australia, April 4, 2005, https://www.rossmckitrick.com/uploads/4/8/0/8/4808045/mckitrick-hockeystick.pdf.
38 Christopher Booker and Richard North, *Scared to Death*, p. 362.
39 This difficulty using tree ring data as a proxy measure of global temperatures was observed by Christopher Booker and Richard North, *Scared to Death*, p. 359.
40 A.W. Montford, *Hiding the Decline: A History of the Climategate Affair* (A.W. Montfort, self-published, 2012), pp. 84-85. See also: "Hide the Decline," JustFacts.com, no date, https://www.justfacts.com/globalwarming.hidethedecline.asp.
41 "Hide the Decline," U.S. Senate Committee on Environment and Public Works," press release, posted by David Lungren, December 14, 2009, https://www.epw.senate.gov/public/index.cfm/press-releases-all?ID=8F16552A-802A-23AD-465F-8858BEB85AC2.
42 Dr. Kelvin Kemm, "Climategate: Ten Years Later," Heartland.org, November 1, 2019, https://www.heartland.org/news-opinion/news/climategate-ten-years-later.
43 Fred Pearce, "The five key leaked emails from UEA's Climatic Research Unit," *The Guardian*, July 7, 2010, https://www.theguardian.com/environment/2010/jul/07/hacked-climate-emails-analysis.
44 John O'Sullivan, "Ten Years After Climategate—Time to Prosecute Michael Mann," *Principia Scientific International*, November 22, 2019, https://principia-scientific.com/ten-years-after-climategate-time-to-prosecute-michael-mann/.
45 Paragraph drawn from the following: Tony Thomas, "Climate Science Proves Scams Don't Die of Exposure," Quadrant.org.au, November 14, 2019, https://quadrant.org.au/opinion/doomed-planet/2019/11/ten-years-after-climategate/.
46 Kelvin Kemm, CFACT editor, "Climategate: Ten years later," CFACT.org, November 1, 2019, https://www.cfact.org/2019/11/01/climategate-ten-years-later/. The YouTube

Endnotes

song "Hide the Decline," was posted on YouTube.com on April 22, 2010, https://www.youtube.com/watch?v=WMqc7PCJ-nc.
47 Marc Morano, *The Politically Incorrect Guide to Climate Change* (Washington, D.C.: Regnery Publishing, 2018), p. 155.
48 Quoted in Marc Morano, "Don't Let Media Whitewash Climategate!" ClimateDepot.com, November 18, 2019, https://www.climatedepot.com/2019/11/18/dont-let-media-whitewash-climategate-read-chapter-excerpt-revealing-the-truth-behind-scandal-10-years-later/.
49 David Rose, "SPECIAL INVESTIGATION: Climate change emails row deepens as Russians admit they DID come from their Siberian Server," *Daily Mail*, December 13, 2009, https://www.dailymail.co.uk/news/article-1235395/SPECIAL-INVESTIGATION-Climate-change-emails-row-deepens--Russians-admit-DID-send-them.html.
50 Michael LePage, "Climate Myths: The 'hockey stick' graph has been proven wrong," *New Scientist*, May 16, 2007, https://www.newscientist.com/article/dn11646-climate-myths-the-hockey-stick-graph-has-been-proven-wrong/.
51 Frédérik Saltré and Corey J.A. Bradshaw, "Climate explained: What was the Medieval Warm Period?" TheConversation.com, April 20, 2021, https://theconversation.com/climate-explained-what-was-the-medieval-warm-period-155294.
52 Don Easterbrook, editor, *Evidence-Based Climate Science: Data Opposing CO_2 Emissions as the Primary Source of Global Warming* (Amsterdam, Netherlands: Elsevier, 2011, second edition, 2016), p. 401.
53 Willie Soon and Sallie Baliunas, "Proxy climatic and environmental changes of the past 1,000 years," *Climate Research*, Volume 23 (2003), pp. 89-110, http://www.int-res.com/articles/cr2003/23/c023p089.pdf.
54 "Project Overview: Medieval Warm Period," CO2science.org, Center for the Study of Carbon Dioxide and Climate, no date, http://co2science.org/data/mwp/description.php; and "Medieval Warm Period: Study Description and Results, CO2science.org, http://www.co2science.org/data/mwp/mwpp.php. See also: S. Fred Singer and Dennis T. Avery, *Unstoppable Global Warming: Every 1,500 Years* (Plymouth, U.K.: Rowman & Littlefield Publishers, Inc., 2007), pp. 105-115.
55 Michael E. Mann, "Medieval Climactic Optimum," in "Volume 1, The Earth system: physical and chemical dimensions of global environmental change," edited by Dr. Michael C. MacCracken and Dr. John S. Perry, in *Encyclopedia of Global Environmental Change*, Thomas Mann, editor in chief (New York: John Wiley & Sons, Ltd., 2002), pp. 514-516, at p. 516, http://www.meteo.psu.edu/holocene/public_html/shared/articles/medclimopt.pdf.
56 Ibid.
57 Don Easterbrook, editor, *Evidence-Based Climate Science: Data Opposing CO_2 Emissions as the Primary Source of Global Warming*, p. 170. See also: J.R. Petit, J. Jouzel, et al., "Climate and atmospheric history of the past 420,000 years from the Vostok ice core, Antarctica," *Nature*, Volume 39 (June 3, 1999), pp. 429-438, http://large.stanford.edu/publications/coal/references/docs/1999.pdf; Manfred Mudelsee, "The phase relations among atmospheric CO_2 content, temperature, and global ice volume over the past 420 ka," *Quaternary Science Reviews*, Volume 20 (2001), pp. 583-589, https://citeseerx.ist.psu.edu/viewdoc/download?doi=10.1.1.582.7554&rep=rep1&type=pdf; and Nicolas Caillon, Jeffrey P. Severinghaus, et al., "Timing of Atmospheric CO_2 and Antarctic Temperature Changes Across Termination III, *Science*, Volume

299, Issue 5613 (March 14, 2003), pp. 1728-1731, https://science.sciencemag.org/content/299/5613/1728.abstract.
58 Hubertus Fischer, Martin Wahlen, et al., "Ice Core Records of Atmospheric CO_2 Around the Last Three Glacial Terminations," *Science*, Volume 283 (March 12, 1999), pp. 1712-1714, https://epic.awi.de/id/eprint/825/1/Fis1999a.pdf.
59 Don Easterbrook, editor, *Evidence-Based Climate Science: Data Opposing CO_2 Emissions as the Primary Source of Global Warming*, p. 403.
60 Ole Humlum, Kjell Stordahl, and Jan-Erik Solheim, "The phase relation between atmospheric carbon dioxide and global temperature," *Global and Planetary Change*, Volume 10 (2013), pp. 51-69, https://www.researchgate.net/profile/J-E-Solheim/publication/257343053_The_phase_relation_between_atmospheric_carbon_dioxide_and_global_temperature/links/56e4581508ae68afa1106148/The-phase-relation-between-atmospheric-carbon-dioxide-and-global-temperature.pdf.
61 Don Easterbrook, editor, *Evidence-Based Climate Science: Data Opposing CO_2 Emissions as the Primary Source of Global Warming*, p. 403-404.
62 Michael E. Mann, *The Hockey Stick and the Climate Wars: Dispatches from the Front Lines* (New York: Columbia University Press, 2012), "Prologue," p. xi.
63 Ibid.
64 Ibid., pp. 73-74.
65 Ibid., p. 66.
66 Edward J. Wegman, David W. Scott, and Yasmin H. Said, *Ad Hoc Committee Report on the 'Hockey Stick' Global Climate Reconstruction*, U.S. House Committee on Energy and Commerce, July 16, 2006, PDF reprinted by the Science & Public Policy Institute, April 26, 2010, "Executive Summary," pp. 2-6, http://scienceandpublicpolicy.org/wp-content/uploads/2010/07/ad_hoc_report.pdf.
67 Michael E. Mann, *The Hockey Stick and the Climate Wars: Dispatches from the Front Lines*, p. 243.
68 Michael E. Mann and Lee R. Kump, *Dire Predictions: Understanding Global Warming, The Illustrated Guide to the Findings of the IPCC* (New York: DK, 2008).
69 Ibid., p. 191.
70 Michael E. Mann, *The New Climate War: The Fight to Take Back Our Planet* (New York: PublicAffairs, Hachette Book Group, 2021).
71 Ibid, "Introduction," pp. 1-2.
72 Ibid., p. 3.
73 Ibid. Emphasis in original.
74 Ibid.
75 Ibid., Chapter 2, "The Climate Wars," p. 41.
76 Ibid., p. 45. Emphasis in original.
77 Ibid., Chapter 4, "It's YOUR Fault," p. 93. Capitalization in original.
78 Ibid., pp. 94-95.
79 Ibid., p. 96.
80 Ibid., Chapter 9, "Meeting the Challenge," p. 230.
81 Ibid., p. 232.
82 Ibid., Chapter 4, "It's YOUR Fault," p. 95. Capitalization in original.
83 Ibid., Chapter 9, "Meeting the Challenge," p. 264.
84 Intergovernmental Panel on Climate Change (IPCC), *Climate Change 2021: The Physical Science Basis*, Working Group contribution, Sixth Assessment Report, August

Endnotes

9, 2021, https://www.ipcc.ch/assessment-report/ar6/. The IPCC Sixth Climate Assessment is the sixth such report issued since 1990.
85 "IPCC report: 'Code Red' for human driven global heating, warns UN chief," United Nations, press release, August 9, 2021, https://news.un.org/en/story/2021/08/1097362.
86 Ibid.
87 Ibid.
88 Ibid.
89 Michael Bastasch, "UN Picks Socialist Politician as New Secretary-General," DailyCaller.com, October 5, 2016, https://dailycaller.com/2016/10/05/un-picks-socialist-politician-as-new-secretary-general/#ixzz4MEueEagr.
90 Zack Colman and Karl Mathiesen, "'Get scared': World's scientists say disastrous climate change is here," *Politico*, August 9, 2021, https://www.politico.com/amp/news/2021/08/09/climate-change-scientists-report-disastrous-502799.
91 Ibid.
92 "Watch: Morano on Fox News Primetime: 'The UN is saying unless we go Marxist, the climate is going to get us!" ClimateDepot.com, August 9, 2021, https://www.climatedepot.com/2021/08/10/watch-morano-on-fox-news-primetime-the-un-is-saying-unless-we-go-marxist-the-climate-is-gonna-get-us-say-hell-no-to-the-un-to-green-new-deal/?mc_cid=cc3ace12d5&mc_eid=84cbea21e4. See also: Fox News Staff, "'Green Fraud' author: Goal of UN climate change report is to 'scare everyone,'" Fox News, August 9, 2021, https://www.foxnews.com/media/green-fraud-un-climate-change-report.
93 "Watch: Morano on Fox News Primetown," ClimateDepot.com, August 10, 2021, https://www.climatedepot.com/2021/08/10/watch-morano-on-fox-news-primetime-the-un-is-saying-unless-we-go-marxist-the-climate-is-gonna-get-us-say-hell-no-to-the-un-to-green-new-deal/.
94 Stephan Farsang, Marion Louvel, et al., "Deep carbon cycle constrained by carbonate solubility," *Nature Communications*, Volume 12, Number 4311 (July 14, 2021), https://www.nature.com/articles/s41467-021-24533-7.
95 "Earth's interior is swallowing up more carbon than thought," *University of Cambridge Research*, press release, July 27, 2021, https://www.cam.ac.uk/research/news/earths-interior-is-swallowing-up-more-carbon-than-thought.
96 Ibid.
97 IPCC Fourth Assessment Report: Climate Change 2007, "Frequently Asked Question 3.2: How Is Precipitation Changing?" https://archive.ipcc.ch/publications_and_data/ar4/wg1/en/faq-3-2.html.
98 Don Easterbrook, editor, *Evidence-Based Climate Science: Data Opposing CO_2 Emissions as the Primary Source of Global Warming*, p. 104.
99 IPCC, "FAQ Chapter 3, FAQ Question 3.1: What are the impacts of 1.5°C and 2.0°C on Warming?" *Special Report on Global Warming of 1.5°C* (SR15), 2018, https://www.ipcc.ch/sr15/faq/faq-chapter-3/.
100 Judith A. Curry and Peter J. Webster, *Thermodynamics of Atmospheres and Oceans*, Volume 65 in the International Geophysics Series (San Diego, CA: Academic Press, 1999), p. 333.
101 Don Easterbrook, editor, *Evidence-Based Climate Science: Data Opposing CO_2 Emissions as the Primary Source of Global Warming*, p. 163.

102 Judith A. Curry and Peter J. Webster, *Thermodynamics of Atmospheres and Oceans*, p. 359.
103 Don Easterbrook, editor, *Evidence-Based Climate Science: Data Opposing CO_2 Emissions as the Primary Source of Global Warming*, pp. 163-164.
104 American Chemical Society, "Water Vapor and Climate Change," ACS.org, no date, https://www.acs.org/content/acs/en/climatescience/climatesciencenarratives/its-water-vapor-not-the-co2.html.
105 Roy W. Spencer, *The Great Global Warming Blunder*, p. 99.
106 Ibid.
107 Ibid., p. 123.
108 Ibid., p. 153.
109 Ibid., p. 100.
110 Ibid., p. 154.
111 Ibid., pp. 154-155.
112 Roy W. Spencer, *The Great Global Warming Blunder*, p. 123.

Chapter 7

1 S. Warren Carey, *The Expanding Earth* (Amsterdam, Netherlands: Elsevier Scientific Publishing Company, 1976), "Face of the Earth," p. 1.
2 Ian Plimer, *Heaven and Earth: Global Warming, the Missing Science* (Lanham, Maryland: Taylor Trade Publishing, an imprint of Rowman & Littlefield Publishing Group, Inc., 2009), p. 10.
3 Gerrit L. Verschuur, *Impact! The Threat of Comets and Asteroids* (New York and Oxford: Oxford University Press, 1996), p. vi.
4 Posted by Samanthi, "Difference Between Uniformitarianism and Catastrophism," DifferenceBetween.com, April 8, 2021, https://www.differencebetween.com/difference-between-uniformitarianism-and-catastrophism/. See also: Marco Romano, "Reviewing the term uniformitarianism in modern Earth Sciences," *Earth-Science Reviews*, Volume 148 (September 2015), pp. 65-76, https://www.sciencedirect.com/science/article/abs/pii/S0012825215000938. See also: David Sepkoski, *Catastrophic Thinking: Extinction and the Value of Diversity from Darwin to the Anthropocene* (Chicago, Illinois: University of Chicago Press, 2020).
5 Gale L. Pooley and Martin L. Tupy, "The Simon Abundance Index: A New Way to Measure Availability of Resources," Cato Institute, *Policy Analysis Number 857*, December 4, 2018, https://www.cato.org/policy-analysis/simon-abundance-index-new-way-measure-availability-resources.
6 Gale L. Pooley and Marian L. Tupy, "The Simon Abundance Index 2020," HumanProgress.org, April 22, 2020, https://www.humanprogress.org/the-simon-abundance-index-2020/. See also: "The Simon Project," HumanProgress.org, no date, https://www.humanprogress.org/simonproject/.
7 Russell McLendon, "How Many Polar Bears Are There?" Treehugger.com, updated July 12, 2021, https://www.treehugger.com/polar-bear-population-4859409.
8 Lawrence M. Krauss, *The Physics of Climate Change* (New York: Post Hill Press, 2021), p. 32.
9 Ibid., p. 31.
10 Ibid., p. 34.
11 Donald J. DePaulo, "Sustainable Carbon Emissions: The Geologic Perspective," *MRS Energy & Sustainability*, Volume 2 (February 2015), pp. 1-16, https://

Endnotes

www.researchgate.net/profile/Donald-Depaolo/publication/281666355_Sustainable_carbon_emissions_The_geologic_perspective_-_CORRIGENDUM/links/56f5a50608ae38d710a0dc7a/Sustainable-carbon-emissions-The-geologic-perspective-CORRIGENDUM.pdf.

12 Ibid., p. 2. Parentheses in original.
13 Ibid. Emphasis in original.
14 Ibid., Abstract, p. 1.
15 Ibid., p. 12.
16 Ibid., Abstract, p. 1.
17 Ibid., p. 13.
18 Ibid., p. 9.
19 Ibid.
20 Ibid.
21 Ibid.
22 Ibid., p. 10.
23 Ibid.
24 "Electricity Generation and Related CO_2 Emissions," *Planete Energies*, an initiative by Total Foundation, December 2, 2016, https://www.planete-energies.com/en/medias/close/electricity-generation-and-related-co2-emissions.
25 Pekka Kauppi, Vilma Sandström, and Antti Lipponen, "Forest resources of nations in relation to human well-being," *PLOS ONE*, Volume 13, Number 5 (May 14, 2018), https://journals.plos.org/plosone/article?id=10.1371/journal.pone.0196248.
26 Ibid.
27 Ibid.
28 David Crisp, Han Dolman, Toste Tanhu, et al., "How Well Do We Understand the Land-Ocean Carbon Cycle?" manuscript submitted to *Reviews of Geophysics*, Earth and Space Science Open Archive, ESSOAr, Washington, D.C., February 23, 2021, https://www.proquest.com/openview/d18bc2ea9f46b87cfd99e29551f5846d/1?pq-origsite=gscholar&cbl=4882998.
29 P. Falkowski, R.J. Scholes, E. Boyle, et al., "The Global Carbon Cycle: A Test of Our Knowledge of Earth as a System," *Science*, Volume 290, Number 5490 (October 13, 2000), pp. 291-296, https://www.jstor.org/stable/3078125.
30 David Crisp, Han Dolman, Toste Tanhu, et al., "How Well Do We Understand the Land-Ocean Carbon Cycle?"
31 Lucian Sfica, Christoph Beck, Andrei-Ion Nita, et al., "Cloud cover changes driven by atmospheric circulation in Europe during the last decades," *International Journal of Climatology* (September 2020), https://rmets.onlinelibrary.wiley.com/doi/full/10.1002/joc.6841.
32 "Cloudy Earth," NASA Earth, no date, https://earthobservatory.nasa.gov/images/85843/cloudy-earth.
33 Alix Martichoux, "NOAA predicts 70% chance of La Niña winter. Here's what that means where you live," Nexstar Media Wire, as reported by WFLA TV, News Channel 8, Tampa, Florida, August 14, 2021, https://www.wfla.com/weather/noaa-predicts-70-chance-of-la-nina-winter-heres-what-that-means-where-you-live/.
34 "The History of Life on Earth," Northern Arizona University, no date, https://www2.nau.edu/lrm22/lessons/timeline/24_hours.html. See also: "Human Evolution: A Timeline of Early Hominids," last updated June 12, 2021, EarthHow, https://earthhow.com/human-evolution-timeline/.

35 Germit L. Verschuur, *Impact! The Threat of Comets and Asteroids* (London and New York: Oxford University Press), pp. 20-21.
36 Luis W. Alvarez, Walter Alverez, et al., "Extraterrestrial Cause for the Cretaceous-Tertiary Extinction: Experimental results and theoretical interpretation," *Science*, Volume 208, Issue 4448 (June 6, 1980), pp. 1095-1108, https://science.sciencemag.org/content/208/4448/1095.abstract.
37 Ibid., p. 1103.
38 Ibid., p. 1102.
39 Ibid., p. 1104.
40 Ibid., p. 1105.
41 David B. Weinreb, "Catastrophic Events in the History of Life: Toward a New Understanding of Mass Extinctions in the Fossil Record—Part 1," *Journal of Young Investigators* (March 23, 2002), https://www.jyi.org/2002-march/2017/10/23/catastrophic-events-in-the-history-of-life-toward-a-new-understanding-of-mass-extinctions-in-the-fossil-record-part-i
42 Ibid.
43 Ibid.
44 Stephen Marshak, *The Essentials of Geology* (New York and London: W. W. Norton & Company Ltd., sixth edition, 2019), p. 385.
45 Robert A. DePalma, Jan Smit, David A. Burnham, et al., "A seismically induced onshore surge deposit at the KPg boundary, North Dakota," *Proceedings of the National Academy of the United States*, Volume 116, Number 17 (April 1, 2019), https://www.pnas.org/content/116/17/8190.
46 Jonathan Amos, "Chicxulub asteroid impact: Stunning fossils record dinosaurs' demise," *BBC News*, March 29, 2019, https://www.bbc.com/news/science-environment-47755275.
47 Steven Goderis, Honami Sato, Ludovic Ferrière, et al., "Globally distributed iridium layer preserved within Chicxulub impact structure," *Science Advances*, Volume 7, Number 9 (February 24, 2021), https://advances.sciencemag.org/content/7/9/eabe3647. See also: David Bressan, "Scientists Find Traces of Asteroid That Wiped Out the Dinosaurs," *Forbes*, February 24, 2021, https://www.forbes.com/sites/davidbressan/2021/02/24/scientists-find-traces-of-asteroid-that-wiped-out-the-dinosaurs/?sh=61dc1d44584d.
48 Ibid.
49 Mara Johnson-Groh, "Origin of dinosaur-ending asteroid possibly found. And it's dark," *Live Science*, August 9, 2021, https://www.livescience.com/dinosaur-killing-asteroid-origin-dark.html. See also: Isabella James, "Origin of the Asteroid that Killed Dinosaurs May Have Been Discovered Using a Computer Model," *Tech Times*, August 10, 2021, https://www.techtimes.com/articles/263948/20210810/origin-asteroid-killed-dinosaurs-discovered-using-computer-model.htm. For an additional journalistic report of the research, see: Mike Wall, "Asteroid that killed the dinosaurs: Likely origin and what we know about the famous space rock," Space.com, August 16, 2021, https://www.space.com/dinosaur-impactor-origin. For the original article, see: David Nesvorný, William F. Bottke, and Simone Marchi, "Dark primitive asteroids account for a large share of K/Pg-scale impacts on Earth," *Icarus*, Volume 38 (November 1, 2021), https://www.sciencedirect.com/science/article/abs/pii/S0019103521002840#!.
50 Ibid. Nesvorný quote found in Mara Johnson-Groh, "Origin of dinosaur-ending asteroid possibly found. And it's dark." Also see: Southwest Research Institute, "Dark

Endnotes

Primitive Asteroids: Zeroing In on Source of the Impactor that Wiped Out the Dinosaurs," SciTechDaily.com, July 28, 2021, https://scitechdaily.com/dark-primitive-asteroids-zeroing-in-on-source-of-the-impactor-that-wiped-out-the-dinosaurs/.

51 Ibid.
52 Amir Siraj and Abraham Loeb, "Breakup of a long-period comet as the origin of the dinosaur extinction," *Scientific Reports*, Volume 11, Number 3803 (February 15, 2021), https://www.nature.com/articles/s41598-021-82320-2.
53 Both quotes sourced from the following: Juan Siliezar, Harvard staff writer, "The cataclysm that killed the dinosaurs," *Harvard Gazette*, February 15, 2021, https://news.harvard.edu/gazette/story/2021/02/new-theory-behind-asteroid-that-killed-the-dinosaurs/.
54 Steve Desch, Alan Jackson, Jessica Noviello, and Ariel Anbar, "The Chicxulub impactor: comet or asteroid?" *Astronomy & Geophysics*, Volume 62, Issue 3 (June 2021), pp. 3.34-3.37, https://academic.oup.com/astrogeo/article/62/3/3.34/6275176.
55 Gerta Keller, "Deccan volcanism, the Chicxulub impact, and the end-Cretaceous mass extinction: Coincidence? Cause and Effect?" Geological Society of America, 2014; published in Gerta Keller and Andrew C. Kerr, *Volcanism, Impacts, and Mass Extinctions: Causes and Effects* (Boulder Colorado: Geological Society of America, Special Paper 505, 2014), pp. 57-89.
56 David P.G. Bond and Paul B. Wignall, "Large igneous provinces and mass extinctions: An update," Geological Society of America, 2014; published in Gerta Keller and Andrew C. Kerr, *Volcanism, Impacts, and Mass Extinctions: Causes and Effects*, pp. 29-55.
57 Richard V. Fisher, volcanologist and former professor of geology at University of California, Santa Barbara, "Effects of Volcanic Gas," UCSB.edu, 1997, https://volcanology.geol.ucsb.edu/gas.htm.
58 David P.G. Bond and Paul B. Wignall, "Large igneous provinces and mass extinctions: An update," p. 32.
59 Ibid., p. 33.
60 "Scientists zero in on the role of volcanoes in the demise of dinosaurs," Advanced Science Research Center, Graduate Center, City University of New York, news release, March 29, 2021, https://www.eurekalert.org/news-releases/586463. For the published research paper, see: Andres Hernandez Nava, Benjamin A. Black, et al., "Reconciling early Deccan Traps CO_2 outgassing and pre-KPB global climate," *Proceedings of the National Academy of Sciences of the United States*, Volume 118, Number 14 (April 6, 2021), https://www.pnas.org/content/118/14/e2007797118.
61 "Scientists zero in on the role of volcanoes in the demise of dinosaurs," Advanced Science Research Center, Graduate Center, City University of New York, news release, March 29, 2021.
62 Paul Voosen, "Did volcanic eruptions help kill off the dinosaurs?" *Science*, February 21, 2019, https://www.sciencemag.org/news/2019/02/did-volcanic-eruptions-help-kill-dinosaurs. For the timeline of Deccan volcanism in relation to the C-T extinction, see: Courtney J. Sprain, Paul R. Renne, et al., "The eruptive tempo of Deccan volcanism in relation to the Cretaceous-Paleogene boundary," *Science*, Volume 363, Issue 6429 (February 22, 2019), pp. 866-870, https://science.sciencemag.org/content/363/6429/866.
63 Paul Voosen, "Did volcanic eruptions help kill off the dinosaurs?" *Science*.
64 Paul R. Renne, Courtney J. Sprain, et al., "State shift in Deccan volcanism at the Cretaceous-Paleogene boundary, possibly induced by impact," *Science*, Volume 350,

Issue 6256 (October 2, 2015), pp. 76-78, https://science.sciencemag.org/content/sci/350/6256/76.full.pdf.
65 Manabu Sakamoto, Michael J. Benton, and Chris Venditti, "Dinosaurs in decline tens of millions of years before their final extinction," *Proceedings of the National Academy of Sciences*, Volume 113, Number 18 (May 3, 2016), pp. 5036-5040, https://www.pnas.org/content/pnas/113/18/5036.full.pdf.
66 Ibid., p. 5039.
67 Joseph A. Bonsor, Paul M. Barrett, et al., "Dinosaur diversification rates were not in decline prior to the K-Pg boundary," *Royal Society Open Science*, Volume 7, Issue 11 (November 18, 2020), https://royalsocietypublishing.org/doi/pdf/10.1098/rsos.201195.
68 Fabien L. Condamine, Guillaume Guinot, Michael J. Benton, and Philip J. Currie, "Dinosaur biodiversity declined well before the asteroid impact, influenced by ecological and environmental pressures," *Nature Communications*, Volume 12, Number 3833 (June 29, 2021), https://www.nature.com/articles/s41467-021-23754-0.
69 Riley Black, "Why Birds Survived, and Dinosaurs Went Extinct, After an Asteroid Hit Earth," *Smithsonian Magazine*, September 15, 2020, https://www.smithsonianmag.com/science-nature/why-birds-survived-and-dinosaurs-went-extinct-after-asteroid-hit-earth-180975801/.
70 Mirzam Abdurrachman, Aswan, Yahdi Zaim, "5 periods of mass extinction on Earth. Are we entering the sixth?" The Conversation, January 30, 2018, https://theconversation.com/5-periods-of-mass-extinction-on-earth-are-we-entering-the-sixth-57575.
71 Douglas H. Erwin, *Extinction: How Life on Earth Nearly Ended 250 Million Years Ago* (Princeton and Oxford: Princeton University Press, 2006), pp. 215-216.
72 Ibid., p. 216.
73 Ibid., pp. 161-162.
74 Ibid., p. 162.
75 Björn Baresel, Hugo Bucher, Borhan Bagherpour, Morgane Brosse, Kuang Guodun, and Urs Schaltegger, "Timing of global regression and microbial bloom linked with the Permian-Triassic boundary mass extinction: implications for driving mechanisms," *Scientific Reports*, Volume 7, Number 43630 (March 6, 2017), https://www.nature.com/articles/srep43630.
76 "The cold exterminated all of them," Université de Genève, press release, March 6, 2017, https://www.unige.ch/communication/communiques/en/2017/cdp060317en/.
77 Ibid.
78 Richard V. Fisher, volcanologist, and former professor of geology at University of California, Santa Barbara, "Effects of Volcanic Gas."
79 John Borland, "Lunar Eclipse Prompts Climate Change Debate," Wired.com, March 3, 2008, https://www.wired.com/2008/03/lunar-eclipse-p/.
80 Ian Plimer, *Heaven and Earth: Global Warming, the Missing Science* (Lanham, Maryland: Taylor Trade Publishing, an imprint of Rowman & Littlefield Publishing Group, Inc., 2009), p. 165.
81 Ibid., p. 191.
82 Ibid., pp. 194-195.
83 Marc Morano, "Greenpeace Co-Founder Dr. Patrick Moore's testimony to Congress on UN Species Report: UN is using 'extinction as a fear tactic to scare the public into compliance,'" ClimateDepot.com, May 22, 2019, https://www.climatedepot.

Endnotes

com/2019/05/22/greenpeace-co-founder-dr-patrick-moore-testimony-to-congress-on-un-species-report-un-is-using-extinction-as-a-fear-tactic-to-scare-the-public-into-compliance/.
84. Stephen Marshak, *The Essentials of Geology*, p. 189. Italics in original.
85. "How much of the Earth is volcanic?" U.S. Geological Survey, no date, https://www.usgs.gov/faqs/how-much-earth-volcanic?qt-news_science_products=0#qt-news_science_products.
86. Stephen Marshak, *The Essentials of Geology*, p. 157.
87. "What is the 'Ring of Fire'?" USGS, no date, https://www.usgs.gov/faqs/what-ring-fire?qt-news_science_products=0#qt-news_science_products.
88. Luigi Foschini, "A solution for the Tunguska event," *Astronomy and Astrophysics*, Volume 342 (1999), L1-L4, https://arxiv.org/abs/astro-ph/9808312.
89. Paul Scott Anderson and Kelly Kizer Whitt, "Today in Science: The Tunguska Explosion," Earth Sky.org, June 30, 2021, https://earthsky.org/space/what-is-the-tunguska-explosion/.
90. Elizabeth Howell, "Chelyabinsk Meteor: A Wake-Up Call for Earth," Space.com, 2019, https://www.space.com/amp/33623-chelyabinsk-meteor-wake-up-call-for-earth.html.
91. Mike Wall, "Russian Meteor's Origin and Size Pinned Down," Space.com, February 27, 2013, https://www.space.com/19974-russian-meteor-explosion-origin-size.html.
92. Eddie Irizarry and Deborah Byrd, "Asteroid 2021 SG came from the sun's direction," EarthSky.org, September 20, 2021, https://earthsky.org/space/asteroid-2021-sg-closest-to-earth-sep21-2021/.
93. Mashable News Staff, India, "NASA Asteroid Simulation Shows We Cannot Stop Asteroid from Hitting Earth and We Need to Be Better Prepared!" *Mashable India*, May 4, 2021, https://in.mashable.com/science/22031/nasa-asteroid-simulation-shows-we-cannot-stop-asteroid-from-hitting-earth-and-we-need-to-be-better-p.
94. NASA, "Planetary Defense Frequently Asked Questions," NASA.gov, no date, https://www.nasa.gov/planetarydefense/faq.
95. NASA Jet Propulsion Laboratory, California Institute of Technology, Center for Near Earth Object Studies (CNEOS), "Discovery Statistics: Cumulative Totals," checked August 20, 2021, NASA.gov, https://cneos.jpl.nasa.gov/stats/totals.html.
96. Gerrit L. Verschuur, *Impact! The Threat of Comets and Asteroids*, p. 33.
97. Ibid. Italics in original.
98. Ian Plimer, *Heaven and Earth*, p. 43.
99. "The Younger Dryas," National Centers for Environmental Information, National Oceanic and Atmospheric Administration (NOAA), no date, https://www.ncdc.noaa.gov/abrupt-climate-change/The%20Younger%20Dryas.
100. Richard B. Alley, "The Younger Dryas cold interval as viewed from central Greenland," *Quaternary Science Reviews*, Volume 19, Issues 1-5 (January 2000), pp. 213-226, https://www.sciencedirect.com/science/article/abs/pii/S0277379199000621?via%3Dihub#! See also: Jørgen Peder Steffensen, Katrine K. Anderson, Matthias Bigler, et al., "High-Resolution Greenland Ice Core Data Show Abrupt Climate Change Happens in Few Years," *Science*, Volume 321, Number 680 (August 1, 2008), pp. 680-684, https://epic.awi.de/id/eprint/17919/1/Ste2007b.pdf.
101. David Sepkoski, *Catastrophic Thinking: Extinction and the Value of Diversity from Darwin to the Anthropocene* (Chicago: University of Chicago Press, 2020), p. 299.

102 Peter Brannen, "Earth Is Not in the Midst of a Sixth Mass Extinction," *The Atlantic*, June 13, 2017, https://www.theatlantic.com/science/archive/2017/06/the-ends-of-the-world/529545/.
103 Gerrit L. Verschuur, *Impact! The Threat of Comets and Asteroids*, p. vi.
104 Ibid., p. 211.

Chapter 8

1 Edward N. Lorenz, "Predictability: Does the Flap of a Butterfly's Wings in Brazil Set Off a Tornado in Texas," presented before the American Association for the Advancement of Science, December 29, 1972, https://mathsciencehistory.com/wp-content/uploads/2020/03/132_kap6_lorenz_artikel_the_butterfly_effect.pdf.
2 Steven E. Koonin, *Unsettled: What Climate Science Tells Us and What It Doesn't and Why It Matters* (Dallas, Texas: BenBella Books, Inc., 2021), p. 79.
3 Reid A. Bryson, "Simulating Past and Forecasting Future Climates," *Environmental Conservation*, Volume 20, Issue 4 (Winter 1993), pp. 339-346, https://www.cambridge.org/core/journals/environmental-conservation/article/abs/simulating-past-and-forecasting-future-climates/179C8A227F1DE9D1AE638D6606404686.
4 Ian Plimer, *Heaven and Earth: Global Warming, the Missing Science* (Lanham, Maryland: Taylor Trade Publishing, an imprint of Rowman & Littlefield Publishing Group, Inc., 2009), p. 234.
5 John von Neumann, member, Atomic Energy Commission, "Can We Survive Technology?" Fortune, June 1955. Also see: Norman Macrae, *John von Neumann: The Scientific Genius Who Pioneered the Modern Computer, Game Theory, Nuclear Deterrence, and Much More* (New York: Pantheon Books, 1992), p. 16.
6 J.G. Charney, R. Fjörtoft, J. von Neumann, "Numerical Integration of the Barotropic Vorticity Equation," *Tellus*, Volume 2, Number 4 (November 1950), pp. 237-254, https://link.springer.com/chapter/10.1007/978-1-944970-35-2_15.
7 S. Ulam, "John von Neumann: 1903-1957," in "John von Neumann: 1903-1957," edited by Oxtoby et al., whole issue of the *Bulletin of the American Mathematical Society*, Volume 64, Number 3, Part 2 (May 1958), 129 pages, pp. 1-49, at p. 30.
8 Ibid.
9 John von Neumann, "Can We Survive Technology?" *Fortune*.
10 Ibid. Italics in original.
11 Peter Dizikes, "When the Butterfly Effect Took Flight," *MIT News Magazine*, February 22, 2011, https://www.technologyreview.com/2011/02/22/196987/when-the-butterfly-effect-took-flight/.
12 Edward N. Lorenz, MIT, "Deterministic Nonperiodic Flow," *Journal of the Atmospheric Sciences*, Volume 20, Issue 2 (March 1, 1963), pp. 130-141, https://journals.ametsoc.org/view/journals/atsc/20/2/1520-0469_1963_020_0130_dnf_2_0_co_2.xml?tab_body=pdf.
13 Ibid., p. 130.
14 Amirmohammad Ketabchi, "Chaos Theory," a PowerPoint presentation, no date.
15 Edward Lorenz, *The Essence of Chaos* (Seattle, Washington: University of Washington Press, 1993), p. 13.
16 Edward N. Lorenz, MIT, "Deterministic Nonperiodic Flow," Abstract p. 130.
17 Ibid., p. 142.
18 Peter Dizikes, "When the Butterfly Effect Took Flight," *MIT News Magazine*.

Endnotes

19 Quoted in the following: "Edward Lorenz, father of chaos theory and butterfly effect, dies at 90," *MIT News*, April 16, 2008, https://news.mit.edu/2008/obit-lorenz-0416.
20 Edward N. Lorenz, "Predictability: Does the Flap of a Butterfly's Wings in Brazil Set Off a Tornado in Texas."
21 James Gleick, *Chaos: Making a New Science* (New York: Viking Penguin, Inc., 1987).
22 Edward Lorenz, *The Essence of Chaos*, p. 15.
23 Ibid.
24 Heinz-Otto Peitgen, Hartmut Jürgens, and Dietmar Saupe, *Chaos and Fractals: New Frontiers of Science* (New York and Berlin: Springer, 1992), p. 48.
25 Ibid.
26 James Gleick, *Chaos: Making a New Science*, pp. 95-96.
27 Benoit B. Mandelbrot, *The Fractal Nature of Nature* (New York: W.H. Freeman and Company, 1977, updated and augmented, 1983), p. 26.
28 Ibid.
29 James Gleick, *Chaos: Making a New Science*, p. 96.
30 Heinz-Otto Peitgen, Hartmut Jürgens, and Dietmar Saupe, *Chaos and Fractals: New Frontiers of Science*, p. 49.
31 Steven E. Koonin, *Unsettled: What Climate Science Tells Us and What It Doesn't and Why It Matters*.
32 "How Reliable Are Weather Forecasts?" SciJinks.com, no date, https://scijinks.gov/forecast-reliability/.
33 Fuqing Zhang, Y. Qiang Sun, et al., "What Is the Predictability Limit of Midlatitude Weather?" *Journal of the Atmospheric Sciences*, Volume 76, Issue 4 (April 1, 2019), pp. 1077-1091, https://journals.ametsoc.org/view/journals/atsc/76/4/jas-d-18-0269.1.xml.
34 James Gleick, *Chaos: Making a New Science*, pp. 20-21, at p. 21.
35 Quoted in William F. Jasper, "Computer Models vs. Climate Reality," *New American*, Volume 31, Number 8 (April 20, 2015), https://thenewamerican.com/computer-models-vs-climate-reality/.
36 Kevin E. Trenberth, "Predictions of climate," Blogs.Nature.com, posted on June 4, 2007, http://blogs.nature.com/climatefeedback/2007/06/predictions_of_climate.html.
37 Vincent Gray, *The Greenhouse Delusion: A Critique of "Climate Change 2001"* (Brentwood, Essex, U.K.: Multi-Science Publishing Co., Ltd., 2004).
38 "IPCC Report slammed as 'dangerous nonsense,'" New Zealand Climate Science Coalition, Scoop.co.nz, press release, April 10, 2007, https://www.scoop.co.nz/stories/SC0704/S00023.htm.
39 Quoted in the following: U.S. Senate Environment and Public Works Committee, Minority Staff Report, Senator James Inhofe, "More Than 650 International Scientists Dissent Over Man-Made Global Warming Claims. Scientists Continue to Debunk 'Consensus' in 2008," December 11, 2008, p. 129, https://www.epw.senate.gov/public/_cache/files/8/3/83947f5d-d84a-4a84-ad5d-6e2d71db-52d9/01AFD79733D77F24A71FEF9DAFCCB056.senateminorityreport2.pdf. All quotations in this paragraph are drawn from the following source: Marc Morano, "UN IPCC Scientist: 'No convincing scientific arguments to support claim that increases in greenhouse gases are harmful to the climate,'" ClimateDepot.com, May 5, 2009, https://www.climatedepot.com/2009/05/05/un-ipcc-scientist-no-convincing-scientific-arguments-to-support-claim-that-increases-in-greenhouse-gases-are-harmful-to-the-climate/.

40 "Global Effects of Mount Pinatubo," NASA Earth Observatory, no date, https://earthobservatory.nasa.gov/images/1510/global-effects-of-mount-pinatubo. See also: Cindy Evans, "Astronauts photograph Mt. Pinatubo," *NASA Earth Observatory*, June 14, 2001, https://earthobservatory.nasa.gov/features/AstronautPinatubo. See also: "Monitoring the Eyjafjallajökull Eruption," Climate.gov, NOAA, April 22, 2010, https://www.climate.gov/news-features/featured-images/monitoring-eyjafjallaj%C3%B6kull-eruption.
41 Henrik Svensmark, "Cosmic Rays, Clouds and Climate," *Euro Physic News*, 2015, pp. 26-29, https://www.europhysicsnews.org/articles/epn/pdf/2015/02/epn2015462 p26.pdf.
42 Reported by Glick in *Chaos: Making a New Science*, p. 21.
43 Ibid., p. 20.
44 Edward Lorenz, *The Essence of Chaos*, p. 85.
45 Ibid., p. 86.
46 Ibid., p. 50.
47 Ross McKitrick, "Checking for model consistency in optimal fingerprinting: a comment," *Climate Dynamics*, August 10, 2021, pp. https://link.springer.com/article/10.1007%2Fs00382-021-05913-7. See also: Nathan Worcester, "Statistical Method Used to Link Climate Change to Greenhouse Gases Challenged," *The Epoch Times*, September 6, 2021, https://www.theepochtimes.com/statistical-method-used-to-link-climate-change-to-greenhouse-gases-challenged_3983949.html.
48 See: M.R. Allen and S.F.B. Tett, "Checking for model consistency in optimal fingerprinting," *Climate Dynamics*, Volume 15 (June 1999), pp. 419-434, https://link.springer.com/article/10.1007%2Fs003820050291. See also: F.B. Simon, Peter A. Stott, et al., "Causes of twentieth-century temperature change near the Earth's surface," *Nature*, Volume 399 (June 10, 1999), pp. 569-572, https://www.nature.com/articles/21164.
49 McKitrick, as quoted in Nathan Worcester, "Statistical Method Used to Link Climate Change to Greenhouse Gases Challenged."
50 Nathan Worcester, "Statistical Method Used to Link Climate Change to Greenhouse Gases Challenged."
51 M.R. Allen and S.F.B. Tett, "Checking for model consistency in optimal fingerprinting."
52 Ross McKitrick, "Checking for model consistency in optimal fingerprinting: a comment," Abstract.
53 Andrew Rothman, "OLS Regression, Gauss-Markov, BLUE, and understanding the math," *Toward Data Science*, June 3, 2020, https://towardsdatascience.com/ols-linear-regression-gauss-markov-blue-and-understanding-the-math-453d7cc630a5.
54 McKitrick, quoted in Nathan Worcester, "Statistical Method Used to Link Climate Change to Greenhouse Gases Challenged."
55 Ibid.
56 McKitrick, quoted in Nathan Worcester, "Statistical Method Used to Link Climate Change to Greenhouse Gases Challenged." For an example of Hasselmann's analsyis, see the following: K. Hasselmann, "Multi-pattern fingerprint method for detection and attribution of climate change," *Climate Dynamics*, Volume 13 (September 1997), pp. 601-611, https://link.springer.com/article/10.1007%2Fs003820050185.
57 Stephen Marshak, *The Essentials of Geology* (New York and London: W. W. Norton & Company Ltd., sixth edition, 2019), p. 355.

Endnotes

58 Ibid., p. 377.
59 Ibid.
60 John J.W. Rogers and M. Santosh, *Continents and Subcontinents* (Oxford and New York: Oxford University Press, 2004), pp. 3-7, at p. 3.
61 Ibid., p. 13.
62 "Plate Tectonics," *National Geographic*, last updated June 10, 2020, https://www.nationalgeographic.org/encyclopedia/plate-tectonics/.
63 Stephen Marshak, *The Essentials of Geology*, p. 67, emphasis in original.
64 Stefan Cwojdzinski, "History of a discussion: selected aspects of the Earth expansion v. plate tectonics theories," from *History of Geoscience: Celebrating 50 Years of INHIGEO*, edited by W. Mayer, R.M. Clary, et al., Geological Society (London: Special Publications, 442), pp. 93-104, at p. 93.
65 Dr. James Marlow, "Global Expansion Tectonics: A Significant Challenge for Physics," *Proceedings of the NPA*, Albuquerque, New Mexico, 2012, reprinted in *Infinite Energy*, Issue 117 (August 28, 2014), http://www.infinite-energy.com/images/pdfs/Max.pdf.
66 Katie Bo Williams, "Why Do Planets Rotate?" *Discover*, August 26, 2015, https://www.discovermagazine.com/the-sciences/why-do-planets-rotate.
67 "Unbalance—The Common Cause of Vibration & Premature Bearing Destruction within Rotating Machinery Equipment," IRD Balancing, no date, https://shop.irdproducts.com/blog/unbalance-cause-of-vibration/.
68 "Rotation about a moving axis," Britannica.com, no date, https://www.britannica.com/science/mechanics/Rotation-about-a-moving-axis.
69 Sybille Hildebrandt, "The earth has lost a quarter of its water," ScienceNordic.com, March 13, 2012, https://sciencenordic.com/chemistry-climate-denmark/the-earth-has-lost-a-quarter-of-its-water/1462713.
70 S. Warren Carey, *The Expanding Earth* (Amsterdam, Oxford, New York: Elsevier Scientific Publishing Company, 1976), pp. 1-2.
71 Ibid., p. 6.
72 S. Warren Carey, *Theories of the Earth and Universe: A History of Dogma in the Earth Sciences* (Stanford, California: Stanford University Press, 1988), pp. 93-94, at 93.
73 Ibid.
74 S. Warren Carey, *The Expanding Earth*, pp. 2-3.
75 Ibid., p. 3.
76 Ibid., p. 9.
77 Ibid., p. 125. Emphasis in original.
78 S. Warren Carey, *Theories of the Earth and Universe: A History of Dogma in the Earth Sciences*, p. 321.
79 Ibid., p. 321.
80 J.B.S. Haldane, "On Being the Right Size," in James R. Newman, *The World of Mathematics* (New York, NY: Simon & Schuster, 1956), Volume 2, pp. 952-957.
81 Galileo Galilei, The Discourses and Mathematical Demonstrations Relating to Two New Sciences, published in 1638. Quoted in the following source: Timothy Paul Smith, *How Big Is Big and How Small Is Small: The Sizes of Everything and Why* (Oxford University Press, 2014), p. 24.
82 Timothy Paul Smith, *How Big Is Big and How Small Is Small: The Sizes of Everything and Why*, p. 25.
83 Stephen Hurrell, *Dinosaurs and the Expanding Earth: Solving the Mystery of the Dinosaurs' Gigantic Size* (One-Off Publishing, 1994, third edition, 2011).

84 S. Warren Carey, *Earth, Universe, Cosmos* (Hobart: University of Tasmania, second edition, 2000), p. 131.
85 Stephen Hurrell, "Ancient Life's Gravity and Its Implications for the Expanding Earth," in The Earth Expansion Evidence—A Challenge for Geology, Geophysics and Astronomy: Selected Contributions to the Interdisciplinary Workshop of the 37th International School of Geophysics, EMFSC, held in Erice, Sicily (October 4-9, 2011), edited by G. Scalera, E. Boschi, and S. Cwojdzínski, pp. 307-325, at p. 308.
86 Ibid., p. 311.
87 Valentin Sapunov, "On the nature of gravity and possible change of Earth mass in geological time." EGU General Assembly, held April 12-17, 2015, in Vienna, Austria, https://ui.adsabs.harvard.edu/abs/2015EGUGA..17..300S/abstract.
88 S. Warren Carey, *Earth, Universe, Cosmos*, p. 131.
89 Paolo Sudiro, "The Earth expansion theory and its transition from scientific hypothesis to pseudoscientific belief," *History of Geo- and Space Sciences*, Volume 5, Issue 1 (2014), pp. 135-148, at p. 141, https://hgss.copernicus.org/articles/5/135/2014/.
90 S. Warren Carey, *Theories of the Earth and Universe: A History of Dogma in the Earth Sciences*, p. 364.
91 Alan Buis, NASA's Jet Propulsion Laboratory, "Milankovitch (Orbital) Cycles and Their Role in Earth's Climate," *NASA News*, February 27, 2020, https://climate.nasa.gov/news/2948/milankovitch-orbital-cycles-and-their-role-in-earths-climate/. See also: John Imbrie and Katherine Palmer Imbrie, *Ice Ages: Solving the Mystery* (Cambridge, Massachusetts: Harvard University Press, 1979), Chapter 8, "Through Distant Worlds and Times," pp. 97-111.
92 The discussion of eccentricity, obliquity, and precession is drawn heavily from the following two sources: (1) Alan Buis, NASA's Jet Propulsion Laboratory, "Milankovitch (Orbital) Cycles and Their Role in Earth's Climate"; and (2) Michel van Biezen, "Astronomy—Ch. 2: Understanding the Night Sky (23 of 23) How Milankovitch Cycles Affect Weather," YouTube.com, September 11, 2014, https://www.youtube.com/watch?v=95CyWye7XcU. In this subchapter on Milankovitch Cycles, the discussion is closely paraphrased from the footnoted sources. When sentences could not be paraphrased without complications, sentences from the original footnoted sources were drawn at times virtually word for word. Reference to the footnoted sources should make this clear.
93 Alan Buis, NASA's Jet Propulsion Laboratory, "Milankovitch (Orbital) Cycles and Their Role in Earth's Climate."
94 Michel van Biezen, "Astronomy—Ch. 2: Understanding the Night Sky (3 of 23) Why Do We Have Seasons?" YouTube.com, August 29, 2014, https://www.youtube.com/watch?v=gAZZjAjBrGA.
95 "Are We on the Brink of a 'New Little Ice Age,'" Woods Hole Oceanographic Institution, no date, https://www.whoi.edu/know-your-ocean/ocean-topics/climate-ocean/abrupt-climate-change/are-we-on-the-brink-of-a-new-little-ice-age/. See also a report commissioned by the National Academy of Sciences' Committee on Abrupt Climate Change. Contributors: National Research Council; Ocean Studies Board; Polar Research Board; Board on Atmospheric Sciences and Climate; Division on Earth and Life Studies. *Abrupt Climate Change: Inevitable Surprises* (Washington, D.C.: National Academy Press, 2002), https://www.nap.edu/catalog/10136/abrupt-climate-change-inevitable-surprises.

Endnotes

96 "About Our Seasons," LPI Education, no date, https://www.lpi.usra.edu/education/skytellers/seasons/.
97 Ibid.
98 Ibid.
99 "Milutin Milankovitch (1879-1958)," NASA Earth Observatory, https://earthobservatory.nasa.gov/features/Milankovitch/milankovitch_2.php. See also: Deanna Connors, "Why Earth Has 4 Seasons," SkyEarth.org, September 22, 2020, https://earthsky.org/earth/can-you-explain-why-earth-has-four-seasons/.
100 Erik Gregersen, senior editor of the *Encyclopedia Britannica*, "Precession of the equinoxes," no date, https://www.britannica.com/science/precession-of-the-equinoxes. See also: "Precession of the Equinoxes," University of British Colombia, no date, https://personal.math.ubc.ca/~cass/courses/m309-01a/tsang/precession.html.
101 Matt Williams, "Precession of the Equinoxes," *Universe Today*, November 7, 2010, https://www.universetoday.com/77640/precession-of-the-equinoxes/.
102 "Precession of the Equinoxes," University of British Colombia, no date.
103 N.C. Brunswick, "Planetarium: Is it true that Vega will be our next North Star?" University of Southern Maine, no date, https://usm.maine.edu/planet/it-true-vega-will-be-our-next-north-star.
104 Alan Buis, NASA's Jet Propulsion Laboratory, "Milankovitch (Orbital) Cycles and Their Role in Earth's Climate."
105 NASA, Archive of Dr. Magneto's Questions and Answers—IMAGE, "Will the equinoxes and solstices switch places in 13,000 years because of the precession of the Earth's rotation axis?" NASA.gov, no date, https://image.gsfc.nasa.gov/poetry/ask/q1795.html.
106 Sal Khan, "Apsidal precession (perihelion precession) and Milankovitch cycles," Kahn Academy, YouTube.com, June 21, 2018, https://www.youtube.com/watch?v=FYewEn-RmZs.
107 Alan Buis, NASA's Jet Propulsion Laboratory, "Milankovitch (Orbital) Cycles and Their Role in Earth's Climate."
108 Mark Landsbaum, "Epidemic of false claims on global warming?" *Orange County Register*, June 2, 2011, https://www.ocregister.com/2011/06/02/epidemic-of-false-claims-on-global-warming/. See also: Philip Scott, "Global Warming Is Not a Crisis," Opinion, ABC News, March 12, 2007, https://abcnews.go.com/International/story?id=2938762&page=1.
109 Hendrik Tennekes, "Some Fresh Air in the Climate Debate," op ed, *Amsterdam De Volkskrant*, March 28, 2007, https://pielkeclimatesci.wordpress.com/2007/03/28/some-fresh-air-in-the-climate-debate-an-op-ed-by-hendrik-tennekes/. The quotes in this paragraph are draw from the following source: Marc Morano, "My Viral Climate Change Video Was Smeared as Fake News. Here Are the Facts," The Daily Signal, August 22, 2018, https://www.dailysignal.com/2018/08/22/my-viral-climate-change-video-was-smeared-as-fake-news-here-are-the-facts/. Morano's article was reprinted on ClimateDepot.com, August 22, 2018, https://www.climatedepot.com/2018/08/22/moranos-point-by-point-rebuttal-to-uk-guardians-nuccitellis-attacks-moranos-facebook-video-continues-to-go-viral-with-over-8-million-views/.
110 John Imbrie and Katherine Palmer Imbrie, *Ice Ages: Solving the Mystery*, pp. 103-109, at p. 105.
111 History.com editors, "Ice Age," History.com, last updated July 15, 2021, https://www.history.com/topics/pre-history/ice-age.

112 Kylie Andrews, "What causes an ice age and what would happen if the Earth endured another one?" ABC Science, June 14, 2006, https://www.abc.net.au/news/science/2016-06-15/what-is-an-ice-age-explainer/7185002.
113 Ian Plimer, *Heaven and Earth: Global Warming, the Missing Science* (Lanham, Maryland: Taylor Trade Publishing, an imprint of Rowman & Littlefield Publishing Group, Inc., 2009), pp. 233-234.
114 Alexander J. Dickson, Christopher J. Beer, Ciara Dempsey, et al., "Oceanic forcing of the Marine Isotope Stage 11 interglacial," *Nature Geoscience*, Volume 2 (May 24, 2009), pp. 428-433, https://www.nature.com/articles/ngeo527.
115 Ibid., p. 234.
116 Ibid.
117 Ibid., p. 240.
118 "Glacial-Interglacial Cycles," National Centers for Environmental Information of the National Oceanic and Atmospheric Administration (NOAA), no date, https://www.ncdc.noaa.gov/abrupt-climate-change/Glacial-Interglacial%20Cycles.
119 Ibid.
120 Ibid.
121 Alan Buis, "Why Milankovitch (Orbital) Cycles Can't Explain Earth's Current Warming," NASA Global Climate Change, February 27, 2020, https://climate.nasa.gov/blog/2949/why-milankovitch-orbital-cycles-cant-explain-earths-current-warming/.
122 A. Ganopolski, R. Winkelmann, and H.J. Schellnhuber, "Critical insolation—CO_2 relation for diagnosing past and future glacial inception," a letter published in *Nature*, Volume 529 (January 13, 2016), pp. 200-203, https://www.nature.com/articles/nature16494.
123 Toby Tyrrell, John G. Shepherd, and Stephanie Castle, "The long-term legacy of fossil fuels," *Tellus B: Chemical and Physical Meteorology*, Volume 59 (August 2007), pp. 664-672, https://www.tandfonline.com/doi/pdf/10.1111/j.1600-0889.2007.00290.x. See also: "Next Ice Age Delayed by Rising Carbon Dioxide Levels," *ScienceDaily*, August 30, 2007, https://www.sciencedaily.com/releases/2007/08/070829193436.htm.
124 Andrew A. Lacis, Gavin A. Schmidt, et al., "Atmospheric CO_2: Principal Control Knob Governing Earth's Temperature," *Science*, Volume 330, Number 6002 (October 15, 2020), pp. 356-359, https://www.science.org/doi/10.1126/science.1190653.
125 Harrison H. Schmitt and William Happer, "In Defense of Carbon Dioxide," *Wall Street Journal*, Opinion, May 8, 2013, https://www.wsj.com/articles/SB10001424127887323528404578452483656067190.
126 William Happer, "Data or Dogma? Promoting Open Inquiry in the Debate over the Magnitude of Human Impact on Earth's Climate," testimony before the U.S. Senate Subcommittee on Space, Science, and Competitiveness, December 8, 2015, https://www.commerce.senate.gov/services/files/c8c53b68-253b-4234-a7cb-e4355a6edfa2.
127 The Physics arXiv Blog, "Earth's Magnetic North Pole Has Been Racing Towards Siberia," *Discover Magazine*, October 28, 2020, https://www.discovermagazine.com/planet-earth/earths-magnetic-north-pole-has-begun-racing-towards-siberia.
128 "True north and magnetic north: what's the difference?" Royal Museums Greenwich, no date, https://www.rmg.co.uk/stories/topics/true-north-magnetic-north-whats-difference. See also: Deborah Byrd, "Why is Earth's magnetic north pole drifting so rapidly?" EarthSky.org, May 19, 2020, https://earthsky.org/earth/magnetic-north-rapid-drift-blobs-flux/.

Endnotes

129 The Physics arXiv Blog, "Earth's Magnetic North Pole Has Been Racing Towards Siberia."
130 "Tracking Changes in Earth's Magnetic Poles," National Centers for Environmental Information of the National Oceanic and Atmospheric Administration (NOAA), last updated June 17, 2021, https://www.ncei.noaa.gov/news/tracking-changes-earth-magnetic-poles.
131 "Historical Magnetic Declination," National Centers for Environmental Information of the National Oceanic and Atmospheric Administration (NOAA), last updated October 2, 2020, https://maps.ngdc.noaa.gov/viewers/historical_declination/.
132 Alan Buis, NASA's Jet Propulsion Laboratory, "Flip Flop: Why Variations in Earth's Magnetic Field Aren't Causing Today's Climate Change," NASA Global Climate Change, August 3, 2021, https://climate.nasa.gov/blog/3104/flip-flop-why-variations-in-earths-magnetic-field-arent-causing-todays-climate-change/.
133 "2012: Magnetic Pole Reversal Happens All the (Geologic) Time," NASA.gov, November 30, 2011, https://www.nasa.gov/topics/earth/features/2012-poleReversal.html.
134 Alan Buis, NASA's Jet Propulsion Laboratory, "Flip Flop: Why Variations in Earth's Magnetic Field Aren't Causing Today's Climate Change." On the NASA website, the last sentence in the quotation is written in bold type for emphasis.
135 Ibid.
136 E. Pallé, C.J. Butler, and K. O'Brien, "The possible connection between ionization in the atmosphere by cosmic rays and low level clouds," *Journal of Atmospheric and Solar-Terrestrial Physics*, Volume 66, Issue 18 (December 2014), pp. 1779-1790, https://www.sciencedirect.com/science/article/abs/pii/S1364682604002056. See also: F. Fluteau, V. Courtillot, Y. Gallet, et al., "Does the Earth's Magnetic Field Influence Climate?" a presentation at the American Geophysical Union, Fall Meeting, December 2006, abstract id: GP51B-02, https://ui.adsabs.harvard.edu/abs/2006AGUFMGP51B..02F/abstract.
137 R. Wang, Y. Balkanski, O. Boucher, et al., "Sources, transport and deposition of iron in the global atmosphere," *Atmospheric Chemistry and Physics*, Volume 15 (June 8, 2015), pp. 6247-6270, https://acp.copernicus.org/articles/15/6247/2015/acp-15-6247-2015.pdf.
138 Louise M.A. Hawkins, J. Michael Grappone, Courtney J. Sprain, et al., "Intensity of the Earth's magnetic field: Evidence for a Mid-Paleozoic dipole low," *Proceedings of the National Academy of the United States of America*, Volume 118, Number 34 (August 24, 2021), https://www.pnas.org/content/118/34/e2017342118.short. See also: University of Liverpool, "New Evidence of 200 Million-Year Cycle for the Earth's Magnetic Field," SciTechDaily.com, September 23, 2021, https://scitechdaily.com/new-evidence-of-200-million-year-cycle-for-earths-magnetic-field/. For an earlier study, see: A.J. Biggin, B. Steinberger, J. Aubert, et al., "Possible links between long-term geomagnetic variations and whole-mantle convection processes," *Nature Geoscience*, Volume 5 (July 29, 2012), pp. 526-522, https://www.nature.com/articles/ngeo1521.
139 John E.A. Marshall, Jon Lakin, et al., "UV-B radiation was the Devonian-Carboniferous boundary terrestrial extinction kill mechanism," *Science Advances*, Volume 6, Issue 22 (May 27, 2020), https://www.science.org/doi/10.1126/sciadv.aba0768.
140 René Thom, Structural Stability and Morphogenesis: An Outline of a General Theory of Models (Reading, Pennsylvania: W.A. Benjamin, 1975).

141 Alexander Woodcock and Monte Davis, *Catastrophe Theory* (Toronto and Vancouver: Clark, Irwin & Company, 1978), p. 32.
142 Jan C. Schmidt, "Challenged by Instability and Complexity: Questioning Classic Stability Assumptions and Presumptions in Scientific Methodology," an article in *Handbook of the Philosophy of Sciences*, Volume 10, *Philosophy of Complex Systems*, series volume edited by Cliff Hooker (Amsterdam: Elsevier, first edition, 2011), Part I, "General Foundations," pp. 223-254, at p. 224.
143 John J.W. Rogers and M. Santosh, *Continents and Supercontinents*, p. 113.
144 Intergovernmental Panel on Climate Change (IPCC), *Climate Change 2021: The Physical Science Basis*, Working Group contribution, Sixth Assessment Report, August 9, 2021, https://www.ipcc.ch/assessment-report/ar6/. The IPCC Sixth Climate Assessment is the sixth such report issued since 1990.
145 Antonio Marco Martínez, "Man is the measure of all things," Antiquitatem.com, June 3, 2013, http://www.antiquitatem.com/en/plato-protagoras-philosophy-sophist/. Martínez's discussion correctly completes his thoughts with reference to Plato writing his *Laws* and properly reworking Protagoras's observation to substitute "God" for human beings being the measure of all things.

Chapter 9

1 Hollis D. Hedberg, "Geologic Aspects of Origin of Petroleum," research article, GeoScienceWorld, *AAPG Bulletin 1964*, Volume 48, Number 11 (November 1, 1964), pp. 1755-1803, https://pubs.geoscienceworld.org/aapgbull/article-abstract/48/11/1755/35065/Geologic-Aspects-of-Origin-of-Petroleum1?redirectedFrom=fulltext.
2 Dmitri Mendeleev, *L'Origine du pétrole*, *Revue Scientific*, Second Series, VIII, 1877, pp. 409-416.
3 Vladimir G. Kutcherov, Division of Heat and Power Technology, Royal Institute of Technology, Stockholm, Sweden, and Vladilen A. Krayushkin, Laboratory of Inorganic Petroleum Origin, Institute of Geological Sciences, National Academy of Sciences, Kiev, Ukraine, "Deep-seated abiogenic origin of petroleum: From geological assessment to physical theory," *Reviews of Geophysics*, Volume 48, Issue 2010 (March 12, 2010), https://agupubs.onlinelibrary.wiley.com/doi/full/10.1029/2008RG000270.
4 I have written two previous books on abiotic oil and numerous published articles between 2005–2015, when I was a senior staff reporter for WorldNetDaily. The first book was published in 2006: Jerome R. Corsi and Craig R. Smith, *Black Gold Stranglehold: Why Does Gasoline Cost So Much* (Washington, D.C.: WorldNetDaily Press, 2005). The second book was published in 2012: Jerome R. Corsi, *The Great Oil Conspiracy: How the U.S. Government Hid the Nazi Discovery of Abiotic Oil from the American People* (New York: Skyhorse Publishing, 2012). Certain materials from these prior publications are reprinted here without footnotes. I have taken from these previous writings sparingly, and I have paraphrased the repeated selections whenever possible. When paraphrasing added unnecessary complications, I have reused some selections word for word. While throughout the book I have been fastidious about footnoting, I do not consider revisiting my own previous writings to be plagiarism.
5 Thomas S. Kuhn, *The Structure of Scientific Revolutions* (Chicago: University of Chicago Press, 1962).
6 Dmitri Mendeleev, *The Principles of Chemistry*, Volume 1, second edition translated from the sixth Russian edition (Collier: New York, 1902), p. 552.

Endnotes

7   Clifford Walters, University of Texas at Austin, "The Origin of Petroleum," in https://en.wikipedia.org/wiki/Dmitri_Mendeleev *Practical Advances in Petroleum Processing*, by Chang S. Hsu and Paul Robinson (Berlin: Springer, 2006), Chapter 2, pp. 79-101, https://www.researchgate.net/publication/226441668_The_Origin_of_Petroleum.

8   Michael D. Gordin, *A Well-Ordered Thing: Dmitrii Mendeleev and the Shadow of the Periodic Table* (New York: Basic Books, 2004), p. 153. Gordin spelled Mendeleev's first name as "Dmitrii" throughout the book. The parenthetical suggestion is in the original.

9   Mark A. Sephton and Robert M. Hazen, "On the Origins of Deep Hydrocarbons," *Mineralogy and Geochemistry*, research article, Volume 75, Number 1 (January 1, 2013), pp. 449-455, from the Abstract at p. 449, https://pubs.geoscienceworld.org/msa/rimg/article-abstract/75/1/449/140972/iOn-the-Origins-of-Deep-Hydrocarbons?redirectedFrom=fulltext.

10 Cecil G. Lalicker, *Principles of Geology* (New York: Appleton-Century-Crofts, Inc., 1949), pp. 59-60.

11 Ibid., p. 58.

12 Ibid., p. 59.

13 A.I. Levorsen, *Geology of Petroleum* (San Francisco: W.H. Freeman and Company, 1954), p. 491.

14 Kenneth K. Landes, *Petroleum Geology* (New York: John Wiley & Sons, Inc., 1951), pp. 132-133. References in the quote are to the following: J. McConnell Sanders, "The Microscopical Examination of Crude Petroleum," *Journal of the Institution of Petroleum Technologists*, Volume 23 (1937), pp. 525-573. See also: W.A. Waldschmidt, "Progress Report on Microscopic Examination of Permian Crude Oils," Program 26th Annual Convention, American Association of Petroleum Geologists, 1941, p. 23.

15 Ibid., p. 133.

16 Richard C. Selley and Stephen A. Sonnenberg, *Elements of Petroleum Geology* (Amsterdam: Academic Press, an imprint of Elsevier, third edition, 2015), p. 199.

17 Ibid.

18 Ibid., p. 295.

19 Ibid.

20 George A. Olah, Árpád Molnár, and G.K. Surya Prakash, *Hydrocarbon Chemistry* (Hoboken, New Jersey: John Wiley & Sons, Inc., third edition, 2018), Volume I, pp. 4-5.

21 Ibid., p. 5.

22 Marc A. Shampo, Ph.D., and Robert A. Kyle, M.D., "Early German Physician First to Synthesize Urea," *Mayo Clinic Proceedings*, Volume 60 (October 1985), p. 662, https://www.mayoclinicproceedings.org/article/S0025-6196(12)60740-X/pdf#:~:text=Friedrich%20W%C3%B6hler%20was%20the%20first,silver%20cyanate%20to%20ammonium%20chloride.

23 "What is organic chemistry?" American Chemical Society, ACS.org, no date, https://www.acs.org/content/acs/en/careers/chemical-sciences/areas/organic-chemistry.html.

24 Ann E. Robinson, "Mendeleev's Periodic Table," Ohio State University and Miami University website entitled "Origins: Current Events in Historic Perspective," Origins. OSU.edu, March 2019, https://origins.osu.edu/milestones/mendeleev-periodic-table-UN-chemistry-radioactivity-noble-gases.

25 Anthony N. Stranges, Department of History, Texas A&M University, prepared for presentation at the AIChE 2003 Spring National Meeting, New Orleans, LA, March 30-April 3, 2001, unpublished.
26 Henry H. Storch, Ph.D., chief, Research and Development Branch; Norma Golumbic, M.S., technical assistant, Research; and Robert B. Anderson, Ph.D., physical chemist, Research and Development Branch, Office of Synthetic Liquid Fuels, Bureau of Mines, U.S. Department of Interior, Pittsburgh, Pennsylvania, *The Fischer-Tropsch and Related Syntheses* (New York: John Wiley & Sons, Inc., 1951), pp. 1-3, and 11. Again, this chapter will closely footnote all sources. When possible, discussions from the sources will be paraphrased if not quoted directly. When paraphrasing adds undue complexity to the text, word-for-word sections will be used here, as sparingly as possible.
27 Ibid.
28 G.A. Somorjai, "The Catalytic Hydrogenation of Carbon Monoxide. The Formation of C1 Hydrocarbons," *Science and Engineering*, Volume 23, Issue 1-2 (1981; published online December 5, 2006), pp. 189-202, https://www.tandfonline.com/doi/abs/10.1080/03602458108068075.
29 "The German Document Retrieval Project," Center for Energy & Mineral Resources," Texas A&M University, September 20, 1977.
30 Paul Schubert, Steve LeViness, and Kym Arcuri, Syntroleum Corporation, Tulsa, Oklahoma, and Anthony Stranges, Texas A&M University, "Fischer-Tropsch Process and Product Development During World War II," April 2, 2011, unpublished paper, at Fischer-Tropsch.org, under "Primary Documents/Presentations."
31 *The United States Strategic Bombing Survey.* The European War report was the first completed, published by the Government Printing Office on September 30, 1945. This report as originally issued can be read on the Internet at the following URL: http://www.anesi.com/ussbs02.htm#page1.
32 Sean McMeekin, *Stalin's War: A New History of World War II* (New York: Basic Books, 2021), pp. 134 and 153-155.
33 Ibid., p. 154.
34 Ibid., p. 155.
35 John Dodaro, "Fischer-Tropsch Process," Stanford University coursework, Stanford.edu, December 11, 2015, http://large.stanford.edu/courses/2015/ph240/dodaro1/.
36 "Core," *National Geographic*, no date, https://www.nationalgeographic.org/encyclopedia/core/. See also: Structure of Earth," GeologyScience.com, no date, https://geologyscience.com/general-geology/structure-of-earth/. See also: Jijo Sudarsan, "Interior of the Earth: Crust, Mantle, and Core," ClearIAS.com, last updated on July 10, 2016, https://www.clearias.com/interior-of-the-earth/.
37 Bin Chen, Zeyu Li, Dongzhou Zhang, et al., "Hidden carbon in Earth's inner core revealed by sear softening in dense Fe_7C_3," *Proceedings of the National Academy of Sciences*, Volume 111, Number 50 (December 16, 2014), pp. 17755-17758, https://www.pnas.org/content/111/50/17755.
38 Kei Hirose, Bernard Wood, and Lidunka Vočadlo, "Light elements in the Earth's core," *Nature Reviews Earth and Environment*, Volume 2 (August 24, 2021), pp. 645-658, https://www.nature.com/articles/s43017-021-00203-6?proof=t%3B.
39 University of Tokyo, "There may be up to 70 more times hydrogen in Earth's core than in the oceans," Phys.org, May 12, 2021, https://phys.org/news/2021-05-hydrogen-earth-core-oceans.html.

Endnotes

40. Suraj K. Bajgain, Mainak Mookherjee, and Rajdeep Dasgupta, "Earth's core could be the largest terrestrial carbon reservoir," *Communications Earth & Environment*, Volume 2, Number 165 (August 19, 2021), https://www.nature.com/articles/s43247-021-00222-7. See also: Daisy Dobrijevic, "Most of Earth's carbon may be locked in our planet's outer core," Space.com, August 23, 2021, https://www.space.com/earth-outer-core-carbon-reservoir.
41. Florida State University, "Researchers refine estimate of amount of carbon in Earth's outer core," ScienceDaily, August 19, 2021, https://www.sciencedaily.com/releases/2021/08/210819113059.htm.
42. Carnegie Institution, "Earth's core deprived of oxygen," ScienceDaily, November 24, 2011, https://www.sciencedaily.com/releases/2011/11/111123133137.htm. The ScienceDaily article reported on the following scientific study: Haijun Huang, Yingwei Fei, et al., "Evidence for an oxygen-depleted liquid outer core of Earth," *Nature*, Volume 479 (November 23, 2011), pp. 513-516, https://www.nature.com/articles/nature10621.
43. Christopher J. Davies, Monica Pozzo, et al., "Transfer of oxygen to Earth's core from a long-lived magma ocean," *Earth and Planetary Science Letters*, Volume 538 (May 15, 2020), https://www.sciencedirect.com/science/article/pii/S0012821X20301515#!.
44. Ibid.
45. Stephanie Pappas, "Earth's core is a billion years old," LiveScience.com, August 26, 2020, https://www.livescience.com/earth-core-billion-years-old.html. The LiveScience.com article reported on the following study: Youjung Zang, Miqiang Hou, et al., "Reconciliation of Experiments and Theory on Transport Properties of Iron and the Geodynamo," *Physical Review Letters*, APS Physics, Journals.APS.org, August 13, 2020, https://journals.aps.org/prl/abstract/10.1103/PhysRevLett.125.078501. We will provide later in this chapter a more complete description of the diamond anvil apparatus, where the understanding of how the apparatus works is more essential for understanding the point of science being made.
46. "Mantle," National Geographic, no date, https://www.nationalgeographic.org/encyclopedia/mantle/. See also: Structure of the Earth's interior," GeologyScience.com, no date. See also: Jijo Sudarsan, "Interior of the Earth: Crust, Mantle, and Core," ClearIAS.com, last updated on July 10, 2016.
47. David D. Pollard and Stephen J. Martel, *Structural Geology: A Quantitative Introduction* (Cambridge, U.K., and New York, NY: Cambridge University Press, 2020), p. 5.
48. Terry Plank and Craig E. Manning, "Subducting Carbon," *Nature*, Volume 574 (October 16, 2019), pp. 343-352, https://www.nature.com/articles/s41586-019-1643-z?proof=t.
49. University of Notre Dame, "Key indicator of carbon sources in Earth's mantle," ScienceDaily, November 9, 2016, https://www.sciencedaily.com/releases/2016/11/161109132623.htm. The ScienceDaily article reported on the following study: Samuel R.W. Hulett, Antonio Simonetti, et al., "Recycling of subducted crustal components into carbonatite melts revealed by boron isotope," *Nature Geoscience*, Volume 9 (November 7, 2016), pp. 904-908, https://www.nature.com/articles/ngeo2831.
50. "Peridotite," *Encyclopedia Britannica*, no date, https://www.britannica.com/science/peridotite.
51. Alessandro Gualtieri, Carlotta Giacobbe, and Cecilia Viti, "The dehydroxylation of serpentine group minerals," *American Mineralogist*, Volume 97, Number 4 (March

2012), pp. 666-680, https://www.researchgate.net/publication/241685023_The_dehydroxylation_of_serpentine_group_minerals.

52 Enrico Bonatti, James R. Lawrence, and Noris Morandi, "Serpentinization of ocean peridotites: temperature dependency of mineralogy and boron content," *Earth and Planetary Science*, Volume 70, Issue 1 (September 1984), pp. 88-94, https://www.sciencedirect.com/science/article/abs/pii/0012821X84902115.

53 Qingyang Hu, Duck Young Kim, Jin Liu, et al., "Dehydrogenation of goethite in Earth's deep lower mantle," *Proceedings of the National Academy of Sciences of the United States of America*, Volume 114, Issue 7 (February 14, 2017), pp. 1498-1501, https://carnegiesscience.edu/news/freeing-hydrogen-earth%E2%80%99s-lower-mantle. See also: "Freeing Hydrogen in Earth's Lower Mantle," *Carnegie Science*, February 2, 2017, https://carnegiescience.edu/news/freeing-hydrogen-earth%E2%80%99s-lower-mantle.

54 Yanhao Lin and Wim van Westrenen, "Oxygen as a catalyst in the Earth's interior?" *National Science Review*, Volume 8, Issue 4 (April 2021), https://academic.oup.com/nsr/article/8/4/nwab009/6102554.

55 "Scientists: Deep Sea Mineral Acts as a Oxygen Reservoir, Stopping the Earth from Becoming a Barren Planet," UnderwaterTimes News Service, September 26, 2007, https://www.underwatertimes.com/news.php?article_id=16278531040. The scientific study can be found here: Arno Rohrbach, Chris Ballhaus, et al., "Metal saturation in the upper mantle," *Nature*, Volume 449, Number 7161 (September 27, 2007), pp. 456-458, https://pubmed.ncbi.nlm.nih.gov/17898766/.

56 Eglantine Boulard, Alexandre Gloter, Alexandre Corgne, et al., "New host for carbon in the deep Earth," *Proceedings of the National Academy of Sciences of the United States of America*, Volume 108, Number 13 (March 29, 2011), pp. 5184-5187, https://www.pnas.org/content/108/13/5184.short.

57 Andy Coghlan, "There's as much water in Earth's mantle as in all the oceans," *New Scientist*, June 7, 2017, https://www.newscientist.com/article/2133963-theres-as-much-water-in-earths-mantle-as-in-all-the-oceans/. See also: Andy Coghlan, "Massive 'ocean discovered towards Earth's core," *New Scientist*, June 12, 2014, https://www.newscientist.com/article/dn25723-massive-ocean-discovered-towards-earths-core/. The study in question is the following: Brandon Schmandt, Steven D. Jacobsen, et al., "Dehydration melting at the top of the lower mantle," *Science*, Volume 344, Issue 6189 (June 13, 2014), pp. 1265-1268, https://www.science.org/doi/abs/10.1126/science.1253358.

58 Shawn E. McGlynn, Jennifer B. Glass, Kristin Johnson-Finn, et al., "Hydrogenation reactions of carbon on Earth: Linking methane, margarine, and life," *American Mineralogist*, Volume 105, Number 5 (April 29, 2020), https://www.degruyter.com/document/doi/10.2138/am-2020-6928CCBYNCND/html.

59 N.G. Holm, C. Oze, O. Mousis et al., "Serpentinization and the Formation of H_2 and CH_4 on Celestial Bodies (Planets, Moons, Comets)," *Astrobiology*, Volume 15, Number 7 (July 1, 2015), pp. 587-600, https://www.ncbi.nlm.nih.gov/pmc/articles/PMC4523005/. See also: Céline Martin, Kennet E. Flores, Alberto Vitale-Brovarone, et al., "Deep mantle serpentinization in subduction zones: Insight from in situ B isotopes in slab and mantle wedge serpentinites," *Chemical Geology*, Volume 545 (July 5, 2020), https://www.sciencedirect.com/science/article/abs/pii/S0009254120301765. See also: "Serpentinization," Oxford University Press, Encyclopedia.com, no date, https://www.encyclopedia.com/science/dictionaries-thesauruses-pictures-and-press-releases/

Endnotes

serpentinization. See also: Steve Drury, "What's happening at the core-mantel boundary?" WileyEarthPages.com, June 16, 2014, https://wileyearthpages.wordpress.com/tag/d-zone/. See also: Terry Colins Assoc., "Rewriting the textbook on fossil fuels: New technologies help unravel nature's methane recipes," American Association for the Advancement of Science, news release, April 22, 2019, https://www.eurekalert.org/news-releases/657159.

60 "How Plants Can Change Our Climate," NASA.gov, no date, https://earthobservatory.nasa.gov/features/LAI/LAI2.php.

61 Ibid.

62 For an explanation of "memory holes" in George Orwell's dystopian novel *1984*, see: Liz Breazeale, "Memory Hole in 1984," Study.com, no date, https://study.com/academy/lesson/memory-hole-in-1984.html.

63 Samson Reiny, "Carbon Dioxide Fertilization Greening Earth, Study Finds," NASA.gov, April 26, 2016, last updated March 27, 2019, https://www.nasa.gov/feature/goddard/2016/carbon-dioxide-fertilization-greening-earth. The scientific study can be found here: Zaichun Ahu, Shilong Piao, et al., "Greening of Earth and its drivers," *Nature Climate Change*, Volume 6 (April 25, 2016), pp. 791-795, https://www.nature.com/articles/nclimate3004.

64 Patrick Moore, Ph.D., "Twelve Invisible Eco-Catastrophes and Threats of Doom That Are Actually Fake," WattsUpWithThat.com, August 2, 2018, https://wattsupwiththat.com/2018/08/03/twelve-invisible-eco-catastrophes-and-threats-of-doom-that-are-actually-fake/.

65 Institute of Physics of the Earth, Russian Academy of Sciences, Linkedin.com, no date, https://www.linkedin.com/company/institute-of-physics-of-the-earth-russian-academy-of-sciences/.

66 This chapter subsection is heavily drawn from J.F. Kenney's website as well as his published articles. See: "Introduction," GasResources.net, no date, https://www.gasresources.net/introduction.htm.

67 History.com editors, "March 25, 1946: Soviets announce withdrawal from Iran," History.com, last updated March 23, 2021, https://www.history.com/this-day-in-history/soviets-announce-withdrawal-from-iran.

68 V.I. Anikeev, Y. Yermakova, B.L. Moroz, Boreskov Institute of Catalysis, Novosibirsk, Russia, "The State of Studies of the Fischer-Tropsch Process in Russia," unpublished paper supported by Syntroleum Corporation, Tulsa, Oklahoma.

69 Vladimir Kutcherov and Vladilen Krayushkin, "Deep-Seated Abiogenic Origin of Petroleum: From Geological Assessment to Physical Theory," *Reviews of Geophysics*, Volume 48, Issue 1 (March 12, 2010), https://agupubs.onlinelibrary.wiley.com/doi/full/10.1029/2008RG000270.

70 Henry P. Scott, Russell J. Hemley, Ho-kwang Mao, Dudley R. Herschbach, Laurence E. Fried, W. Michael Howard, and Sorin Bastea, "Generation of methane in the Earth's mantle: *In situ* high pressure-temperature measurements of carbonate reduction," *Proceedings of the National Academy of Sciences of the United States of America*, Volume 101, Number 39 (September 28, 2004), pp. 14023-14026, http://www.ncbi.nlm.nih.gov/pmc/articles/PMC521091/. Also, see: http://www.pnas.org/content/101/39/14023.full.pdf+html.

71 Quoted in: Erin O'Donnell, "Rocks into Gas," *Harvard Magazine*, March-April 2005, http://harvardmagazine.com/2005/03/rocks-into-gas.html.

72 Quoted in: Lawrence Livermore National Laboratory, "Methane in deep earth: A possible new source of energy," llnl.gov, press release, September 13, 2004, https://www.llnl.gov/news/methane-deep-earth-possible-new-source-energy.
73 Anton Kolesnikov, Vladimir G. Kutcherov, and Alexander F. Goncharov, "Methane-derived hydrocarbons produced under upper-mantle conditions," *Nature Geoscience*, Volume 2 (July 26, 2009), pp. 566-570, https://www.nature.com/articles/ngeo591.
74 Vetenskapspsrådet (The Swedish Research Council), "Fossils from Animals and Plants Are Not Necessary for Crude Oil and Natural Gas, Swedish Researchers Find," ScienceDaily, September 12, 2009, https://www.sciencedaily.com/releases/2009/09/090910084259.htm.
75 Ibid.
76 There is a long list of tributes to Thomas Gold, including those listed here. (1) Cornell news release, "Thomas Gold, Astronomer and Brilliant Scientific Gadfly, Dies at 84," June 22, 2004, http://www.spaceref.com/news/viewpr.html?pid=14439. (2) "Thomas Gold (1920–2004)," Bulletin of the AAS (American Astronomical Society), Volume 36, Issue 4 (December 1, 2004), https://baas.aas.org/pub/thomas-gold-1920-2004/release/1. (3) "The Art of Genius: Thomas Gold: Lived 1920-2004," FamousScientists.org, no date, https://www.famousscientists.org/thomas-gold/. (4) Hermann Bondi, Biographical Memoirs of Fellows of the Royal Society, "Thomas Gold. 22 May 1920–22 June 2004, *The Royal Society Publishing*, December 1, 2006, https://royalsocietypublishing.org/doi/10.1098/rsbm.2006.0009. (5) Geoffrey Burbidge and Margaret Burbidge, "Thomas Gold 1920-2004," National Academy of Sciences, *Biographical Memoirs*, Volume 88 (2006), http://nasonline.org/publications/biographical-memoirs/memoir-pdfs/gold-thomas.pdf. (6) Daniel R. Colman, Saroj Poudel, et al., "The deep, hot biosphere: Twenty-five years of retrospection," *Proceedings of the National Academy of Sciences of the United States of America*, Volume 114, Number 27 (July 3, 2017), pp. 6895-6903, https://www.pnas.org/content/114/27/6895. See also for a discussion of his life and a long list of his most important publications: "Thomas Gold," Wikipedia.org, last edited on September 22, 2021, https://en.wikipedia.org/wiki/Thomas_Gold.
77 Thomas Gold, *The Deep Hot Biosphere: The Myth of Fossil Fuels* (New York: Springer-Verlag New York, Inc., 1999), p. 44.
78 Thomas Gold, *Power from Earth: Deep Earth Gas—Energy for the Future* (London: J.M. Dent & Sons Ltd., 1987).
79 Ibid., p. 125.
80 Ibid., p. 126.
81 Ibid., p. 125.
82 Ibid., p. 133.
83 Ibid., p. 147.
84 Ibid., p. 173.
85 Thomas Gold, *The Deep Hot Biosphere*.
86 Ibid., p. 17.
87 Ibid., p. 19.
88 Ibid., p. 22. Emphasis in original.
89 Ibid., p. 5.
90 Ibid. Emphasis in original.

Endnotes

91   "Methane hydrates," WorldOceanReview.com, no date, https://worldoceanreview.com/en/wor-1/energy/methane-hydrates/.
92   Thomas Gold, *The Deep Hot Biosphere*, p. 26.
93   Ibid. Parentheses in original.
94   Ibid.
95   Ibid., p. 27.
96   Ibid.
97   Thomas Gold, *The Deep Hot Biosphere*, p. 11.
98   Ibid., p. 12.
99   Deborah S. Kelley, Jeffrey A. Karson, et al., "A Serpentinite-Hosted Ecosystem: The Lost City Hydrothermal Field," *Science*, Volume 307, Issue 5714 (March 4, 2005), https://www.science.org/doi/abs/10.1126/science.1102556.
100 Ibid., "Abstract."
101 Giora Proskurowski, Marvin D. Lilley, Jeffery S. Seewald, Gretchen L. Früh-Green, Eric J. Olson, John E. Lupton, Sean P. Sylva, and Deborah S. Kelley, "Abiogenic Hydrocarbon Production at Lost City Hydrothermal Field," *Science* Volume 319, Number 5863 (February 1, 2008), pp. 604-607, https://www.science.org/doi/abs/10.1126/science.1151194.
102 Ibid., p. 604.
103 Ibid.
104 Ibid.
105 Ibid., Abstract.
106 "Understanding Carbon-14 Analysis," Beta Analytic Testing Laboratory, no date, BetaLabServices.com, https://www.betalabservices.com/biobased/carbon14-dating.html.
107 Olaf Kniemeyer, Florin Musat, Stefan M. Sievert, et al., "Anaerobic oxidation of short-chain hydrocarbons by marine sulfate-reducing bacteria," *Nature*, Volume 449, Number 18 (October 18, 2007), pp. 898-90, https://www.nature.com/articles/nature06200.
108 "Anaerobic bacteria," *MedicinePlus*, U.S. National Library of Medicine, National Institute of Health (NIH), MedLinePlus.gov, no date, https://medlineplus.gov/ency/article/003439.htm.
109 Lonny Lippsett, "Some Things New Under the Sea...and other recent findings by WHOI deep-ocean researchers," *Oceanus*, Woods Hole Oceanographic Institute, February 22, 2008, https://www.whoi.edu/oceanus/feature/some-things-new-under-the-sea/.
110 Sarah K. Hu, Erica L. Herrera, Amy R. Smith, et al., "Protistan grazing impacts microbial communities and carbon cycling at deep-sea hydrothermal vents," *Proceedings of the National Academy of Sciences of the United States of America*, Volume 118, Number 29 (July 20, 2021), https://www.biorxiv.org/content/10.1101/2021.02.08.430233v1.abstract.
111 Aparna Vidyasagar, "What Are Protists?" LiveScience.com, March 30, 2016, https://www.livescience.com/54242-protists.html.
112 Woods Hole Oceanographic Institute, "Microbial Predators at Hydrothermal Vents Play Important Role in Deep-Sea Carbon Cycling," SciTechDaily.com, July 20, 2021, https://scitechdaily.com/microbial-predators-at-hydrothermal-vents-play-important-role-in-deep-sea-carbon-cycling/amp/.
113 Thomas Gold, *The Deep Hot Biosphere*, p. 64.
114 Ibid., pp. 6-7.

115 Guy Ourisson, Pierre Albrecht, and Michael Rohmer, "The Microbial Origin of Fossil Fuels," *Scientific American*, Volume 251, Number 2 (August 1984), pp. 44-51, https://www.jstor.org/stable/24969433.
116 Thomas Gold, *The Deep Hot Biosphere*, p. 83.
117 Thomas Gold, "Letters," *Scientific American*, Volume 251, Number 5 (November 1984), pp. 6-10, https://www.jstor.org/stable/24969468.
118 Thomas Gold, *The Deep Hot Biosphere*, p. 83.
119 Ibid., p. 84.
120 Ibid.
121 Ibid.
122 Ibid.
123 Ibid., p. 86.
124 Ibid.
125 Thomas Gold, "The deep, hot biosphere," *Proceedings of the National Academy of Sciences of the United States of America*, Volume 89, Number 13 (July 1, 1992), pp. 6045-6049, https://www.pnas.org/content/89/13/6045.short.
126 Richard C. Selley and Stephen A. Sonnenberg, *Elements of Petroleum Geology*, p. 207.
127 Kenneth K. Landes, *Petroleum Geology* (New York: John Wiley & Sons, Inc, 1951), p. 168.
128 H.D. Klemme and G.F. Ulmishek, "Effective Petroleum Source Rocks of the World: Stratigraphic Distribution and Controlling Depositional Factors," *AAAP Bulletin*, Volume 50 (1991), pp. 1809-1851, reprinted in part by AAPG Datapages, Inc., Search and Discovery Article #30003, 1999, https://www.searchanddiscovery.com/documents/animator/klemme2.htm.
129 Seppo A. Korpela, "Oil Depletion in the United States and the World," a working paper for a talk to Ohio Petroleum Marketers Association at their annual meeting in Columbus, Ohio, May 1, 2002, https://www.academia.edu/25749980/Oil_Depletion_in_the_United_States_and_the_World.
130 Seppo A. Korpela, "Oil depletion in the world," *Current Science*, Volume 91, Number 9 (November 10, 2006), pp. 1148-1152, https://www.jstor.org/stable/24094091.
131 Ker Than, "The Mysterious Origin and Supply of Oil," *Live Science*, October 10, 2005, at http://www.livescience.com/9404-mysterious-origin-supply-oil.html.
132 Richard C. Selley and Stephen A. Sonnenberg, *Elements of Petroleum Geology*, pp. 208-209.
133 Ibid., p. 211.
134 Ibid.
135 Thomas Gold, *The Deep Hot Biosphere*, p. 85.
136 Ibid.
137 The discussion of carbon isotope fractionation is drawn heavily from the following: Thomas Gold, *Power from Earth: Deep Earth Gas—Energy for the Future*, Appendix I, pp. 175-188.
138 Thomas Gold, *The Deep Hot Biosphere*, p. 72.
139 Ibid.
140 "Rock Cycle—Past Life," msnucleus.org, no date, https://www.msnucleus.org/membership/html/k-6/rc/pastlife/5/rcpl5_2a.html.
141 "What Is a Fossil?" FossilEra.com, no date, https://www.fossilera.com/fossils. Scroll down to text.
142 Ibid.

Endnotes

143 Matthew R. Simmons, *Twilight in the Desert: The Coming Saudi Oil Shock and the World Economy* (Hoboken, New Jersey: John Wiley & Sons, Inc., 2005), p. 231.
144 Ibid.
145 Ibid., p. 281.
146 Ibid., p. 170.
147 Tim Kennedy, "Saudi Oil Is Secure and Plentiful, Say Officials," *Arab News*, April 29, 2004, https://www.arabnews.com/node/248483.
148 Matthew R. Simmons, *Twilight in the Desert: The Coming Saudi Oil Shock and the World Economy*, p. 155.
149 U.S. Energy Information Administration (EIA), "Saudi Arabia," last updated December 2, 2021, https://www.eia.gov/international/overview/country/SAU.
150 H.S. Edgell, "Basement Tectonics of Saudi Arabia as Related to Oil Field Structures," in *Basement Tectonics 9: Australia and Other Regions*, edited by M.J. Rickard, H.J. Harrington, and P.R. Williams, Proceedings of the Ninth International Conference on Basement Tectonics, held in Canberra, Australia, July 1990 (Dordrecht/Boston/London: Kluwer Academic Publishers, 1992), pp. 169-193.
151 Ibid., p. 169.
152 Ibid.
153 Ibid., p. 170.
154 Scott Weeden, senior editor, "Meteoric History of Cantarell Field Continues for Pemex," HartEnergy.com, May 1, 2015, https://www.hartenergy.com/ep/exclusives/meteoric-history-cantarell-field-continues-pemex-175129.
155 Alan R. Hildebrand, Glen T. Penfield, David A. Kring, Mark Pilkington, Antonio Carmargo Z., Stein B. Jacobsen, and William V. Boynton, "Chicxulub Crater: A possible Cretaceous/Tertiary boundary impact crater on the Yucatán Peninsula, Mexico," *Geology*, Volume 19, Number 9 (September 1991), p. 867-871, https://pubs.geoscienceworld.org/gsa/geology/article-abstract/19/9/867/205322/Chicxulub-Crater-A-possible-Cretaceous-Tertiary.
156 José Manuel Grajales-Nishimura, Estaban Cedillo-Pardo, Ricardo Martínez-Ibarra, Jesús García-Hernández-García, and Rafael Penaloza-Romero, "The Stratigraphic Architecture of the K/T Boundary Carbonate Breccia Sedimentary Succession in the Cantarell Oil Field: The Most Important Oil-Producing Horizon in Offshore Campeche," 2004 AAPG International Conference and Exhibition. See also: Xu Shunshan, Ángel Francisco Nieto-Samaniego, et al., "Factors influencing the fault displacement-length relationship: an example from the Cantarell oilfield, Gulf of Mexico," *Geofísica Internacional*, Volume 50, Number 3 (July/September 2011), Ciudad de México, Scielo.org.mx, http://www.scielo.org.mx/scielo.php?pid=S0016-71692011000300003&script=sci_arttext&tlng=en.
157 Roger Barton, Ken Bird, Jesús García Hernández, et al., "High Impact Reservoirs," *Oilfield Review*, July/September 2011, Volume 21, Number 4 (Winter 2009/2010), pp. 14-29, https://www.researchgate.net/profile/Oliver-Schenk/publication/287920314_High-impact_reservoirs/links/5977982845851570a1b31775/High-impact-reservoirs.pdf.
158 Ibid.
159 Russell Gold, "In Gulf of Mexico, Industry Closes In on New Oil Source," *Wall Street Journal*, September 5, 2006, at http://online.wsj.com/article/SB115742365939953524.html?mod=home_whats_news_us.

160 "Mexico discovers 'huge' oil field," BBC News, BBC.co.uk, March 15, 2006, http://news.bbc.co.uk/2/hi/americas/4808466.stm.
161 Associated Press, "Offshore oil discovery could help make Brazil major petroleum exporter," *International Herald Tribune*, November 8, 2007. See also: Alexei Barrionuevo, "Underwater oil discovery to transform Brazil into a major exporter," *New York Times*, January 11, 2008, https://www.nytimes.com/2008/01/11/business/worldbusiness/11iht-oil.1.9147825.html. See also: Jeb Blount, "Brazil to announce massive offshore oil discovery on October 23," Reuters, October 3, 2013, https://www.reuters.com/article/brazil-oil-discovery/brazil-to-announce-massive-offshore-oil-discovery-on-oct-23-idUSL1N0HT1IS20131003.
162 "How Far Do We Drill to Find Oil?" Petro-Online.com, November 5, 2014, https://www.petro-online.com/news/fuel-for-thought/13/breaking-news/how-far-do-we-drill-to-find-oil/32357.
163 Ibid.
164 Thomas Gold, *The Deep Hot Biosphere: The Myth of Fossil Fuels* (New York: Copernicus Books, an imprint of Springer-Verlag New York, Inc., paperback edition, 2001).
165 Goddard Space Flight Center, "Titan's Mysterious Methane Comes from Inside, Not the Surface," SpaceRef.com, November 30, 2005, at http://www.spaceref.com/news/viewpr.html?pid=18410.
166 H.B. Niemann, S.K. Atreya, S.J. Bauer, et al., "The abundances of constituents of Titan's atmosphere from the GCMS instrument on the Huygens probe," *Nature*, Volume 438 (November 30, 2005), pp. 779-784, https://www.nature.com/articles/nature04122.
167 Donald E. Jennings, Paul N. Romani, Gordon L. Bjoraker, et al., "$^{12}C/^{13}C$ Ratio in Ethane on Titan and Implications for Methane's Replacement," *Journal of Physical Chemistry*, Volume 113, Number 42 (June 24, 2009), pp. 11101-11106, https://pubs.acs.org/doi/10.1021/jp903637d.
168 "Titan's surface organics surpass oil reserves on Earth," European Space Agency (ESA) Space Science, February 13, 2008, at http://www.esa.int/esaSC/SEMCSUUHJCF_index_0.html.
169 Ibid.
170 Christopher R. Glein, "Noble gases, nitrogen, and methane from the interior to the atmosphere of Titan," *Icarus*, Volume 250 (April 2015), pp. 570-586, https://www.sciencedirect.com/science/article/abs/pii/S0019103515000032. See also: Goddard Space Flight Center, "Titan's Mysterious Methane Comes from Inside, Not the Surface," SpaceRef.com, press release, November 30, 2005, http://www.spaceref.com/news/viewpr.html?pid=18410.
171 Sebastian Seibold, Werner Rammer, Torsten Hothorn, et al., "The contribution of insects to global forest deadwood composition," *Nature*, Volume 597 (September 1, 2021), pp. 77-81, https://www.nature.com/articles/s41586-021-03740-8.
172 Australian National University, "Deadwood Releasing 10.9 Gigatons of Carbon Every Year—More Than All Fossil Fuel Emissions Combined," SciTechDaily.com, September 20, 2021, https://scitechdaily.com/deadwood-releasing-10-9-gigatons-of-carbon-every-year-more-than-all-fossil-fuel-emissions-combined/.
173 Ibid.
174 "Importance of Methane," United States Environmental Protection Agency (EPA), no date, https://www.epa.gov/gmi/importance-methane.

Endnotes

175 Adam Voiland, "Methane Matters: Scientists Work to Quantify the Effects of a Potent Greenhouse Gas," NASA Earth Observatory, March 8, 2016, https://earthobservatory.nasa.gov/features/MethaneMatters.

176 Joe Schwarcz, "Why isn't the carbon dioxide from breathing a concern for global warming?" McGill University Office for Science and Society, March 20, 2017, https://www.mcgill.ca/oss/article/environment-quirky-science-you-asked/humans-and-animals-exhale-carbon-dioxide-every-breath-why-not-considered-be-problem-far-global.

177 Ibid.

178 Clyde Spencer, "Contribution of Anthropogenic CO_2 Emissions to Changes in Atmospheric Concentrations," WattsUpWithThat.com, June 11, 2021, https://wattsupwiththat.com/2021/06/11/contribution-of-anthropogenic-co2-emissions-to-changes-in-atmospheric-concentrations/.

179 Jamie Shutler, associate professor, University of Exeter, and Andy Watson, Royal Society research professor, University of Exeter, "The oceans are absorbing more carbon than previously thought," CarbonBrief.org, September 28, 2020, https://www.carbonbrief.org/guest-post-the-oceans-are-absorbing-more-carbon-than-previously-thought.

180 Carolyn W. Snyder, Michael D. Mastrandrea, and Stephen H. Schneider, "The Complex Dynamics of the Climate System: Constraints on Our Knowledge, Policy Implications, and the Necessity of Systems Thinking," an article in *Handbook of the Philosophy of Sciences*, Volume 10, *Philosophy of Complex Systems*, series volume edited by Cliff Hooker (Amsterdam: Elsevier, first edition, 2011), Part V, "Climatology," pp. 467-505.

181 Ibid., p. 498.

182 Ibid., p. 493.

183 Ibid.

184 Daniel R. Coleman, Saroj Poudel, Blake W. Stamps, et al., "The deep, hot biosphere: Twenty-five years of retrospection," *Proceedings of the National Academy of Sciences of the United States of America*, Volume 114, Number 27 (July 3, 2017), pp. 6895-6903, https://www.pnas.org/content/114/27/6895.

185 Ibid.

Chapter 10

1 Marc Morano, *Green Fraud: Why the Green New Deal Is Even Worse Than You Think* (Washington, D.C.: Regnery Publishing, 2021), p. 1.

2 Stephen Moore and Kathleen Hartnett White, *Fueling Freedom: Exposing the Mad War on Energy* (Washington, D.C.: Regnery Publishing, 2016), p. 169.

3 Michael Shellenberger, *Apocalypse Never: Why Environmental Alarmism Hurts Us All* (New York: Harper, 2020), p. xiii.

4 U.S. Energy Information Administration, U.S. Department of Energy, "U.S. energy facts explained," Energy.gov, last updated May 14, 2021, https://www.eia.gov/energyexplained/us-energy-facts/.

5 See: Linda Hunt, Secret Agenda: The United States Government, Nazi Scientists, and Project Paperclip, 1945 to 1990 (New York: St. Martin's Press, 1991); and Tom Bower, *The Paperclip Conspiracy: The Hunt for the Nazi Scientists* (Boston: Little, Brown and Company, 1987).

6 Burton H. Davis, Center for Applied Energy Research, University of Kentucky, "An Overview of Fischer-Tropsch Synthesis at the U.S. Bureau of Mines," prepared for

presentation at the AIChE 2003 Spring National Meeting, New Orleans, Louisiana, March 30–April 3, 2001, unpublished.
7 Ibid.
8 "German Document Retrieval Project," Texas A&M University, Center for Energy and Mineral Resources, no date, http://www.fischer-tropsch.org/DOE/germ_doc_ret_proj/report.pdf.
9 Francis Menton, "Texas Starts Waking Up to the Issue of the Full Cost of 'Renewables,'" *Manhattan Contrarian Blog*, Manhattan.contrarian.com, June 20, 2021, https://www.manhattancontrarian.com/blog/2021-6-20-texas-starts-waking-up-to-the-issue-of-the-full-costs-of-renewables.
10 U.S. Energy Information Administration, U.S. Department of Energy, "What is U.S. electricity generation by energy source?" Energy.gov, last updated March 5, 2021, https://www.eia.gov/tools/faqs/faq.php?id=427&t=3.
11 U.S. Energy Information Administration, "Annual Energy Outlook 2021 with projections to 2050," EIA.gov, February 2021, p. 14, https://www.eia.gov/outlooks/aeo/pdf/AEO_Narrative_2021.pdf.
12 Aaron Larson, "The Solar and Wind Power Cost-Value Conundrum," *Power Magazine*, August 2, 2021, https://www.powermag.com/the-solar-and-wind-power-cost-value-conundrum/. For the IRENA report, see: International Renewable Energy Agency (IRENA), Renewable Power Generation Costs in 2020, IRENA.org, 2021, https://www.irena.org/-/media/Files/IRENA/Agency/Publication/2021/Jun/IRENA_Power_Generation_Costs_2020.pdf.
13 Svitlana Kolosok, Iuliia Myroshnychenko, et al., "Renewable energy innovation in Europe: Energy efficiency analysis," International Conference on Innovation, Modern Applied Science and Environment Studies (ICIES2020), *E3S Web of Conferences*, Volume 234, Number 21 (February 2, 2021), p. 4, https://www.e3s-conferences.org/articles/e3sconf/abs/2021/10/e3sconf_icies2020_00021/e3sconf_icies2020_00021.html.
14 Ibid., pp. 4-5.
15 Peter Ferrara, "Renewable Fuel Standard Follies," *The American Spectator*, November 1, 2017, https://spectator.org/renewable-fuel-standard-follies/.
16 U.S. Energy Information Agency, "State Profile: Texas," EIA.gov, no date, https://www.eia.gov/state/?sid=TX.
17 Robert Brice, "Why was $66 billion spent on renewables before the Texas blackouts?" CFact.org, June 24, 2021, https://www.cfact.org/2021/06/24/why-was-66-billion-spent-on-renewables-before-the-texas-blackouts-big-wind-and-solar-got-22-billion-in-subsidies/. See also: Dick Law Firm, "Winter Storm Uri Causes Texas $200 Billion in Damages," DickLawFirm.com, March 27, 2021, https://www.dicklawfirm.com/Blog/2021/March/Winter-Storm-Uri-Causes-Texas-200-Billion-in-Dam.aspx. See also: Hobby School of Public Affairs, University of Houston, "The Winter Storm of 2021," UH.edu, no date, https://uh.edu/hobby/winter2021/storm.pdf.
18 Erin Douglas and Mitchell Ferman, "Texas Legislature approves bills to require power plants to 'weatherize,' among other measures to overhaul electric grid," *Texas Tribune*, May 26, 2021, https://www.texastribune.org/2021/05/26/texas-power-grid-reform-legislature/.
19 "How Long Does It Take to Charge an Electric Car?" Pod-Point.com, last updated November 11, 2021, https://pod-point.com/guides/driver/how-long-to-charge-an-electric-car#article-block-9.

Endnotes

20 Eric Peters, "What They Don't Tell You About Electric Cars," *The American Spectator*, December 22, 2020, https://spectator.org/electric-vehicle-battery/.
21 Aarian Marshall, "Biden Wants More EVs on Road. What About the Charging Stations?" *Wired*, August 10, 2021, https://www.wired.com/story/biden-wants-more-evs-charging-stations/.
22 Mark Kane, "Check Electric Cars Listed by Weight Per Battery Capacity (kWh)," InsideEvs.com, August 23, 2021, https://insideevs.com/news/528346/ev-weight-per-battery-capacity/.
23 Peter Valdes-Dapena, CNN Business, "Why electric cars are so much heavier than regular cars," CNN.com, June 7, 2021, https://www.cnn.com/2021/06/07/business/electric-vehicles-weight/index.html.
24 Ibid.
25 Rachael Nealer, David Reichmuth, and Don Anair, "Cleaner Cars from Cradle to Grave: How Electric cars Beat Gasoline Cars on Lifetime Global Warming Emissions," Union of Concerned Scientists, November 2015, https://www.ucsusa.org/sites/default/files/attach/2015/11/Cleaner-Cars-from-Cradle-to-Grave-full-report.pdf.
26 Dave Quast, "Myth Busting: 'Zero Emissions' EVs Actually Run on Natural Gas," EnergyInDepth.com, September 16, 2018, https://www.energyindepth.org/myth-busting-zero-emissions-evs-actually-run-on-natural-gas/.
27 Jude Clement, "More Electric Cars Mean More Coal and Natural Gas," *Forbes*, January 24, 2018, https://www.forbes.com/sites/judeclemente/2018/01/24/more-electric-vehicles-mean-more-coal-and-natural-gas/?sh=15ac5cba2a37.
28 Rory Carroll, "Five Myths about electric vehicles. No, they aren't only for rich people," *Washington Post*, September 24, 2021, https://www.washingtonpost.com/outlook/five-myths/electric-vehicles-five-myths/2021/09/23/e554b6cc-1bf2-11ec-a99a-5fe-a2b2da34b_story.html.
29 Jessie Lin and Yusin Hu, "Lithium price skyrockets as EV sales double," *DigiTimes Asia*, September 30, 2021, https://www.digitimes.com/news/a20210930PD207.html.
30 Aaron Turpen, "EV charger 101: It's not just plug-and-play," NewAtlas.com, September 30, 2021, https://newatlas.com/automotive/how-to-install-electric-car-charger-ev/.
31 Dave Quast, "Myth Busting: 'Zero Emissions' EVs Actually Run on Natural Gas."
32 Luke Richardson, "How many sun hours do I need? Calculating peak sun," EnergySage.com, May 17, 2017, https://news.energysage.com/many-sunlight-hours-need-calculating-peak-sun-hours/. See also: "What is a peak sun hour? What are peak sun hour numbers for your state?" SolarReviews.com, updated September 24, 2021, https://www.solarreviews.com/blog/peak-sun-hours-explained.
33 Timothy Paul Smith, *How Big Is Big and How Small Is Small: The Sizes of Everything and Why* (Oxford University Press, 2014), p. 99. See also: "Solar Panel Efficiency—Pick the Most Efficient Solar Panels," Solar.com, 2021, https://www.solar.com/learn/solar-panel-efficiency/.
34 RS Components, "How much land is needed to power the world's major cities by Solar?" AltEnergMag.com, September 26, 2017, https://www.altenergymag.com/article/2017/09/how-much-land-is-needed-to-power-the-worlds-major-cities-by-solar/27181.
35 Office of Energy Efficiency and Renewable Energy, U.S. Department of Energy, Solar Energy Technology Office, "Solar-Plus-Storage 101," Energy.gov, March 11, 2019, https://www.energy.gov/eere/solar/articles/solar-plus-storage-101.

36 Anuradha Varanasi, "Windiest States in America," Stacker.com, December 27, 2019, https://stacker.com/stories/3809/windiest-states-america.

37 Paul Denholm, Jacob Nunemaker, Pieter Gagnon, and Wesley Cole, "The Potential for Battery Energy Storage to Provide Peaking Capacity in the United States," National Renewable Energy Laboratory (NREL), Office of Energy Efficiency & Renewable Energy, U.S. Department of Energy, *Technical Report NREL/TP-6A20-74184*, June 2019, https://www.nrel.gov/docs/fy19osti/74184.pdf.

38 William J. Korchinsky, senior fellow, Reason Foundation, "The Limits of Wind Power," Reason.org, October 4, 2012, https://reason.org/policy-study/the-limits-of-wind-power/.

39 Timothy Paul Smith, *How Big Is Big and How Small Is Small: The Sizes of Everything and Why*, p. 105.

40 "Backgrounders: Generating Electricity: Fossil Fuels," LetsTalkScience.ca, no date, https://letstalkscience.ca/educational-resources/backgrounders/generating-electricity-fossil-fuels.

41 Timothy Paul Smith, *How Big Is Big and How Small Is Small: The Sizes of Everything and Why*, p. 104.

42 Liter of Light USA staff, "Solar and Wind Energy: Everything There Is to Know," LiterOfLight.org, no date, https://www.literoflightusa.org/solar-and-wind-energy/.

43 Office of Energy Efficiency and Renewable Energy, U.S. Department of Energy, Wind Energy Technologies Office, "How Do Wind Turbines Work?" Energy.gov, no date, https://www.energy.gov/eere/wind/how-do-wind-turbines-work.

44 John A. Dutton, e-Education Institute, "Alternative Fuels from Biomass Sources: How Corn Is Processed to Make Ethanol," College of Earth and Mineral Sciences, Penn State, no date, https://www.e-education.psu.edu/egee439/node/673.

45 Taxpayers for Common Sense, "Understanding U.S. Corn Ethanol and Other Corn-Based Biofuels," May 2021, https://www.taxpayer.net/wp-content/uploads/2021/05/TCS-Biofuels-Subsidies-Report.pdf.

46 Stephanie Kelly and Jarrett Renshaw, "Biden administration mulls big cuts to biofuel mandates in win for oil industry," Reuters, September 22, 2021, https://www.reuters.com/business/energy/exclusive-us-epa-considering-cuts-biofuel-blending-obligations-2020-2021-2022-2021-09-22/.

47 C. Ford Runge, "The Case Against More Ethanol: It's Simply Bad for Environment," *Yale Environment 360*, published at the Yale School of the Environment, May 25, 2016, https://e360.yale.edu/features/the_case_against_ethanol_bad_for_environment.

48 Ann Davis and Russell Gold, "U.S. Biofuel Boom Running on Empty," *Wall Street Journal*, August 27, 2009, https://www.wsj.com/articles/SB125133578177462487.

49 Susan S. Lang, "Cornell ecologist's study finds that producing ethanol and biodiesel from corn and other crops is not worth the energy," *Cornell Chronicle*, July 5, 2005, https://news.cornell.edu/stories/2005/07/ethanol-biodiesel-corn-and-other-crops-not-worth-energy.

50 John Kemp, "Worldwide energy shortage shows up in surging coal, gas, and oil prices," Reuters, September 26, 2021, https://www.reuters.com/business/energy/worldwide-energy-shortage-shows-up-surging-coal-gas-oil-prices-kemp-2021-09-24/. See also: Stine Jacobsen and Christopher Steitz, "Lighter winds slow progress at offshore firms Orsted, RWE," Reuters, August 12, 2021, https://www.reuters.com/business/sustainable-business/weaker-winds-slow-progress-offshore-firms-orsted-rwe-2021-08-12/.

Endnotes

51. Rachel Morison and Todd Gillespie, *Bloomberg News*, "Three More U.K. Power Suppliers Collapse as Energy Crisis Deepens," *Financial Post*, September 29, 2021, https://financialpost.com/pmn/business-pmn/three-more-u-k-power-suppliers-collapse-as-energy-crisis-deepens.
52. Alexander Smith, "Covid is at the center of world's energy crunch, but a cascade of problems is fueling it," NBC News, October 8, 2021, https://www.nbcnews.com/news/amp/rcna2688.
53. Ibid.
54. Ibid.
55. Alfred Cang, "China Orders Top Energy Firms to Secure Supplies at All Costs," *Bloomberg*, September 30, 2021, https://www.bloomberg.com/news/articles/2021-09-30/china-orders-top-energy-firms-to-secure-supplies-at-all-costs.
56. Tyler Durden, "Millions of Chinese Residents Lose Power After Widespread, 'Unexpected' Blackouts; Power Company Warns This Is 'New Normal,'" ZeroHedge.com, September 27, 2021, https://www.zerohedge.com/markets/millions-chinese-residents-lose-power-after-widespread-unexpected-blackouts-power-company.
57. Charles Kennedy, "China Bets on Shale to Raise Its Oil Production," OilPrice.com, September 30, 2021, https://oilprice.com/Energy/Crude-Oil/China-Bets-On-Shale-To-Raise-Its-Oil-Production.html.
58. David Rose, "Beijing's dirtiest secret: With 1,000 coal-fired power stations (and climbing) China's energy pollution mocks the world's bid to combat climate change," *Daily Mail*, September 24, 2021, https://www.dailymail.co.uk/news/article-10026335/Chinas-dirtiest-secret-1-000-coal-fired-power-stations-climbing.html.
59. David Stanway, "China energy crunch triggers shutdowns, pleas for more coal," Reuters, September 28, 2021, https://www.reuters.com/world/china/china-energy-crunch-triggers-alarm-pleas-more-coal-2021-09-28/.
60. Elena Mazneva and Anna Shiryaevskaya, "The Surge in Gas Prices Is the Equivalent to a $190 Oil Shock," *Bloomberg News*, October 1, 2021, https://www.bloomberg.com/news/articles/2021-10-01/the-surge-in-gas-prices-is-the-equivalent-to-a-190-oil-shock.
61. Harry Robertson, "Legendary gas trader says Europe's energy crisis could spread to the U.S. if it ditches fossil fuels too fast," Business Insider, September 28, 2021, https://africa.businessinsider.com/markets/legendary-gas-trader-says-europes-energy-crisis-could-spread-to-the-us-if-it-ditches/bkglefg.
62. Andrew Freedman, "Fossil Fuels to Dominate World Energy Use Through 2040," ClimateCentral.org, July 25, 2013, https://www.climatecentral.org/news/fossil-fuels-to-dominate-world-energy-use-through-2040-16284.
63. U.S. Energy Information Administration, Office of Energy Analysis, U.S. Department of Energy, *International Energy Outlook 2019, with projections to 2050*, September 24, 2019, https://www.eia.gov/outlooks/ieo/pdf/ieo2019.pdf. See also: U.S. Energy Information Administration, "EIA projects nearly 50 percent increase in world energy usage by 2050, led by growth in Asia," EIA.gov, September 24, 2019, https://www.eia.gov/todayinenergy/detail.php?id=41433.
64. Alex Longley, "Someone is betting that oil will soar to a record US$200 per barrel," *Bloomberg News*, September 30, 2021, https://www.bnnbloomberg.ca/someone-is-betting-that-oil-will-soar-to-a-record-us-200-a-barrel-1.1659692.
65. Tim De Chant, "China's carbon pollution now surpasses all developed countries combined," ArsTechnica.com, May 6, 2021, https://arstechnica.com/tech-policy/2021/05/chinas-carbon-pollution-now-surpasses-all-developed-countries-combined/.

66 Ibid.
67 Environmental Defense Fund, "Why China is at the center of our climate strategy," EDF. org, no date, https://www.edf.org/climate/why-china-center-our-climate-strategy.
68 "Synfuels China demonstrates First Fischer-Tropsch products from the new Shenhua Ningxia Coal-to-Liquids plant," SynGasChem.com, December 12, 2016, https://www.syngaschem.com/2016/12/12/synfuels-china-demonstrates-first-fischer-tropsch-products-from-the-new-shenhua-ningxia-coal-to-liquids-plant/.
69 David Rose, "Beijing's dirtiest secret: With 1,000 coal-fired power stations (and climbing) China's energy pollution mocks the world's bid to combat climate change."
70 Madison Gesiotto, opinion contributor, "Why do environmental Democrats ignore massive Chinese pollution?" *The Hill*, October 28, 2019, https://thehill.com/opinion/energy-environment/467785-why-do-environmental-democrats-ignore-massive-chinese-pollution.
71 Xiaoying You, "China's 2060 climate pledge is 'largely consistent' with 1.5°C goal, study finds," CarbonBrief.com, April 22, 2021, https://www.carbonbrief.org/chinas-2060-climate-pledge-is-largely-consistent-with-1-5c-goal-study-finds. The study in question can be found here: Hongbo Duan, Sheng Zhou, et al., "Assessing China's efforts to pursue the 1.5°C warming limit," *Science*, Volume 372, Number 6540 (April 23, 2021), pp. 378-385, https://www.science.org/doi/abs/10.1126/science.aba8767.
72 United States Environmental Protection Agency, "Greenhouse Gas Emissions Continue to Decline as the American Economy Flourishes Under the Trump Administration," EPA.gov, press release, November 9, 2020, https://www.epa.gov/newsreleases/greenhouse-gas-emissions-continue-decline-american-economy-flourishes-under-trump-0.
73 Devin Dwyer and Sarah Herndon, "US greenhouse gas emissions drop under Trump, but climate experts aren't celebrating," ABC News, December 22, 2020, https://abcnews.go.com/Politics/us-greenhouse-gas-emissions-drop-trump-climate-experts/story?id=74848440.
74 White House Briefing Room, "FACT SHEET: President Biden Sets 2030 Greenhouse Gas Pollution Reduction Target Aimed at Creating Good-Paying Union Jobs and Securing U.S. Leadership on Clean Energy Technologies," April 22, 2021, WhiteHouse.gov, https://www.whitehouse.gov/briefing-room/statements-releases/2021/04/22/fact-sheet-president-biden-sets-2030-greenhouse-gas-pollution-reduction-target-aimed-at-creating-good-paying-union-jobs-and-securing-u-s-leadership-on-clean-energy-technologies/. See also: U.S. Department of State, "Leaders Summit on Climate," State.gov, no date, https://www.state.gov/leaders-summit-on-climate/.

Conclusion

1 John Brooke, Michael Bevis, and Steve Rissing, "How Understanding the History of the Earth's Climate Can Offer Hope Amid Crisis," *Time*, September 20, 2019, updated September 23, 2019, https://time.com/5680432/climate-change-history-carbon/. *Time* credits for the article noted the following: "John Brooke, Michael Bevis, and Steve Rissing teach History, Geophysics, and Biology at The Ohio State University, where they team-teach a general education course on climate change."
2 Gary Anderson, "There Are Conservative Solutions to the Climate Problem," *The American Spectator*, August 20, 2021, https://spectator.org/climate-change-2/. *American Spectator* credits for the article noted the following: "Gary Anderson lectures on

Endnotes

Alternative Analysis at the George Washington University's Elliott School of International Affairs."
3 Pope Francis, *Encyclical Letter LAUDATO SI' of the Holy Father Francis on Care for Our Common Home*, Vatican Press, May 24, 2015, https://www.vatican.va/content/dam/francesco/pdf/encyclicals/documents/papa-francesco_20150524_enciclica-laudato-si_en.pdf.
4 Pat McCloskey, OFM, "Is Every Encyclical Infallible?" Franciscan Media, no date, https://www.franciscanmedia.org/ask-a-franciscan/is-every-encyclical-infallible.
5 Pope Francis, *Encyclical Letter LAUDATO SI' of the Holy Father Francis on Care for Our Common Home*, p. 18.
6 George F. Wills, "Pope Francis's fact-free flamboyance," *Washington Post*, Opinions, September 18, 2015, https://www.washingtonpost.com/opinions/pope-franciss-fact-free-flamboyance/2015/09/18/7d711750-5d6a-11e5-8e9e-dce8a2a2a679_story.html.
7 Ibid.
8 Paul R. Ehrlich and John Harte, "Biophysical limits, women's rights and the climate encyclical," *Nature Climate Change*, Volume 5 (September 24, 2014), pp. 904-905, https://www.nature.com/articles/nclimate2795.
9 Suzanne Goldenberg, "Pope's climate push is 'raving nonsense' without population control, says top U.S. scientist," *Guardian*, September 24, 2015, https://www.theguardian.com/world/2015/sep/24/popes-climate-stance-is-nonsense-rejects-population-control-says-top-us-scientist.
10 Paul R. Ehrlich, Anne H. Ehrlich, and John P. Holdren, *Ecoscience: Population, Resources, Environments* (San Francisco: W.H. Freeman and Company, 1977).
11 Ibid., p. 839.
12 Ibid., p. 760. Emphasis in original.
13 Ibid.
14 Ibid., p. 839.
15 Ibid.
16 Ibid., p. 715.
17 Timothy Paul Smith, *How Big Is Big and How Small Is Small: The Sizes of Everything and Why*, pp. 106-107.
18 World Nuclear Association, "Generation IV Nuclear Reactors," World-Nuclear.org, updated December 2020, https://world-nuclear.org/information-library/nuclear-fuel-cycle/nuclear-power-reactors/generation-iv-nuclear-reactors.aspx.
19 James E. Hanley, "Will We Accept Nuclear Fusion When It Comes?" American Institute for Economic Research (AIER), September 24, 2021, https://www.aier.org/article/will-we-accept-nuclear-fusion-when-it-comes/.
20 Office of Science, U.S. Department of Energy, "DOE Explains…Nuclear Fusion Reactions," Energy.gov, https://www.energy.gov/science/doe-explainsnuclear-fusion-reactions.
21 Timothy Paul Smith, *How Big Is Big and How Small Is Small: The Sizes of Everything and Why*, p. 144.
22 Jeff Tollefson, "U.S. achieves laser-fusion record: what it means for nuclear-weapons research," *Nature*, August 27, 2021, https://www.nature.com/articles/d41586-021-02338-4.
23 James Hanley, "Will We Accept Nuclear Fusion When It Comes?"

24. Ian Duncan, "Biden administration sets goals of replacing all jet fuel with sustainable alternatives by 2050," *Washington Post*, September 9, 2021, https://www.washingtonpost.com/transportation/2021/09/09/jets-sustainable-aviation-fuel-goal/.
25. White House Briefing Room, "FACT SHEET: Biden Administration Advances the Future of Sustainable Fuels in American Aircraft," WhiteHouse.gov, September 9, 2021, https://www.whitehouse.gov/briefing-room/statements-releases/2021/09/09/fact-sheet-biden-administration-advances-the-future-of-sustainable-fuels-in-american-aviation/.
26. Trevor English, "What's the Difference between Jet Fuel and Gasoline?" Interesting Engineering.com, May 31, 2020, https://interestingengineering.com/whats-the-difference-between-jet-fuel-and-gasoline.
27. Allen Herbert, "The Differences Between AvGas, Jet Fuel, Auto Fuel, and Diesel," AirplaneAcademy.com, no date, https://airplaneacademy.com/the-differences-between-avgas-jet-fuel-auto-fuel-and-diesel/.
28. Office of Energy Efficiency and Renewable Energy, U.S. Department of Energy, "Sustainable Aviation Fuel: Review of Technical Pathways," DOE/EE-2041, Energy.gov, September 2020, https://www.energy.gov/sites/prod/files/2020/09/f78/beto-sust-aviation-fuel-sep-2020.pdf.
29. Ibid., "Executive Summary," p. vi.
30. Ibid., p. vii.
31. Ibid., p. vi.
32. Rael Jean Isaac, "How to End Biden's Fake Climate Apocalypse," *The American Spectator*, March 15, 2021, https://spectator.org/climate-change-global-warming/.
33. Andrea Shalal, Timothy Gardner, and Steve Holland, "U.S. waives sanctions on Nord Stream 2 as Biden seeks to mend Europe ties," Reuters, May 19, 2021, https://www.reuters.com/business/energy/us-waive-sanctions-firm-ceo-behind-russias-nord-stream-2-pipeline-source-2021-05-19/.
34. "Russia's new pipeline bypasses Ukraine in pumping gas to Europe, Kyiv says," EuroNews.com, October 1, 2021, https://www.euronews.com/2021/10/01/russia-s-new-pipeline-bypasses-ukraine-in-pumping-gas-to-europe-kyiv-says.
35. Valerie Volcovici, "Explainer: How Biden could use his whole government to take on climate change," Reuters, January 19, 2021, https://www.reuters.com/article/us-usa-biden-climate-policy-explainer-idUSKBN29O260.
36. Timothy Paul Smith, *How Big Is Big and How Small Is Small: The Sizes of Everything and Why*, p. 143.
37. Ibid., p. 145.

About the Author

Since 2004, Jerome R. Corsi has published twenty-five books on economics, history, and politics, including six *New York Times* bestsellers, two at #1. He co-authored his first #1 NYT bestseller in 2004 with John O'Neill: *Unfit for Command: Swift Boat Veterans Speak out Against John Kerry*. Dr. Corsi authored his second #1 *NYT* bestseller in 2008: *The Obama Nation: Leftist Politics and the Cult of Personality*. From 2004–2016, Dr. Corsi was a senior editor at WorldNetDaily.com, where he authored hundreds of articles.

In 2019, the Mueller investigation targeted him for prosecution for "lying" to federal authorities in the Russian collusion investigation. Dr. Corsi refused a plea agreement, and the Mueller prosecutors declined prosecution. His 2018 book *Killing the Deep State: The Fight to Save President Trump* was on the *New York Times* bestseller list for several weeks. He authored two books on the Trump administration and his experience with the Mueller prosecutors: *Silent No More: How I Became a Political Prisoner of Mueller's "Witch Hunt"* (2019) and *Coup d'État: Exposing Deep State Treason and the Plan to Re-Elect President Trump* (2020).

The Truth about Energy, Global Warming, and Climate Change: Exposing Climate Lies in an Age of Disinformation will be his third book on energy, climate change, and global warming. His first book on this subject was *Black Gold Stranglehold: The Myth of Scarcity and the Politics of Oil* (2005), co-authored with Craig R. Smith. In that book, Dr. Corsi correctly predicted that oil would hit one hundred dollars a barrel when oil industry experts thought a price that high would never happen. His second book was *The Great Oil Conspiracy: How the US Government Hid the Nazi Discovery of Abiotic Oil from the American People* (2012). That book was the first in which Dr. Corsi argued that German chemists in the Weimar Republic formulated the chemical equations describing how abiogenic hydrocarbons are created in the mantle of the Earth.

Currently, Dr. Corsi is working to assist his wife, Monica, the founder of Corstet LLC, and her new website https://corstet.com. Dr. Corsi has retired from active involvement in current politics to devote full-time to writing books. He resides with his family in New Jersey.

Made in the USA
Columbia, SC
14 May 2023